Student Supplement

to accompany

Calculus
with Analytic Geometry

Second Edition

by   **Howard E. Campbell and
Paul F. Dierker**

University of Idaho

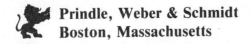 **Prindle, Weber & Schmidt
Boston, Massachusetts**

CONTENTS

Exercises 1.3: Functions (page 8)

1. $f(3) = 3 + 5 = 8$

3. $k(0) = (0)^2 - 2(0) + 1 = 1$

5. $k(a + 3) = (a + 3)^2 - 2(a + 3) + 1 = a^2 + 6a + 9 - 2a - 6 + 1 = a^2 + 4a + 4$

7. $(f \circ g)(x) = f(g(x)) = f(x^4) = x^4 + 5$

9. $(f \circ g)(2) = f(g(2)) = f(2^4) = 2^4 + 5 = 16 + 5 = 21$

11. $(k \circ f)(x) = k(f(x)) = k(x + 5) = (x + 5)^2 - 2(x + 5) + 1$
$$= x^2 + 10x + 25 - 2x - 10 + 1 = x^2 + 8x + 16$$

13. $k(y^3) = (y^3)^2 - 2(y^3) + 1 = y^6 - 2y^3 + 1$

15. $g(x) + f(x) = x^4 + x + 5$

17. $g(x)k(x) = x^4(x^2 - 2x + 1) = x^6 - 2x^5 + x^4$

19. $k(-3)f(-1) = [(-3)^2 - 2(-3) + 1][-1 + 5] = (9 + 6 + 1)(4) = (16)(4) = 64$

21. $f(x + h) - f(x) = (x + h + 5) - (x + 5) = h$

23. $\dfrac{k(x + h) - k(x)}{h} = \dfrac{[(x + h)^2 - 2(x + h) + 1] - [x^2 - 2x + 1]}{h}$

$$= \frac{x^2 + 2hx + h^2 - 2x - 2h + 1 - x^2 + 2x - 1}{h}$$

$$= \frac{2hx + h^2 - 2h}{h} = 2x + h - 2$$

25. (a) The only value x cannot have is -2 so the domain is the set of all real numbers except -2. (b) The only value the fraction $\dfrac{1}{x + 2}$ does not assume for values of x in the domain is 0. Thus, the range is the set of all real numbers except 0.

27. (a) The function $f(x) = \sqrt{x + 2}$ is real for all values of x for which $x + 2 \geq 0$ and only these values. Thus, the domain is the set of all $x \geq -2$. (b) Since $\sqrt{}$ means the positive square root or zero, the range is the set of all non-negative real numbers.

29. (a) Since the cube root of any real number is real, the domain of $\sqrt[3]{\dfrac{1}{x + 2}}$ is the set of all real numbers $\neq -2$. (b) Since $\dfrac{1}{x + 2}$ is never zero and cube roots can be positive or negative, the range is the set of all real numbers except 0.

35. Since a total of 600 yd of fencing is available, the situation may be pictured as

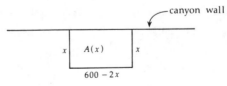

Thus, $A(x) = x(600 - 2x)$.

37. The volume of a right circular cone with radius r and height h is $V = \dfrac{1}{3} \pi r^2 h$. In this case $r = 2h$ so on substitution we get $V(h) = \dfrac{1}{3} \pi (2h)^2 h = \dfrac{4}{3} \pi h^3$.

1. $|3 - 7| = |-4| = 4$

3. $|4 - (-6)| = |4 + 6| = |10| = 10$

5. $|-3 - 8| = |-11| = 11$

9. $|x - (-6)| = |x + 6|$

23. $|x - (-6)| > 8$ so $|x + 6| > 8$

27. From the figure we see that
    $\tan \pi/6 = \dfrac{1}{\sqrt{3}}$

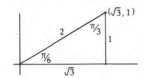

29. From the figure we see that
    $\cos 5\pi/6 = \dfrac{-\sqrt{3}}{2} = -\dfrac{\sqrt{3}}{2}$

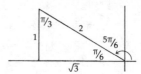

33. Since $\log_2 16$ is the power to which 2 must be raised to obtain 16 and $2^4 = 16$, we see that $\log_2 16 = 4$.

35. Since $\log_3 (\frac{1}{9})$ is the power to which 3 must be raised to obtain $\frac{1}{9}$ and $3^{-2} = \frac{1}{9}$, we see that $\log_3 (\frac{1}{9}) = -2$.

37. Since $\log_5 .04$ is the power to which 5 must be raised to obtain .04 and $.04 = \frac{1}{25} = 5^{-2}$, we see that $\log_5 .02 = -2$.

39. Since $.25 = \frac{1}{4}$ and $2^{-2} = \frac{1}{4}$, we see that $\log_2 .25 = -2$.

41. Since $\cos (-x) = \cos x$ and $(-x)^2 = x^2$, $f(x) = \cos x$ or $f(x) = x^2$ have the property that $f(-x) = f(x)$ and so are even functions. There are many others.

45. Let $f(x) = kx$ where k is a constant. Then
    $f(ax + by) = k(ax + by) = a(kx) + b(ky) = af(x) + bf(y)$
    so $f(x) = kx$ is a linear function.

47. Let $f(x) = \log_a x$. Then $f(xy) = \log_a (xy) = \log_a x + \log_a y$ so $f(x) = \log_a x$ is such a function.

53. By definition $|x| = -x$ when $x < 0$. Also $|0| = 0 = -0$. Therefore, $|x| = -x$ for all x such that $x \leq 0$.

55. Since $|-x| = |x|$ for all x, the given equation is equivalent to $|x| = x$. Thus, $|-x| = x$ for all x such that $x \geq 0$.

Exercises 1.5: Graphs of Functions and Equations (page 18)

5. The graph of $y = 3x$ is the set of all points with y-coordinate 3 times the x-coordinate. Using the indicated table of values as an aid we obtain the indicated graph.

x	y
-2	-6
-1	-3
0	0
1	3
2	6
3	9

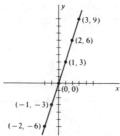

The graph has no breaks.

9. Using the indicated table and the fact that $y = 2|x|$ we get

x	y
-3	6
-2	4
-1	1
0	0
1	1
2	4
3	6

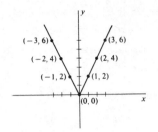

The graph has no breaks.

11. Using $y = x + |x|$ we get

x	y
-3	0
-1	0
0	0
1	2
3	6

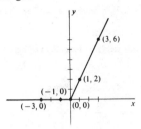

The graph has no breaks.

25. Using $y = f(x) = \dfrac{x^2 + 2x}{x} = \begin{cases} x + 2 \text{ if } x \neq 0 \\ \text{no value if } x = 0 \end{cases}$ we get

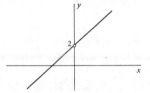

There is a break at $x = 0$.

Exercises 1.6: Lines (page 26)

1. $m = \dfrac{-1 - 5}{6 - 2} = -\dfrac{6}{4} = -\dfrac{3}{2}$

3. Since the x-coordinates of the given points are the same, the line is vertical and thus has no slope.

5. Using the equation $y - y_1 = m(x - x_1)$ we obtain

$$y - (-6) = \tfrac{2}{5}(x - 1).$$

This can be put in the form $5y + 30 = 2x - 2$ or $2x - 5y = 32$.

7. Here the slope $m = \dfrac{7 - 3}{4 + 1} = \dfrac{4}{5}$ so an equation of the line is

$y - 7 = \tfrac{4}{5}(x - 4)$, or $5y - 35 = 4x - 16$, or $4x - 5y + 19 = 0$.

9. Since the y-coordinates of the given points (1,6) and (5,6) are equal, the line is horizontal and has equation $y = 6$.

11. Since $x = 3$ is vertical, the line has equation $x = -1$.

13. To find the slope of the line $3x + 5y = 2$, we solve for y to get

$$y = \frac{3}{5}x + \frac{2}{5}.$$

Then the coefficient of x, $-\frac{3}{5}$, is the slope of the given line. Since non-vertical parallel lines have equal slopes, the slope of the line in question is also $-\frac{3}{5}$. Thus, it has an equation $y + 1 = -\frac{3}{5}(x - 2)$ or $3x + 5y = 1$.

15. The given equation can be put in the form $y = \frac{1}{2}x - \frac{5}{2}$ and so it has slope $\frac{1}{2}$.

But a line perpendicular to the given one must have slope m where $\frac{1}{2}m = -1$, or $m = -2$. Thus, an equation of the line in question is $y + 2 = -2(x - 1)$ or $y = -2x$.

19. Since $x = -3$ is vertical, a perpendicular line must be horizontal and thus an equation of the line in question is $y = -5$.

21. Since $y = 5$ is horizontal, a perpendicular line must be vertical. Thus, an equation of the line in question is $x = 6$.

25. The equations of the lines can be put in the forms

$$y = \frac{2}{3}x - \frac{5}{9} \text{ and } y = -\frac{3}{2}x + \frac{7}{10}.$$

Thus, their slopes are $\frac{2}{3}$ and $-\frac{3}{2}$. The product of their slopes is $\frac{2}{3}\left(-\frac{3}{2}\right) = -1$, so the lines are perpendicular.

27. The line $y + 3 = 0$ or $y = -3$ is horizontal and the line $x - 5 = 0$ or $x = 5$ is vertical. Thus, the lines are perpendicular.

29. The point $(2,3)$ is on the line if and only if it satisfies the equation, that is if and only if $5(2) + k(3) = 7$. Thus, $3k = -3$ and $k = -1$.

43. Let F and C denote Fahrenheit and Celsius temperatures, respectively. Since the relationship between C and F is linear, we know that $C - C_1 = m(F - F_1)$. Moreover, the graph must pass through the points $(212,100)$ and $(32,0)$. Thus,

$$m = \frac{100}{212 - 32} = \frac{5}{9}.$$

Using $(F_1, C_1) = (32,0)$ we then get

$$C = \frac{5}{9}(F - 32).$$

Exercises 1.7: Distance, Circles, Ellipses, Hyperbolas, and Parabolas
(page 35)

1. $D((4,7), (1,3)) = \sqrt{(4 - 1)^2 + (7 - 3)^2} = \sqrt{9 + 16} = \sqrt{25} = 5$.

3. Since these points are on the same vertical line, the distance between them is $|5 - (-3)| = |5 + 3| = |8| = 8$.

9. The distance between (x,x) and $(2x,0)$ is

$$\sqrt{(x - 2x)^2 + (x - 0)^2} = \sqrt{2x^2} = \sqrt{2}\,|x|.$$

Since this distance must be one, $\sqrt{2}\,|x| = 1$. Thus, $|x| = 1/\sqrt{2}$ or $x = \pm 1/\sqrt{2}$.

11. Using the formula $(x - h)^2 + (y - k)^2 = r^2$ we get

$$x^2 + (y - 5)^2 = 36.$$

13. Here the radius is $D((2,-4), (1,3)) = \sqrt{(2-1)^2 + (-4-3)^2} = \sqrt{50}$. Thus, an equation of the circle is $(x-2)^2 + (y+4)^2 = 50$.

15. Since the radius is 5, an equation of the circle is
$$(x - h)^2 + (y - k)^2 = 25.$$
Since it passes through the points $(3,-4)$ and $(0,5)$, these points must satisfy the equation. That is
$$(3 - h)^2 + (-4 - k)^2 = 25 \text{ and } (0 - h)^2 + (5 - k)^2 = 25.$$
Multiplying we get the pair of equations
$$9 - 6h + h^2 + 16 + 8k + k^2 = 25, \quad h^2 + 25 - 10k + k^2 = 25.$$
Subtraction gives $9 - 6h - 9 + 18k = 0$ or $h = 3k$. Substituting in (2) we get $9k^2 + 25 - 10k + k^2 = 25$ or $10k^2 - 10k = 0$. Cancelling the 10's and factoring we get $k(k - 1) = 0$, so $k = 0$ or $k = 1$. From (3) $h = 0$ when $k = 0$ and $h = 3$ when $k = 1$ so we get two circles
$$x^2 + y^2 = 25 \text{ and } (x - 3)^2 + (y - 1)^2 = 25.$$

17. To satisfy these conditions, a circle must look like the figure indicated:

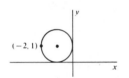

Thus, if the radius is r, the circle must have $h = -r$ and $k = r$ so its equation must have the form $(x + r)^2 + (y - r)^2 = r^2$, and since it passes through $(-2,1)$ we have $(-2 + r)^2 + (1 - r)^2 = r^2$. So $4 - 4r + r^2 + 1 - 2r + r^2 = r^2$ or $r^2 - 6r + 5 = 0$, $r = 5$ or $r = 1$. Thus, there are two such circles $(x + 5)^2 + (y - 5)^2 = 25$ and $(x + 1)^2 + (y - 1)^2 = 1$.

35. Since $|x| = \begin{cases} -x \text{ if } x < 0 \\ x \text{ if } x \geq 0 \end{cases}$ the equation can be expressed as
$$y = \begin{cases} -x^2 \text{ if } x < 0 \\ x^2 \text{ if } x \geq 0 \end{cases}$$

Thus, the graph is

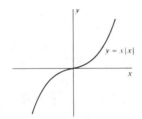

$y = x|x|$

37.

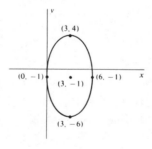

The distance between any point (x,y) and the line $y = 1$ is $|1 - y|$.

Thus, a point (x,y) is on the graph if and only if
$$|1 - y| = \sqrt{(x + 1)^2 + (y + 1)^2}$$
or squaring both sides, if and only if
$$1 - 2y + y^2 = (x + 1)^2 + (y + 1)^2.$$
Simplifying this equation we get
$$1 - 2y + y^2 = x^2 + 2x + 1 + y^2 + 2y + 1 \text{ or } x^2 + 2x + 1 + 4y = 0.$$

39. A point (x,y) is equidistant from $(1,-3)$ and $(-2,5)$ if and only if
$$\sqrt{(x - 1)^2 + (y + 3)^2} = \sqrt{(x + 2)^2 + (y - 5)^2}.$$
Squaring both sides and multiplying out we obtain
$$x^2 - 2x + 1 + y^2 + 6y + 9 = x^2 + 4x + 4 + y^2 - 10y + 25 \text{ or } 16y - 6x = 19 \text{ as an}$$
equation of the set.

Exercises 1.8: Translation of Axes (page 41)

17. Factoring out the coefficients of x^2 and y^2 we get
$$25(x^2 - 6x) + 9(y^2 + 2y) = -9.$$
Completing the squares in x and y we obtain
$$25(x^2 - 6x + 9) + 9(y^2 + 2y + 1) = -9 + 225 + 9$$
$$\text{or } 25(x - 3)^2 + 9(y + 1)^2 = 225.$$
Dividing by 225 produces
$$\frac{(x - 3)^2}{9} + \frac{(y + 1)^2}{25} = 1.$$

6

23. We obtain $\sqrt{(x + 3)^2 + y^2} + \sqrt{(x - 3)^2 + y^2} = 10$

   or $\sqrt{(x + 3)^2 + y^2} = 10 - \sqrt{(x - 3)^2 + y^2}$.

   Squaring both sides gives

   $(x + 3)^2 + y^2 = 100 - 20\sqrt{(x - 3)^2 + y^2} + (x - 3)^2 + y^2$

   or $3x - 25 = -5\sqrt{(x - 3)^2 + y^2}$.

   Squaring again produces

   $9x^2 - 150x + 625 = 25x^2 - 150x + 225 + 25y^2$

   or $16x^2 + 25y^2 = 400$, and we get $\dfrac{x^2}{25} + \dfrac{y^2}{16} = 1$.

Exercises 1.9: Parametric Equations (page 44)

5. Since $y = v - 1$, we have $v = y + 1$. Substitution into $x = 3v + 7$ gives
   $x = 3(y + 1) + 7$ or $x = 3y + 10$ or $x - 3y = 10$. The graph is a straight line.

9. Since $x = \sin t$ and $y = \cos t$, we have
   $$x^2 + y^2 = \sin^2 t + \cos^2 t = 1.$$
   Thus, $x^2 + y^2 = 1$. However, since $0 \le t \le \pi$, $x = \sin t \ge 0$. Consequently,
   the graph is that portion of the graph of $x^2 + y^2 = 1$ for which $x \ge 0$. That
   is, the graph is the right half of the circle of radius one centered at the
   origin.

11. Since $x = 4 \sin t$ and $y = 3 \cos t$, we have
    $$x/4 = \sin t \text{ and } y/3 = \cos t.$$
    Thus, $(x/4)^2 + (y/3)^2 = \sin^2 t + \cos^2 t = 1$ or,

    (1) $$\left(\frac{x}{4}\right)^2 + \left(\frac{y}{3}\right)^2 = 1.$$

    Note that since $0 \le t \le \pi$, $x = 4 \sin t \ge 0$. Consequently, the graph is that
    portion of the graph of (1) for which $x \ge 0$. That is, the graph is the right
    half of the ellipse $\left(\frac{x}{4}\right)^2 + \left(\frac{y}{3}\right)^2 = 1$.

13. Since $x = 2 + \cos t$, and $y = 1 + \sin t$, we have $x - 2 = \cos t$ and $y - 1 = \sin t$.
    Thus,
    $$(x - 2)^2 + (y - 1)^2 = \cos^2 t + \sin^2 t = 1 \text{ or,}$$
    $$(x - 2)^2 + (y - 1)^2 = 1.$$
    Consequently, the graph is a circle of radius one centered at $(2,1)$.

21. For any angle θ the distance the wheel has rolled is equal to the arc length
    subtended by an angle θ on a circle of radius r. Consequently, the center of
    the wheel must have coordinates $(r\theta, r)$ and we have the following diagram.

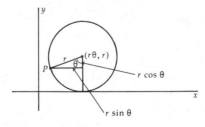

Consequently, the coordinates of P are $(r\theta - r\sin\theta, r - r\cos\theta)$. The desired parametric equations are $x = r\theta - r\sin\theta$ and $r = r - r\cos\theta$.

<u>Technique Review Exercises, Chapter 1 (page 49)</u>

1. (a) $f(x^3) = (x^3)^2 + 3(x^3) = x^6 + 3x^3$

   (b) $(f\circ g)(2) = f(g(2)) = [g(2)]^2 + 3[g(2)]$

   But $g(2) = 1 - 2^2 = -3$ so $(f\circ g)(2) = (-3)^2 + 3(-3) = 9 - 9 = 0$

   (c) $g(x + 1) = 1 - (x + 1)^2 = 1 - (x^2 + 2x + 1) = -x^2 - 2x$.

2. Since $\sqrt{5 - x}$ is real if and only if $5 - x \geq 0$ or $5 \geq x$, the domain is the set of all $x \leq 5$. Since every nonnegative real number is the square root of a number, the range is the set of all nonnegative real numbers.

3. (a) $|x - 5|$ is the distance between x and 5 on the real line.

   (b) $|6 + x| = |x - (-6)|$ is the distance between x and -6 on the real line.

4.

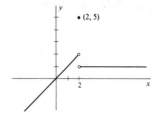

5. The equation $5x + 3y = 7$ can be put in the form $y = -\frac{5}{3}x + \frac{7}{3}$. Thus, the line has slope $-\frac{5}{3}$. A perpendicular line must have slope m where $-\frac{5}{3}m = -1$. Hence, $m = \frac{3}{5}$ and the line in question has an equation $y - 3 = \frac{3}{5}(x + 2)$ or $3x - 5y + 21 = 0$.

6. Since the y-coordinates of these points are the same, the line is horizontal and has equation $y = -2$.

7. $D((-3,5), (6,1)) = \sqrt{(-3 - 6)^2 + (5 - 1)^2} = \sqrt{81 + 16} = \sqrt{97}$.

8. A point (x,y) is on the graph if and only if
$$\sqrt{(x + 1)^2 + (y - 2)^2} = \sqrt{(x - 3)^2 + (y - 5)^2}$$
or $(x + 1)^2 + (y - 2)^2 = (x - 3)^2 + (y - 5)^2$.

   Multiplying out and simplifying we get
   $x^2 + 2x + 1 + y^2 - 4y + 4 = x^2 - 6x + 9 + y^2 - 10y + 25$ or $8x + 6y = 29$.

9. The radius of the circle is $\sqrt{(5 - 7)^2 + (-1 - 2)^2} = \sqrt{4 + 9} = \sqrt{13}$. Thus, an equation of the circle is
$$(x - 5)^2 + (y + 1)^2 = 13.$$

13. We complete the squares to obtain $(x - 2)^2 + (y - 3)^2 = 4$. Thus, the graph is a circle centered at (2,3) with radius 2.

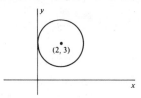

15. Since $y = 5 \sin t$ and $x = -2 \cos t$, we have $y/5 = \sin t$ and $-x/2 = \cos t$. Thus,

$$\left(\frac{y}{5}\right)^2 + \left(-\frac{x}{2}\right)^2 = \sin^2 t + \cos^2 t = 1 \text{ or,}$$

$$\left(\frac{y}{5}\right)^2 + \left(\frac{x}{2}\right)^2 = 1. \text{ The graph is an ellipse.}$$

Additional Exercises, Chapter 1 (page 49)

1. $f(5) = 5^2 + 2(5) = 25 + 10 = 35$

3. $(f \circ g)(x) = f(g(x)) = (x^2 - 3x + 5)^2 + 2(x^2 - 3x + 5)$
    $$= x^4 - 6x^3 + 21x^2 - 36x + 35$$

5. $(x + h)^2 - 3(x + h) + 5 - x^2 + 3x - 5 = 2hx + h^2 - 3h$

7. (a) The only values x cannot have are 2 and −1. Thus, the domain is the set of all real numbers except 2 and −1.

    (b) The only value the function does not assume, for values of x in the domain, is zero. Therefore, the range is the set of all real numbers except 0.

27. Here $m = \frac{3 + 2}{1 - 5} = -\frac{5}{4}$. Thus, an equation of the line is
    $$y - 3 = -\frac{5}{4}(x - 1) \text{ or } 5x + 4y = 17.$$

29. The equation $7x + 3y = 1$ can be put in the form $y = -\frac{7}{3}x + \frac{1}{3}$. Thus, the slope of both lines is $-\frac{7}{3}$. Therefore, an equation of the line in question is
    $$y - 2 = -\frac{7}{3}(x + 1) \text{ or } 7x + 3y = -1.$$

33. The given equations can be put in the forms
    $$y = \frac{2}{3}x - \frac{1}{6} \text{ and } y = \frac{2}{3}x + \frac{7}{15}.$$
    Therefore, both lines have the same slope $\frac{2}{3}$ and so they are parallel.

35. $D((-4,2),(-6,-3)) = \sqrt{(-4 + 6)^2 + (2 + 3)^2} = \sqrt{4 + 25} = \sqrt{29}$.

37. The radius is $D((-2,7), (1,-3)) = \sqrt{(-2 - 1)^2 + (7 + 3)^2} = \sqrt{109}$, so an equation of the circle is $(x + 2)^2 + (y - 7)^2 = 109$.

41. The distance between a point (x,y) and the line x = 2 is $|x - 2|$. The distance between (x,y) and (6,1) is $\sqrt{(x - 6)^2 + (y - 1)^2}$. Thus, we get
    $$|x - 2| = \sqrt{(x - 6)^2 + (y - 1)^2}$$

or $x^2 - 4x + 4 = (x - 6)^2 + (y - 1)^2$. Simplifying we obtain

$$x^2 - 4x + 4 = x^2 - 12x + 36 + y^2 - 2y + 1$$

or $y^2 - 2y - 8x + 33 = 0$.

57. Since $x = -\cos 3t$ and $y = 2 \sin 3t$, we have

$$-x = \cos 3t \text{ and } y/2 = \sin 3t. \quad \text{Thus,}$$

$$(-x)^2 + \left(\frac{y}{2}\right)^2 = \cos^2 3t + \sin^2 3t = 1 \text{ or,}$$

$$x^2 + \left(\frac{y}{2}\right)^2 = 1. \quad \text{The graph is an ellipse.}$$

59. Since $x = 1 + \cos t$ and $y = 2 \sin t$, we have

$$x - 1 = \cos t \text{ and } y/2 = \sin t. \quad \text{Thus,}$$

$$(x - 1)^2 + \left(\frac{y}{2}\right)^2 = \cos^2 t + \sin^2 t = 1 \text{ or,}$$

$$(x - 1)^2 + \left(\frac{y}{2}\right)^2 = 1. \quad \text{The graph is an ellipse.}$$

CHAPTER 2: LIMITS, CONTINUITY, AND DERIVATIVES

Exercises 2.1: The Tangent Problem (page 58)

1. $m(h) = \dfrac{f(3 + h) - f(3)}{h} = \dfrac{(3 + h)^2 - 3^2}{h} = \dfrac{6h + h^2}{h} = 6 + h.$

So $m(h) \to 6$ as $h \to 0$ and $m = 6$. Thus, an equation of the tangent at $(3,9)$ is $y - 9 = 6(x - 3)$ or $y = 6x - 9$.

3. $m(h) = \dfrac{f(0 + h) - f(0)}{h} = \dfrac{(0 + h)^2 - 0^2}{h} = \dfrac{h^2}{h} = h.$

So $m(h) \to 0$ as $h \to 0$ and $m = 0$. An equation of the tangent at $(0,0)$ is $y = 0$.

5. At $(3,1)$ $m(h) = \dfrac{f(3 + h) - f(3)}{h} = \dfrac{[4 - (3 + h)] - [4 - 3]}{h} = \dfrac{-h}{h} = -1.$ So

$m(h) \to -1$ as $h \to 0$ and an equation of the tangent at $(3,1)$ is $y - 1 = -1(x - 3)$ or $y = -x + 4$.

7. At $(1,7)$ $m(h) = \dfrac{f(1 + h) - f(1)}{h} = \dfrac{[5(1 + h) + 2] - [5(1) + 2]}{h} = \dfrac{5h}{h} = 5.$ So

$m(h) \to 5$ as $h \to 0$ and an equation of the tangent at $(-2,6)$ is $y - 7 = 5(x - 1)$ or $y = 5x + 2$.

9. At $(1,0)$ $m(h) = \dfrac{f(1 + h) - f(1)}{h} = \dfrac{[(1 + h)^2 - (1 + h)] - [1^2 - 1]}{h} = \dfrac{h + h^2}{h}$

$= 1 + h.$ So $m(h) \to 1$ as $h \to 0$ and $m = 1$. Thus, an equation of the tangent at $(1,0)$ is $y = x - 1$.

11. $m(h) = \dfrac{f(2 + h) - f(2)}{h} = \dfrac{[(2 + h)^2 - 2(2 + h) + 3] - [2^2 - 2(2) + 3]}{h}$

$= \dfrac{4 + 4h + h^2 - 4 - 2h + 3 - 4 + 4 - 3}{h} = \dfrac{h^2 + 2h}{h} = h + 2.$ So $m(h) \to 2$

as $h \to 0$ and $m = 2$. Thus, an equation of the tangent at $(2,3)$ is $y - 3 = 2(x - 2)$ or $y = 2x - 1$.

17. $f(x) = x^3$; $(2,8)$

$m(h) = \dfrac{f(2 + h) - f(2)}{h} = \dfrac{(2 + h)^3 - 2^3}{h} = 12 + 6h + h^2.$

So $m(h) \to 12$ as $h \to 0$ and $m = 12$.

21. $m(h) = \dfrac{f(3 + h) - f(3)}{h} = \dfrac{\dfrac{1}{3 + h} - \dfrac{1}{3}}{h} = \dfrac{\dfrac{3}{3(3 + h)} - \dfrac{3 + h}{3(3 + h)}}{h}$

$= \dfrac{\dfrac{3 - 3 - h}{3(3 + h)}}{h} = \dfrac{\dfrac{-h}{3(3 + h)}}{h} = \dfrac{-1}{3(3 + h)}.$

So $m(h) \to -\dfrac{1}{9}$ as $h \to 0$ and $m = -\dfrac{1}{9}.$

23. $m(h) = \dfrac{\dfrac{1}{2 + h - 1} - \dfrac{1}{2 - 1}}{h} = \dfrac{\dfrac{1}{1 + h} - 1}{h} = \dfrac{\dfrac{-h}{1 + h}}{h} = \dfrac{-1}{1 + h}.$

So $m(h) \to -1$ as $h \to 0$ and $m = -1$.

25. $m(h) = \dfrac{\sqrt{1 + h} - \sqrt{1}}{h} = \dfrac{\sqrt{1 + h} - 1}{h} \cdot \dfrac{\sqrt{1 + h} + 1}{\sqrt{1 + h} + 1} = \dfrac{1 + h - 1}{h(\sqrt{1 + h} + 1)}$

$= \dfrac{h}{h(\sqrt{1 + h} + 1)} = \dfrac{h}{\sqrt{1 + h} + 1}.$

So $m(h) \to \dfrac{1}{2}$ as $h \to 0$ and $m = \dfrac{1}{2}.$

27. $m(h) = \dfrac{(0 + h) a^{0+h} - 0a^0}{h} = \dfrac{ha^h}{h} = a^h.$ So $m(h) \to 1$ as $h \to 0$ and $m = 1$.

29. $m(h) = \dfrac{(0 + h)^2 \cos (0 + h) - 0^2 \cos 0}{h} = \dfrac{h^2 \cos h}{h} = h \cos h.$

So $m(h) \to 0$ as $h \to 0$ and $m = 0$.

31. $m(h) = \dfrac{(0 + h)^7 - 0^7}{h} = \dfrac{h^7}{h} = h^6.$ So $m(h) \to 0$ as $h \to 0$ and $m = 0$.

33. $m(h) = \dfrac{(2 + h - 2)^2 \sin (2 + h) - (2 - 2)^2 \sin 2}{h} = \dfrac{h^2 \sin (2 + h)}{h}$

$= h \sin (2 + h).$ So $m(h) \to 0$ as $h \to 0$ and $m = 0$.

35. $m(h) = \dfrac{(a + h)^2 + 2(a + h) - a^2 - 2a}{h} = \dfrac{2ah + h^2 + 2h}{h} = 2a + h + 2.$

So $m(h) \to 2a + 2$ as $h \to 0$ and $m = 2a + 2$.

37. $m(h) = \dfrac{3(1 + h)^2 + 1 - 3(1) - 1}{h} = \dfrac{6h + 3h^2}{h} = 6 + 3h.$ So $m(h) \to 6$ as $h \to 0$

and $m = 6$. Thus, the slope of the normal is $-\dfrac{1}{6}$ and an equation of the normal

is $y - 4 = -\dfrac{1}{6} (x - 1)$ or $x + 6y = 25$.

Exercises 2.2: Limits (page 66)

1. $\displaystyle\lim_{x \to 0} \dfrac{x^2 + 5x}{x} = \lim_{x \to 0} (x + 5) = 5$

3. $\displaystyle\lim_{x \to 0} \dfrac{x^2 - 4}{x - 2} = \lim_{x \to 2} \dfrac{(x - 2)(x + 2)}{x - 2} = \lim_{x \to 2} (x + 2) = 4$

5. If $x > 0$, $|x| = x$ so $\frac{x}{|x|} = \frac{x}{x} = 1$ if $x > 0$.

   If $x < 0$, $|x| = -x$ so $\frac{x}{|x|} = \frac{x}{-x} = -1$ if $x < 0$.

   Thus, $\frac{x}{|x|} = \begin{cases} 1 \text{ if } x > 0 \\ -1 \text{ if } x < 0 \end{cases}$ and so $\lim\limits_{x \to 0} \frac{x}{|x|}$ does not exist because it gets arbi-

   trarily close to two different numbers and not a unique number for values of x
   sufficiently close to 0 but not equal to 0.

7. $\lim\limits_{x \to 9} \frac{3 - \sqrt{x}}{9 - x} = \lim\limits_{x \to 9} \frac{3 - \sqrt{x}}{(3 - \sqrt{x})(3 + \sqrt{x})} = \lim\limits_{x \to 9} \frac{1}{3 + \sqrt{x}} = \frac{1}{6}.$

9. $\lim\limits_{x \to 0} \frac{3x}{x^2 - 8x} = \lim\limits_{x \to 0} \frac{3}{x - 8} = -\frac{3}{8}$

11. $\lim\limits_{x \to 3} \frac{x - 3}{x^2 - x - 6} = \lim\limits_{x \to 3} \frac{x - 3}{(x - 3)(x + 2)} = \lim\limits_{x \to 3} \frac{1}{x + 2} = \frac{1}{5}.$

15. $\lim\limits_{x \to 1} f(x) = 1$ since both x^2 for $x < 1$ and x for $x > 1$ can be made arbitrarily

    close to 1 if x is sufficiently close to 1 but not equal to 1.

17. Here $f(x) = \begin{cases} \dfrac{x^2 - 2x}{x} \text{ if } x \neq 0 \\ 1 \text{ if } x = 0 \end{cases} = \begin{cases} x - 2 \text{ if } x \neq 0 \\ 1 \text{ if } x = 0 \end{cases}$

    Since x must be $\neq 0$ in the consideration of $\lim\limits_{x \to 0} f(x)$, we only need to consider

    $f(x) = x - 2$ if $x \neq 0$. Hence, $\lim\limits_{x \to 0} f(x) = -2$.

19. Here f(x) can be made arbitrarily close to both 4 and 6 for values of x suf-
    ficiently close to 2 but not equal to 2. Thus, the limit does not exist because
    f(x) does not get arbitrarily close to a single number.

21. Since x cannot equal −2, the value 5 need not be considered. Since f(x) can be
    made arbitrarily close to both 1 and 9 for values of x sufficiently close to −2
    but not equal to −2, the limit does not exist.

23. Since $-1 \leq \sin\left(\frac{1}{x}\right) \leq 1$, the product $x \sin\left(\frac{1}{x}\right)$ can be made arbitrarily close to
    zero for values of x sufficiently close to zero but not equal to zero. Thus,
    the limit is zero.

25. Since $\sin\left(\frac{1}{x}\right)$ takes <u>all</u> of the values from −1 to 1 for values of x close to
    zero but not equal to zero, it gets arbitrarily close to many different numbers
    rather than a single number and so the limit does not exist.

27. We obtain the table:

x	f(x)		x	f(x)
1.9	5.983287		1.999	5.999833
2.1	6.016621		2.001	6.000167
1.99	5.998333		1.9999	5.999983
2.01	6.001666		2.0001	6.000017

   The limit appears to exist and have the value 6.

31. We obtain the indicated table showing that the limit appears to exist and have the value 0.

x	f(x)
± .5	.244834
± .1	.04996
± .01	.00500
± .001	.00000

33.

x	1 − x	1/x	f(x)
.1	.9	10	2.86796
−.1	1.1	−10	2.59374
.01	.99	100	2.73191
−.01	1.01	−100	2.70472
.001	.999	1000	2.71828
−.001	1.001	−1000	2.71556
.0001	.9999	10000	2.71828
−.0001	1.0001	−10000	2.71828

It appears that the limit exists and has a value near 2.71828.

35. (a) Here we must make $|2x - 6| < \frac{1}{8}$. But $|2x - 6| = |2(x - 3)| = 2|x - 3|$, so we must make $2|x - 3| < \frac{1}{8}$ or $|x - 3| < \frac{1}{16}$. That is, we must restrict x to be within $\frac{1}{16}$ of 3.

(b) Here we must make $|2x - 6| < \varepsilon$. As above, $|2x - 6| = 2|x - 3|$, so we must make $2|x - 3| < \varepsilon$ or $|x - 3| < \varepsilon/2$. That is, we must restrict x to be within $\frac{\varepsilon}{2}$ of 3.

39. Here $m(h) = \dfrac{f(4 + h) - f(4)}{h} = \dfrac{\sqrt{4 + h} - \sqrt{4}}{h} = \dfrac{\sqrt{4 + h} - 2}{h}$

$$= \frac{(\sqrt{4 + h} - 2)(\sqrt{4 + h} + 2)}{h(\sqrt{4 + h} + 2)} = \frac{4 + h - 4}{h(\sqrt{4 + h} + 2)}$$

$$= \frac{h}{h(\sqrt{4 + h} + 2)} = \frac{1}{\sqrt{4 + h} + 2}.$$

Hence, $m = \lim\limits_{h \to 0} m(h) = \lim\limits_{h \to 0} \dfrac{1}{\sqrt{4 + h} + 2} = \dfrac{1}{4}$.

Exercises 2.3: The Limit Theorem (page 70)

1. $\lim\limits_{x \to 2} (x^2 + 3x) = \lim\limits_{x \to 2} x^2 + \lim\limits_{x \to 2} 3x = \left(\lim\limits_{x \to 2} x\right)\left(\lim\limits_{x \to 2} x\right) + 3 \lim\limits_{x \to 2} x = (2)(2) + 3(2) = 10.$

3. Since $\lim\limits_{h \to 1} (h - 2) = \lim\limits_{h \to 1} h + \lim\limits_{h \to 1} (-2) = 1 - 2 = -1 \neq 0$, we may say that that $\lim\limits_{h \to 1} \dfrac{h^3}{h - 2} = \dfrac{\lim\limits_{h \to 1} h^3}{\lim\limits_{h \to 1} (h - 2)}.$

But $\lim\limits_{h \to 1} h^3 = \left(\lim\limits_{h \to 1} h\right)\left(\lim\limits_{h \to 1} h\right)\left(\lim\limits_{h \to 1} h\right) = (1)(1)(1) = 1$, so $\lim\limits_{h \to 1} \dfrac{h^3}{h - 2} = \dfrac{1}{-1} = -1.$

5. $\lim\limits_{h \to 0} \dfrac{h^2 - 5h}{h} = \lim\limits_{h \to 0} (h - 5) = \lim\limits_{h \to 0} h + \lim\limits_{h \to 0} (-5) = 0 - 5 = -5$

7. $\lim\limits_{x \to 0} (3x - 2)(x^2 + 7) = \lim\limits_{x \to 0} (3x - 2) \lim\limits_{x \to 0} (x^2 + 7)$

$= [3 \lim\limits_{x \to 0} x + \lim\limits_{x \to 0} (-2)][(\lim\limits_{x \to 0} x)(\lim\limits_{x \to 0} x) + \lim\limits_{x \to 0} 7]$

$= [3(0) + (-2)][(0)(0) + 7] = (-2)(7) = -14$

13

9. $m = \lim_{h \to 0} \dfrac{f(3+h) - f(3)}{h} = \lim_{h \to 0} \dfrac{(3+h)^2 - 5(3+h) - 3^2 + 5(3)}{h}$

$= \lim_{h \to 0} \dfrac{9 + 6h + h^2 - 15 - 5h - 9 + 15}{h} = \lim_{h \to 0} \dfrac{h + h^2}{h} = \lim_{h \to 0} (1 + h) = 1$

Hence, an equation of the tangent at $(3, -6)$ is

$$y + 6 = 1(x - 3) \text{ or } y = x - 9.$$

13. $m = \lim_{h \to 0} \dfrac{f(2+h) - f(2)}{h} = \lim_{h \to 0} \dfrac{(2+h)^3 - 2(2+h)^2 + 6 - (2)^3 + 2(2)^2 - 6}{h}$

$= \lim_{h \to 0} \dfrac{8 + 12h + 6h^2 + h^3 - 8 - 8h - 2h^2 + 6 - 8 + 8 - 6}{h}$

$= \lim_{h \to 0} \dfrac{4h + 4h^2 + h^3}{h} = \lim_{h \to 0} (4 + 4h + h^2) = 4.$

Thus, an equation of the tangent at $(2,6)$ is $y - 6 = 4(x - 2)$ or $y = 4x - 2$.

Exercises 2.4: Continuity (page 75)

1. Since $7 - \pi x^2$ is a polynomial, it is continuous for all values of x.

3. Since this is a rational function, it is continuous for all values of x except those where the denominator is zero. That is, all values of x except those where

$$x^2 - x - 6 = (x - 3)(x + 2) = 0.$$

Thus, the function is continuous for all values of x except $x = 3$ and $x = -2$.

9. Except for $x = -2$ this function is the same as the function of Exercise 7. For $x = -2$ we get

$\lim_{x \to -2} f(x) = \lim_{x \to -2} \dfrac{x^2 - x - 6}{x^2 + 7x + 10} = \lim_{x \to -2} \dfrac{(x-3)(x+2)}{(x+5)(x+2)} = \lim_{x \to -2} \dfrac{x-3}{x+5} = -\dfrac{5}{3}.$

Since $f(-2) = -\dfrac{5}{3} = \lim_{x \to -2} f(x)$ the function is continuous at $x = -2$.

Thus, $f(x)$ is continuous for all values of x except $x = -5$.

11. Except for $x = 0$ this function is the same as the function of Exercise 5. For $x = 0$ we get

$\lim_{x \to 0} f(x) = \lim_{x \to 0} \dfrac{x}{x^2 - 3x} = \lim_{x \to 0} \dfrac{x}{x(x-3)} = \lim_{x \to 0} \dfrac{1}{x - 3} = -\dfrac{1}{3}.$

But $f(0) = -\dfrac{1}{2} \neq \lim_{x \to 0} f(0)$ so $f(x)$ is not continuous at $x = 0$. Thus, the function is continuous for all values of x except $x = 0$ and $x = 3$.

13. In the interval between any two integers this function is the polynomial 2x and so it is continuous for all values of x that are not integers. If i is any integer then

$\lim_{x \to i} f(x) = 2i$ (since when $x \to i$, x must not equal i). But since i is an integer, $f(i) = 0$. Therefore,

$\lim_{x \to i} f(x) \neq f(i)$ if $i \neq 0$ but $\lim_{x \to 0} f(x) = 2(0) = 0 = f(0).$

Thus, $f(x)$ is continuous for all values of x that are not integers and also at x = 0.

15. $\lim\limits_{x\to 0} \dfrac{x}{x^2 - 3x} = \lim\limits_{x\to 0} \dfrac{x}{x(x - 3)} = \lim\limits_{x\to 0} \dfrac{1}{x - 3} = -\dfrac{1}{3}$, so define $f(0) = -\dfrac{1}{3}$.

19. $\lim\limits_{x\to -1} \dfrac{x^2 - x - 2}{x^2 + 2x + 1} = \lim\limits_{x\to -1} \dfrac{(x + 1)(x - 2)}{(x + 1)(x + 1)} = \lim\limits_{x\to -1} \dfrac{x - 2}{x + 1}$ does not exist so it is

impossible to define the function of -1 so it is continuous at -1.

21. Let $f(x)$ be the speed in miles per hour of the plane at a distance x in miles from its loading gate in Seattle. Then, according to the given information

$f(0) = 0$

$f(s) = 600$ for some s, $0 < s < d$

$f(d) = 0$ where d is the distance to the New York gate.

Also, it is natural to assume here that $f(x)$ is continuous. Thus, by the Intermediate Value Theorem for continuous functions, since $0 < 150 < 600$, there is at least one c_1, $0 < c_1 < p$ and one c_2, $p < c_2 < d$ such that $f(c_1) = f(c_2) = 150$. Thus, the plane must have been traveling at 150 mph for at least two points of its journey.

<u>Exercises 2.5:</u> Graphs of Factored Polynomials (page 82)

1. $\sqrt{x - 3}$ is real when $x - 3 \geq 0$, that is, when $x \geq 3$.

3. We must solve the inequality $x(x^2 + 3x - 10) \geq 0$ or $x(x + 5)(x - 2) \geq 0$.
Sketching the graph of $y = x(x + 5)(x - 2)$ we get

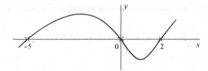

Thus, the square root is real when $y \geq 0$, that is, when $-5 \leq x \leq 0$ or $x \geq 2$.

7. The only values for which this square root is zero are $x = -1$ and $x = 3$. We now solve

$$\dfrac{(x + 1)(x - 3)}{x^2(x - 5)^3} > 0$$

by solving $y = x^2(x - 5)^3(x + 1)(x - 3) > 0$. Sketching, we get

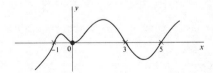

Thus, $y > 0$, and the square root is positive, for $-1 < x < 0$, $0 < x < 3$ or $x > 5$. Combining this with the values for which the square root is zero we get the following for which the square root is real

$$-1 \leq x < 0 \text{ or } 0 < x \leq 3 \text{ or } x > 5.$$

11. Sketching $y = 2x + 11$ we get

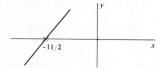

So the solution is $x < -11/2$.

13. Sketching $y = x^2 + 3x + 2 = (x + 2)(x + 1)$ we get

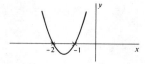

So the solution is $-2 < x < -1$.

15. This is equivalent to $x^2 - 5x + 6 > 0$. Sketching
$$y = x^2 - 5x + 6 = (x - 3)(x - 2) \text{ we get}$$

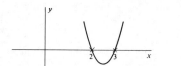

So the solution is $x < 2$ or $x > 3$.

17. Sketching $y = x(x - 1)(x + 2)$ we get

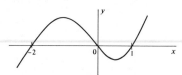

So the solution is $x \leq -2$ or $0 \leq x \leq 1$.

19. The fraction is zero only when $x = 0$ or $x = 3$. To find where it is positive we sketch $y = x(3 - x)^3(x + 1)^2(x - 5)$ to get

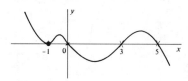

Thus, the fraction is greater than or equal to zero when $x < -1$ or $-1 < x \leq 0$ or $3 \leq x < 5$.

21. To solve this we sketch $y = (3 + x)(7 - x)^3(x + 1)(1 - x)(6 + x)(1 - x)^3$
$$= (3 + x)(7 - x)^3(x + 1)(6 + x)(1 - x)^4 \text{ to get}$$

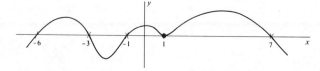

Thus, the solution is $-6 < x < -3$ or $-1 < x < 1$ or $1 < x < 7$.

23. We sketch $y = (3 - x)^3(x + 7)^2(x + 1)$ to obtain

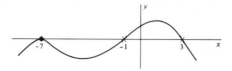

Thus, we have (a) $-1 < x < 3$, (b) $x < -7$ or $-7 < x < -1$ or $x > 3$.

25. We sketch $y = (6 + x)^3(x - 7)^5(x)(2 + x)(1 - x)$ to get

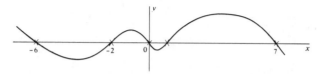

Thus, the solutions are (a) $x < -6$ or $-2 < x < 0$ or $1 < x < 7$

(b) $-6 < x < -2$ or $0 < x < 1$ or $x > 7$.

27. We sketch $y = 4x^3 - x^5 = x^3(4 - x^2) = x^3(2 - x)(2 + x)$ to get

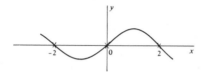

Thus, the solutions are (a) $x < -2$ or $0 < x < 2$

(b) $-2 < x < 0$ or $x > 2$.

Exercises 2.6: The Derivative (page 90)

1. $f'(x) = \lim_{h \to 0} \dfrac{f(x + h) - f(x)}{h} = \lim_{h \to 0} \dfrac{x + h + 7 - x - 7}{h} = \lim_{h \to 0} \dfrac{h}{h} = \lim_{h \to 0} 1 = 1.$

3. $\lim_{h \to 0} \dfrac{f(x + h) - f(x)}{h} = \lim_{h \to 0} \dfrac{(x + h)^2 + 5 - x^2 - 5}{h} = \lim_{h \to 0} \dfrac{x^2 + 2hx + h^2 + 5 - x^2 - 5}{h}$

$= \lim_{h \to 0} \dfrac{2hx + h^2}{h} = \lim_{h \to 0} (2x + h) = 2x.$

7. $f'(x) = \lim_{h \to 0} \dfrac{f(x + h) - f(x)}{h} = \lim_{h \to 0} \dfrac{2(x + h)^2 - 4(x + h) - 2x^2 + 4x}{h}$

$= \lim_{h \to 0} \dfrac{2x^2 + 4xh + 2h^2 - 4x - 4h - 2x^2 + 4x}{h}$

$= \lim_{h \to 0} \dfrac{4xh + 2h^2 - 4h}{h} = \lim_{h \to 0} (4x + 2h - 4) = 4x - 4.$

9. $f'(x) = \lim_{h \to 0} \dfrac{(x + h)^3 - x^3}{h} = \lim_{h \to 0} \dfrac{x^3 + 3x^2h + 3xh^2 + h^3 - x^3}{h}$

$= \lim_{h \to 0} \dfrac{3x^2h + 3xh^2 + h^3}{h} = \lim_{h \to 0} (3x^2 + 3xh + h^2) = 3x^2$

11. $f'(x) = \lim\limits_{h \to 0} \dfrac{\dfrac{1}{x+h} - \dfrac{1}{x}}{h} = \lim\limits_{h \to 0} \dfrac{\dfrac{x}{x(x+h)} - \dfrac{x+h}{x(x+h)}}{h}$

$= \lim\limits_{h \to 0} \dfrac{\dfrac{-h}{x(x+h)}}{h} = \lim\limits_{h \to 0} \dfrac{-1}{x(x+h)} = -\dfrac{1}{x^2}$

13. $f'(x) = \lim\limits_{h \to 0} \dfrac{\sqrt{x+h+1} - \sqrt{x+1}}{h}$

$= \lim\limits_{h \to 0} \dfrac{\sqrt{x+h+1} - \sqrt{x+1}}{h} \cdot \dfrac{\sqrt{x+h+1} + \sqrt{x+1}}{\sqrt{x+h+1} + \sqrt{x+1}}$

$= \lim\limits_{h \to 0} \dfrac{x+h+1-(x+1)}{h(\sqrt{x+h+1} + \sqrt{x+1})}$

$= \lim\limits_{h \to 0} \dfrac{h}{h(\sqrt{x+h+1} + \sqrt{x+1})} = \lim\limits_{h \to 0} \dfrac{1}{\sqrt{x+h+1} + \sqrt{x+1}}$

$= \dfrac{1}{\sqrt{x+1} + \sqrt{x+1}} = \dfrac{1}{2\sqrt{x+1}}.$

15. $f'(x) = \lim\limits_{h \to 0} \dfrac{f(x+h) - f(x)}{h} = \lim\limits_{h \to 0} \dfrac{5-5}{h} = \lim\limits_{h \to 0} \dfrac{0}{h} = 0$

19. $f(x) = \dfrac{x^2 - 1}{x} = x - \dfrac{1}{x}$

$f'(x) = \lim\limits_{h \to 0} \dfrac{f(x+h) - f(x)}{h} = \lim\limits_{h \to 0} \dfrac{x+h - \dfrac{1}{x+h} - x + \dfrac{1}{x}}{h}$

$= \lim\limits_{h \to 0} \dfrac{h + \dfrac{1}{x} - \dfrac{1}{x+h}}{h}$

$= \lim\limits_{h \to 0} \left[1 + \dfrac{1}{h}\left(\dfrac{1}{x} - \dfrac{1}{x+h}\right)\right] = \lim\limits_{h \to 0} \left[1 + \dfrac{1}{h}\left(\dfrac{x+h}{x(x+h)} - \dfrac{x}{x(x+h)}\right)\right]$

$= \lim\limits_{h \to 0} \left[1 + \dfrac{1}{h}\left(\dfrac{h}{x(x+h)}\right)\right] = \lim\limits_{h \to 0} \left[1 + \left(\dfrac{1}{x(x+h)}\right)\right]$

$= 1 + \dfrac{1}{x^2}.$

21. (a) Here $s = f(t) = 4t - 9 - t^2$ so

$v = \dfrac{ds}{dt} = f'(t) = \lim\limits_{h \to 0} \dfrac{f(t+h) - f(t)}{h}$

$= \lim\limits_{h \to 0} \dfrac{4(t+h) - 9 - (t+h)^2 - 4t + 9 + t^2}{h}$

$= \lim\limits_{h \to 0} \dfrac{4t + 4h - 9 - t^2 - 2ht - h^2 - 4t + 9 + t^2}{h}$

$= \lim\limits_{h \to 0} \dfrac{4h - 2ht - h^2}{h} = \lim\limits_{h \to 0} (4 - 2t - h) = 4 - 2t.$ Thus,

when $t = 0$ sec, $s = -9''$ and $v = 4$ in/sec

when $t = 1$ sec, $s = -6''$ and $v = 2$ in/sec

when $t = 6$ sec, $s = -21''$ and $v = -8$ in/sec.

(b) When v = 0, 4 − 2t = 0 so t = 2 sec and s = 4(2) − 9 − 2^2 = −5" so the object is 5" to the <u>left</u> of the starting point when v = 0.

23. Here f(x) = 7x − x^2 − 12 = (4 − x)(x − 3) so a rough sketch of the graph is

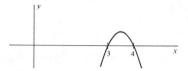

Thus, the largest value of f(x) is somewhere between 3 and 4. Also,

$$f'(x) = \lim_{h \to 0} \frac{f(x + h) - f(x)}{h}$$

$$= \lim_{h \to 0} \frac{7(x + h) - (x + h)^2 - 12 - 7x + x^2 + 12}{h}$$

$$= \lim_{h \to 0} \frac{7x + 7h - x^2 - 2xh - h^2 - 12 - 7x + x^2 + 12}{h}$$

$$= \lim_{h \to 0} \frac{7h - 2xh - h^2}{h} = \lim_{h \to 0} (7 - 2x - h)$$

$$= 7 - 2x = 2(\frac{7}{2} - x).$$

So if x < 7/2, f'(x) > 0 and the tangents to the curve have positive slope. Thus, the graph is increasing to the left of 7/2. If x > 7/2, f'(x) < 0 and the tangents to the curve have negative slope. Thus, the graph is decreasing to the right of 7/2. Therefore, the largest value of the function occurs at 7/2. Since f(7/2) = $7(\frac{7}{2}) - (\frac{7}{2})^2 - 12 = \frac{49}{2} - \frac{49}{4} - 12 = \frac{1}{4}$, the largest value of the function is $\frac{1}{4}$.

27. Here $f'(0) = \lim_{h \to 0} \frac{f(0 + h) - f(0)}{h} = \lim_{h \to 0} \frac{|0 + h| - |0|}{h} = \lim_{h \to 0} \frac{|h|}{h}.$

But $|h| = \begin{cases} -h \text{ if } h < 0 \\ h \text{ if } h > 0 \end{cases}$

so $\frac{|h|}{h} = \frac{-h}{h} = -1$ if h < 0 and $\frac{|h|}{h} = \frac{h}{h} = 1$ if h > 0.

So $f'(0) = \lim_{h \to 0} \frac{|h|}{h}$ does not exist because $\frac{|h|}{h}$ does not approach one unique value.

Technique Review Exercises, Chapter 2 (page 91)

1. (a) $\lim_{x \to -2} \frac{x^2 + x - 2}{x^2 + 5x + 6} = \lim_{x \to -2} \frac{(x + 2)(x - 1)}{(x + 2)(x + 3)} = \lim_{x \to -2} \frac{x - 1}{x + 3} = \frac{-2 - 1}{-2 + 3} = -3$

(b) $\lim_{x \to 3} f(x) = \lim_{x \to 3} \frac{(x + 2)(x - 1)}{(x + 2)(x + 3)} = \lim_{x \to 3} \frac{x - 1}{x + 3} = \frac{3 - 1}{3 + 3} = \frac{1}{3}$

2. (a) $\lim\limits_{x \to 5} f(x) = 10$ because both $2|x|$ for $x < 5$, and $x + 5$ for $x > 5$ can be made

arbitrarily close to 10 for values of x sufficiently close to 5.

(b) Since $f(x) = 2|x|$ if $x < 5$,

$$\lim\limits_{x \to 3} f(x) = \lim\limits_{x \to 3} 2|x| = 2 \lim\limits_{x \to 3} |x| = 6.$$

3. If $x \neq 3$, $f(x) = \dfrac{x^2 - 2x - 3}{x^2 - x - 6} = \dfrac{(x - 3)(x + 1)}{(x - 3)(x + 2)} = \dfrac{x + 1}{x + 2}.$

Thus, $f(x)$ is continuous for all x except $x = -2$ and possibly $x = 3$. For

$x = 3$, $f(3) = 4/5$ and

$$\lim\limits_{x \to 3} f(x) = \lim\limits_{x \to 3} \frac{x + 1}{x + 2} = \frac{3 + 1}{3 + 2} = \frac{4}{5}.$$

Since $\lim\limits_{x \to 3} f(x) = f(3)$, $f(x)$ is continuous at $x = 3$. Therefore, $f(x)$ is con-

tinuous for all x except $x = -2$.

5. The graph crosses the x-axis at 1,4 and -3 and touches without crossing at

$x = 0$. A large positive value of x makes y negative. Thus, we get the follow-

ing sketch:

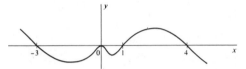

6. This is real when the expression under the radical sign is greater than or

equal to zero. Thus, it is real when $x = 1$, $x = -5$, and when the expression

under the sign is positive, and hence when

$$y = x(3 - x)(x + 2)(x - 1)^3(x + 5)^2$$

is positive. We sketch the graph of this product to get

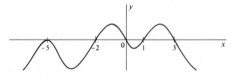

Thus, the square root is real if $x = -5$ or $-2 < x < 0$ or $1 \leq x < 3$.

7. $f'(x) = \lim\limits_{h \to 0} \dfrac{f(x + h) - f(x)}{h} = \lim\limits_{h \to 0} \dfrac{2 - 3(x + h) - (x + h)^2 - 2 + 3x + x^2}{h}$

$= \lim\limits_{h \to 0} \dfrac{-3h - 2hx - h^2}{h} = \lim\limits_{h \to 0} (-3 - 2x - h) = -3 - 2x$

8. To find the slope of the tangent at $(-1,5)$ we find

$m = f'(-1) = \lim\limits_{h \to 0} \dfrac{f(-1 + h) - f(-1)}{h}$

$= \lim\limits_{h \to 0} [\frac{1}{h} (2 - \frac{3}{-1 + h} - 2 + \frac{3}{-1})]$

$$= \lim_{h \to 0} [\tfrac{1}{h} (\tfrac{3}{1-h} - 3)] = \lim_{h \to 0} [\tfrac{1}{h} (\tfrac{3}{1-h} - \tfrac{3-3h}{1-h})]$$

$$= \lim_{h \to 0} [\tfrac{1}{h} \tfrac{3h}{1-h}] = \lim_{h \to 0} \tfrac{3}{1-h} = 3.$$

Thus, an equation of the tangent at $(-1,5)$ is $y - 5 = 3(x + 1)$ or $y = 3x + 8$.

9. Since $s = 100t - t^2$, the velocity v is

$$v = \frac{ds}{dt} = \lim_{h \to 0} \frac{100(t + h) - (t + h)^2 - 100t + t^2}{h}$$

$$= \lim_{h \to 0} \frac{100t + 100h - t^2 - 2th - h^2 - 100t + t^2}{h}$$

$$= \lim_{h \to 0} \frac{100h - 2th - h^2}{h} = \lim_{h \to 0} (100 - 2t - h)$$

$$= 100 - 2t.$$

(a) So when $t = 5$ sec, the velocity is $100 - 2(5) = 90$ ft/sec.

(b) Since $v = 100 - 2t = 2(50 - t)$, we see that $v > 0$ when $t < 50$ and $v = 0$ when $t = 50$. Thus, the object starts to move downward when $t = 50$ sec.

(c) Since the object starts downward when $t = 50$ sec, its highest point occurs at $t = 50$ sec. Here

$$s = 100(50) - 50^2 = 5,000 - 2,500 = 2,500 \text{ ft}$$

so the object goes up to a height of 2,500 ft.

Additional Exercises, Chapter 2 (page 92)

1. $m(h) = \dfrac{3 - (2 + h)^2 - 3 + 2^2}{h} = \dfrac{3 - 4 - 4h - h^2 - 3 + 4}{h} = \dfrac{-4h - h^2}{h} = -4 - h.$

So $m(h) \to -4$ as $h \to 0$ and $m = -4$. Thus, an equation of the tangent at $(2,-1)$ is $y + 1 = -4(x - 2)$ or $4x + y = 7$.

5. $m(h) = \dfrac{3 - 3}{h} = \dfrac{0}{h} = 0$ so $m = 0$. Thus, an equation of the tangent at $(5,3)$ is $y - 3 = 0(x - 5)$ or $y = 3$.

9. $\lim_{x \to 0} \dfrac{x^2 - 6x}{3x} = \lim_{x \to 0} \dfrac{x - 6}{3} = \dfrac{-6}{3} = -2.$

11. $\lim_{x \to 16} \dfrac{\sqrt{x} - 4}{x - 16} = \lim_{x \to 16} \dfrac{\sqrt{x} - 4}{(\sqrt{x} - 4)(\sqrt{x} + 4)} = \lim_{x \to 16} \dfrac{1}{\sqrt{x} + 4} = \dfrac{1}{8}$

13. $\lim_{x \to -4} \dfrac{x + 4}{x^2 + 3x - 4} = \lim_{x \to -4} \dfrac{x + 4}{(x + 4)(x - 1)} = \lim_{x \to -4} \dfrac{1}{x - 1} = -\dfrac{1}{5}$

15. Here $f(x)$ can be made arbitrarily close to both 42 and 48 for values of x sufficiently close to 6 but not equal to 6. Thus, $\lim_{x \to 6} f(x)$ does not exist.

17. Here $f(x) = \begin{cases} \dfrac{1 - x^2}{x - 1} = \dfrac{(1 - x)(1 + x)}{x - 1} = -x - 1 \text{ if } x \neq 1 \\ 2 \text{ if } x = 1 \end{cases}$

Since x must be $\neq 1$ in the consideration of $\lim\limits_{x \to 1} f(x)$, we only need to consider

$f(x) = -x - 1$ for $x \neq 1$. Hence,

$$\lim_{x \to 1} f(x) = -1 - 1 = -2.$$

19. $\lim\limits_{x \to 4} [(x^2 + 3)(x - 1)] = \lim\limits_{x \to 4} (x^2 + 3) \lim\limits_{x \to 4} (x - 1)$

$$= [(\lim_{x \to 4} x)(\lim_{x \to 4} x) + \lim_{x \to 4} 3][\lim_{x \to 4} x - \lim_{x \to 4} 1]$$

$$= [(4)(4) + 3][4 - 1] = 19(3) = 57$$

23. $m = \lim\limits_{h \to 0} \dfrac{f(2 + h) - f(2)}{h} = \lim\limits_{h \to 0} \dfrac{2(2 + h) - (2 + h)^2 - 2(2) + 2^2}{h}$

$$= \lim_{h \to 0} \frac{4 + 2h - 4 - 4h - h^2 - 4 + 4}{h} = \lim_{h \to 0} \frac{-2 - h^2}{h} = \lim_{h \to 0} (-2 - h) = -2.$$

Thus, an equation of the tangent at $(2,0)$ is $y = -2(x - 2)$ or $y = -2x + 4$.

27. Being a rational function, this is continuous for all values of x except those where the denominator is zero. Since $x^2 + 2x - 8 = (x + 4)(x - 2)$, the function is continuous for all values of x except $x = -4$ and $x = 2$.

29. $\lim\limits_{x \to -4} \dfrac{x + 4}{x^2 + 2x - 8} = \lim\limits_{x \to -4} \dfrac{x + 4}{(x + 4)(x - 2)} = \lim\limits_{x \to -4} \dfrac{1}{x - 2} = -\dfrac{1}{6}$, so define

$f(-4) = -\dfrac{1}{6}.$

31. $\lim\limits_{x \to 2} \dfrac{x - 2}{x^2 - 4x + 4} = \lim\limits_{x \to 2} \dfrac{x - 2}{(x - 2)(x - 2)} = \lim\limits_{x \to 2} \dfrac{1}{x - 2}.$

Since this limit does not exist it is impossible to define the function so that it is continuous at $x = 2$.

35. The only values for which the square root is zero are $x = -5$ and $x = 1$. We now solve

$$\frac{(x + 5)(1 - x)}{(1 + x)(x - 3)} > 0$$

by solving $y = (x + 5)(1 - x)(1 + x)(x - 3) > 0$. Sketching we get

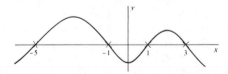

Thus, $y > 0$, and the square root is positive for $-5 < x < -1$ or $1 < x < 3$.

Thus, the square root is real for $-5 \leq x < -1$ or $1 \leq x < 3$.

41. We sketch $y = x^3 + 2x^2 - 8x = x(x^2 + 2x - 8) = x(x + 4)(x - 2)$ to obtain

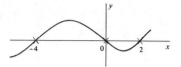

Thus, we have (a) $-4 < x < 0$ or $x > 2$, (b) $x < -4$ or $0 < x < 2$.

45. $f'(x) = \lim_{h \to 0} \dfrac{7(x + h) - 2(x + h)^2 - 5 - 7x + 2x^2 + 5}{h}$

$\quad = \lim_{h \to 0} \dfrac{7x + 7h - 2x^2 - 4xh - 2h^2 - 5 - 7x + 2x^2 + 5}{h}$

$\quad = \lim_{h \to 0} \dfrac{7h - 4xh - 2h^2}{h} = \lim_{h \to 0} (7 - 4x - 2h) = 7 - 4x.$

47. $f'(x) = \lim_{h \to 0} \dfrac{f(x + h) - f(x)}{h} = \lim_{h \to 0} \dfrac{17 - 17}{h} = \lim_{h \to 0} \dfrac{0}{h} = \lim_{h \to 0} 0 = 0$

49. $f'(x) = \lim_{h \to 0} \dfrac{1}{h} [\dfrac{1}{3(x + h)} - \dfrac{1}{3x}] = \lim_{h \to 0} \dfrac{1}{h} [\dfrac{3x - 3(x + h)}{3(x + h)(3x)}]$

$\quad = \lim_{h \to 0} \dfrac{1}{h} [\dfrac{3x - 3x - 3h}{3(x + h)(3x)}] = \lim_{h \to 0} \dfrac{1}{h} [\dfrac{-3h}{3(x + h)(3x)}]$

$\quad = \lim_{h \to 0} \dfrac{-3}{3(x + h)(3x)} = \dfrac{-1}{x(3x)} = - \dfrac{1}{3x^2}.$

CHAPTER 3: THE DERIVATIVE

Exercises 3.1: The Derivative of a Polynomial (page 101)

11. $(x^2 + 8)(x^3 - 2) = x^5 + 8x^3 - 2x^2 - 16$ so the derivative is $5x^4 + 24x^2 - 4x$
15. $x^3(x - 3) = x^4 - 3x^3$ so the derivative is $4x^3 - 9x^2$

Exercises 3.2: The Product and Quotient Rules (page 106)

5. Use of the product rule gives $(x^5 + 3x)(2x) + (x^2 - 1)(5x^4 + 3)$
    $= 7x^6 - 5x^4 + 9x^2 - 3.$
9. We get $(7x - 3x^{-1})(6x^2) + (2x^3 + 4)(7 + 3x^{-2})$
    $= 42x^3 - 18x + 14x^3 + 28 + 6x + 12x^{-2} = 56x^3 - 12x + 12x^{-2} + 28$
11. Since $(x^3 + x)^2 = x^6 + 2x^4 + x^2$ we get $6x^5 + 8x^3 + 2x$ for the derivative.
15. Use of the quotient rule gives

$$\dfrac{(x - 3)(1) - (x + 5)(1)}{(x - 3)^2} = \dfrac{-8}{(x - 3)^2}$$

17. Use of the quotient rule gives

$$\dfrac{(2x^4 + 1)(0) - 1(8x^3)}{(2x^4 + 1)^2} = \dfrac{-8x^3}{(2x^4 + 1)^2}$$

21. We get $\dfrac{(x - 1)(2x + 3) - (x^2 + 3x + 2)(1)}{(x - 1)^2} = \dfrac{x^2 - 2x - 5}{(x - 1)^2}$

23. Since $(x + 2)(x - 3)(x + 5) = (x^2 - x - 6)(x + 5)$ we get

$(x^2 - x - 6)(1) + (x + 5)(2x - 1) = 3x^2 + 8x - 11.$

25. Since $\dfrac{dy}{dx} = 2x + 5$ we see that $\dfrac{dy}{dx} - 2x = 2x + 5 - 2x = 5$, so the equation is satisfied.

29. Since $y' = -kx^{-2} + 3$

$$y + xy' = kx^{-1} + 3x + x(-kx^{-2} + 3)$$
$$= kx^{-1} + 3x - kx^{-1} + 3x = 6x \text{ so the equation is}$$

satisfied.

Exercises 3.3: The Chain Rule (page 111)

1. $\dfrac{dy}{dx} = \dfrac{dy}{du}\dfrac{du}{dx} = (6u - 5)(2x) = [6(x^2 - 1) - 5](2x)$

$= 12x(x^2 - 1) - 10x = 12x^3 - 22x$

3. $\dfrac{dy}{dx} = \dfrac{dy}{du} \cdot \dfrac{du}{dx} = (20u^4 - 7 - 2u^{-3})(1 + 2x)$

$= [20(x + x^2)^4 - 7 - 2(x + x^2)^{-3}](1 + 2x)$

7. $\dfrac{dy}{dx} = \dfrac{dy}{du}\dfrac{du}{dx} = \dfrac{(u^2 - 1)\,3u^2 - (u^3 + 3)(2u)}{(u^2 - 1)^2} \ (3)(x^2 + 4)^2(2x)$

$= 6x(x^2 + 4)^2\,\dfrac{u^4 - 3u^2 - 6u}{(u^2 - 1)^2}$

$= 6x(x^2 + 4)^2\,\dfrac{(x^2 + 4)^{12} - 3(x^2 + 4)^6 - 6(x^2 + 4)^3}{((x^2 + 4)^6 - 1)^2}$

$= 6x(x^2 + 4)^5\,\dfrac{(x^2 + 4)^9 - 3(x^2 + 4)^3 - 6}{((x^2 + 4)^6 - 1)^2}$

15. $\dfrac{dy}{dx} = 4\left(\dfrac{x + 2}{x - 3}\right)^3 \dfrac{d}{dx}\left(\dfrac{x + 2}{x - 3}\right) = 4\left(\dfrac{x + 2}{x - 3}\right)^3\left[\dfrac{(x - 3)(1) - (x + 2)(1)}{(x - 3)^2}\right]$

$= 4\left(\dfrac{x + 2}{x - 3}\right)^3\left[\dfrac{-5}{(x - 3)^2}\right] = \dfrac{-20(x + 2)^3}{(x - 3)^5}$

17. $\dfrac{dy}{dx} = -6\left(\dfrac{x^{-1} + 3x^{-2}}{4x + 5}\right)^{-7}\dfrac{(4x + 5)(-x^{-2} - 6x^{-3}) - (x^{-1} + 3x^{-2})4}{(4x + 5)^2}$

$= -6\,\dfrac{(x^{-1} + 3x^{-2})^{-7}}{(4x + 5)^{-5}}\,(-8x^{-1} - 41x^{-2} - 30x^{-3})$

$= \dfrac{6(4x + 5)^5}{x^3(x^{-1} + 3x^{-2})^7}\,(8x^2 + 41x + 30)$

21. $\dfrac{dy}{dx} = 3[(3x^2 + 5x)(x - 5)]^2[(6x + 5)(x - 5) + 3x^2 + 5x]$

$= 3[(3x^2 + 5x)(x - 5)]^2(9x^2 - 20x - 25)$

23. $\dfrac{dy}{dx} = (x^2 + 2)^4(2)(x^3 - 1)(3x^2) + 4(x^2 + 2)^3(2x)(x^3 - 1)^2$

$= 6x^2(x^2 + 2)^4(x^3 - 1) + 8x(x^2 + 2)^3(x^3 - 1)^2$

27. $\dfrac{dy}{dx} = \dfrac{dy/dt}{dx/dt} = \dfrac{4}{3t^2 - 7}$

31. $\dfrac{dy}{dx} = \dfrac{dy/dt}{dx/dt} = \dfrac{5(u^3 + 4u)^4(3u^2 + 4)}{4(u - 7u^{-2})^3(1 + 14u^{-3})}$

33. Using $[f \circ g]'(x) = [f(g(x))]' = f'(g(x))g'(x)$ we get $[f \circ g]'(2) = f'(g(2))g'(2)$. But $f'(u) = 2u$ so $f'(g(2)) = f'(3) = 2(3) = 6$, and $g'(2) = -1$, so $[f \circ g]'(2) = 6(-1) = -6$.

35. $[f \circ g]'(a) = f'(g(a))g'(a)$. But $f'(u) = -12u^3 - u^{-2}$ so
$[f \circ g]'(a) = [-12(2)^3 - (2)^{-2}](-3) = \dfrac{1155}{4}$.

37. $y' = 4(x^2 + 5x + 2)^3(2x + 5)$ so at $(0,16)$, $m = 4(2)^3(5) = 160$. Thus, an equation of the tangent at $(0,16)$ is $y - 16 = 160x$ or $y = 160x + 16$.

39. $\dfrac{dy}{dx} = \dfrac{dy/dv}{dx/dv} = \dfrac{-(3v^2 - 2v)^{-2}(6v - 2)}{8(4v - 2)}$

Thus, when $v = 1$ we have $\dfrac{dy}{dx} = m = -\dfrac{1}{4}$, that is the tangent line must have slope $-1/4$. Also, since $x = 4$ and $y = 1$ when $v = 1$, the tangent line must pass through the point $(4,1)$. Consequently, an equation for the tangent line is $y - 1 = -\dfrac{1}{4}(x - 4)$ or $x + 4y = 8$.

41. The tangent is horizontal for the values of x for which $y' = 0$. Since $y' = 5(x^2 - 4)^4(2x)$, these values are $x = 0$, $x = 2$, and $x = -2$.

43. $\dfrac{dT(x)}{dt} = \dfrac{dT(x)}{dt}\dfrac{dx}{dt} = (2x - 3)\dfrac{dx}{dt}$ so when $x = 0$ and $\dfrac{dx}{dt} = -5$, the rate of change the temperature with respect to the time is $(-3)(-5) = 15$ units/sec.

<u>Exercises 3.4</u>: Derivatives and Curve Sketching (page 119)

For Exercises 1-6 find y''.

1. Since $y' = 3x^2 - 10x + 8$, $y'' = 6x - 10$.

2. Since $y' = \dfrac{(x + 1)(1) - (x - 2)(1)}{(x + 1)^2} = \dfrac{3}{(x + 1)^2} = 3(x + 1)^{-2}$,
$y'' = -6(x^2 + 1)^{-3}$

7. Since $y' = 2x - 4 = 2(x - 2)$, $y' > 0$ if $x > 2$ and $y' < 0$ if $x < 2$. Thus, we get (a) $x > 2$, (b) $x < 2$. Since $y'' = 2 > 0$ for all x, we get (c) all x, (d) no x. The only critical point occurs at $(2,-2)$. Since y'' is never zero there are no points of inflection. A sketch of the graph is in the answer section of the text.

9. Since $y' = -6 - 2x = -2(x + 3)$, $y' > 0$ if $x < -3$ and $y' < 0$ if $x > -3$. Thus, we get (a) $x < -3$, (b) $x > -3$. Since $y' = -2 < 0$ for all x we get (c) no x, (d) all x. The only critical point is at $(-3,12)$. Since y'' is not zero there are no points of inflection. A sketch of the graph is in the answer section of the text.

11. $y = 12x - x^3 + 1$. Since $y' = 12 - 3x^2 = -3(x^2 - 4) = -3(x - 2)(x + 2)$, $y' > 0$ if $-2 < x < 2$ and $y' < 0$ if $x < -2$ or $x > 2$. Thus, we get (a) $-2 < x < 2$, (b) $x < -2$ or $x > 2$. Since $y'' = -6x$, we have $y'' > 0$ if $x < 0$ and $y'' < 0$ if $x > 0$. Thus, we get (c) $x < 0$, (d) $x > 0$. Critical points are at $(-2,-15)$ and $(2,17)$ and a point of inflection occurs at $(0,1)$. A sketch of

the graph is in the answer section of the text.

15. $y = 3x^4 - 4x^3 + 2$. Since $y' = 12x^3 - 12x^2 = 12x^2(x - 1)$, $y' > 0$ if $x > 1$ and $y' < 0$ if $x < 0$ or $0 < x < 1$. Thus, we get (a) $x > 1$, (b) $x < 0$ or $0 < x < 1$. Since $y'' = 36x^2 - 24x = 36x(x - 2/3)$, $y' > 0$ if $x < 0$ or $x > 2/3$ and $y' < 0$ if $0 < x < 2/3$. Thus, we get (c) $x < 0$ or $x > 2/3$, (d) $0 < x < 2/3$. Critical points occur at $(0,2)$ and $(1,1)$ and points of inflection are at $(0,2)$ and $(2/3, 38/27)$ and a sketch of the graph is in the answer section of the text.

21. $y = \dfrac{x - 2}{x + 1}$. Since $y' = \dfrac{(x + 1)(1) - (x - 2)(1)}{(x + 1)^2} = \dfrac{3}{(x + 1)^2}$, $y' > 0$ for all $x \neq -1$ and we have (a) all $x \neq -1$, (b) no x. Since $y'' = \dfrac{-6}{(x + 1)^3}$, $y'' > 0$ if $x < -1$ and $y'' < 0$ if $x > -1$ so we get (c) $x < -1$, (d) $x > -1$. There are no critical points or inflection points but y is not defined for $x = -1$. A sketch of the graph is in the answer section of the text.

29. We are to show that R is a decreasing function of p if $E < -1$. That is, we are to show that
$$\frac{dR}{dp} < 0 \text{ if } E < -1.$$
Now, since $R = px$, $\dfrac{dR}{dp} = p\dfrac{dx}{dp} + x$.

But if $E < -1$ then $\dfrac{p}{x}\dfrac{dx}{dp} < -1$ or $\dfrac{dx}{dp} < -\dfrac{x}{p}$.

Thus, $\dfrac{dR}{dp} = p\dfrac{dx}{dp} + x < p\left(-\dfrac{x}{p}\right) + x = -x + x = 0$.

That is, if $E < -1$, $\dfrac{dR}{dp} < 0$.

Exercises 3.5: Maximum–Minimum Problems (page 127)

1. Since $f'(x) = 6x - 12 = 6(x - 2)$, $f'(x) = 0$ only at $x = 2$ and there are no endpoints. Thus, $(2,-11)$ is the only point where local maxima or minima can occur. Now, $f''(x) = 6$ so $f''(2) = 6 > 0$ and $(2,-11)$ is a local minimum point and there are no local maximum points.

3. Since $f'(x) = 2x - 4 = 2(x - 2)$, $f'(x) > 0$ and hence $f(x)$ is increasing for $3 \leq x \leq 5$. Thus, $(3,-1)$ is a local minimum and $(5,7)$ is a local maximum, and these are the only local extremes for $3 \leq x \leq 5$.

5. Since $f'(x) = 2x - 4 = 2(x - 2)$ local extrema can occur only at the endpoints $(0,2)$ and $(3,-1)$ or at $(2,-2)$ where $f'(x) = 0$. Also, $f'(x) < 0$ and hence $f(x)$ is decreasing for $x < 2$ and $f'(x) > 0$ and so $f(x)$ is increasing for $x > 2$. Thus, $(2,-2)$ is a local minimum while $(0,2)$ and $(3,-1)$ are local maxima.

9. Since $f'(x) = -4x^3 + 4x = -4(x^2 - 1) = -4x(x - 1)(x + 1)$, a sketch of the graph of $f'(x)$ is

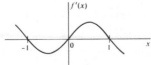

Thus, $f'(x) > 0$, and hence $f(x)$ is increasing if $x < -1$ or $0 < x < 1$ and $f'(x) < 0$ and so $f(x)$ is decreasing if $-1 < x < 0$ or $x > 1$. Thus, $(-3,-62)$, $(0,1)$, and $(3,-62)$ are local minima while $(-1,2)$ and $(1,2)$ are local maxima.

11. $x^{-2} - x^{-1}$, $1 \le x \le 3$. Since $f'(x) = -2x^{-3} + x^{-2} = \dfrac{-2}{x^3} + \dfrac{1}{x^2} = \dfrac{x-2}{x^3}$, local

extrema can occur only at the endpoints $(1,0)$ and $(3,-2/9)$ or at $(2,-\frac{1}{4})$ where $f'(x) = 0$. Also, $f'(x) > 0$ and hence $f(x)$ is increasing for $x < 0$ or $x > 2$ and $f'(x) < 0$ and so $f(x)$ is decreasing for $0 < x < 2$. Thus, $(1,0)$ and $(3,-2/9)$ are local maxima while $(2,-\frac{1}{4})$ is a local minimum point.

13. $x^{-1} + 4$. Since $f'(x) = -x^{-2} = \dfrac{-1}{x^2} \ne 0$, the only possible place where a local extrema could exist is where $f'(x)$ does not exist, namely at $x = 0$. But, $f(x)$ does not exist at $x = 0$ either so the function has no local maximum or minimum.

15. Since $f(x) = |x| = \begin{cases} x & \text{if } x > 0 \\ -x & \text{if } x < 0 \end{cases}$, $f'(x) = \begin{cases} 1 & \text{if } x > 0 \\ -1 & \text{if } x < 0 \end{cases}$.

Thus, the only possible point where a local maximum or minimum could exist is at $(0,0)$ where $f'(x)$ does not exist. Since $f'(x) = 1 > 0$ if $x > 0$ and $f'(x) = -1 < 0$ for $x < 0$, $f(x)$ is increasing if $x > 0$ and decreasing if $x < 0$. Thus, $(0,0)$ must be a local minimum point.

Exercises 3.6: Absolute Maxima and Minima, The Mean Value Theorem
(page 132)

1. Since the function is continuous on the closed interval it must have absolute maximum and minimum points. Each absolute extrema can only occur at a point $(x,f(x))$ where $f'(x) = 0$, $f'(x)$ does not exist or at one of the endpoints. Since $f'(x) = 2x - 6 = 2(x - 3)$ the absolute extrema can occur only at $(2,-9)$, $(3,-10)$, or $(5,-6)$. Thus, the absolute maximum is at $(5,-6)$ and the absolute minimum is at $(3,-10)$.

3. As in Exercise 1, $f'(x) = 2(x - 3)$. So $f'(x) \ne 0$ for $0 \le x \le 2$. Thus, the absolute maximum and minimum points can only occur at $(0,-1)$ or $(2,-9)$. Thus, $(0,-1)$ is the absolute maximum point and $(2,-9)$ is the absolute minimum point.

7. Being continuous for $-4 \le x \le 4$, the function must have absolute extrema. Since $f'(x) = 3x^2 + 12x + 12 = 3(x + 2)^2$, the possible points are $(-4,-16)$, $(-2,-8)$, and $(4,208)$. Thus, the absolute maximum point is at $(4,208)$ and the absolute minimum point is at $(-4,-16)$.

11. This function is not continuous at $x = 0$ and so may not have absolute maximum and minimum points. Since $f'(x) = -2x^{-3}$, absolute extrema can only occur where $x = -1$, $x = 0$, or $x = 3$. Since the function is not defined at $x = 0$, that value of x must be eliminated leaving only the points $(-1,1/2)$ and $(3,1/9)$. Since $f'(x) > 0$ for $x < 0$ and $f'(x) < 0$ for $x > 0$, the function is increasing for $x < 0$ and decreasing for $x > 0$. Thus, $(3,1/9)$ must be the absolute minimum point. There is no absolute maximum point because x^{-2} gets arbitrarily large for values of x near zero.

17. Since the function is not restricted to a closed interval, absolute extrema may not occur. Neither an absolute maximum value nor an absolute minimum value exists because the function assumes both positive and negative values whose absolute values are arbitrarily large.

19. We are to find all values of c such that
$$f'(c) = \frac{f(9) - f(1)}{9 - 1} = \frac{243 - 45 + 7 - 5}{8} = \frac{200}{8} = 25.$$
Since $f'(x) = 6x - 5$, we must have $f'(c) = 6c - 5 = 25$, or $c = 5$.

23. Let s be the distance in miles travelled by the driver from the toll road entrance in t hours. Then $s = f(t)$ where we can assume that f is continuous for the 240 mile interval, that is for $0 \le t \le 8/3$. By the mean value theorem, there must be at least one value c, $0 \le c \le 8/3$ such that
$$f'(c) = \frac{f(8/3) - f(0)}{8/3 - 0} = \frac{240 - 0}{8/3 - 0} = 90.$$
But $f'(c)$ is the velocity of the driver when $t = c$ hours. Thus, there must be at least one point of his trip where he was going at exactly 90 mph.

Exercises 3.7: Applied Maximum-Minimum Problems (page 136)

1. Neither of the numbers can be negative since they must have a positive sum and produce a maximum (and hence positive) product. Neither of the numbers can exceed their sum 30 so let x be one of the numbers where $x \le 30$. Then we have $0 \le x \le 30$ and the other number must be $30 - x$. Their product $p(x)$ is thus $p(x) = x(30 - x) = 30x - x^2$, $0 \le x \le 30$. Being continuous for $0 \le x \le 30$, $p(x)$ must have an absolute maximum value that must occur at a point where $p'(x) = 0$ or $p'(x)$ does not exist or at an endpoint 0 or 30. Since $p'(x) = 30 - 2x = 2(15 - x)$, the possible points occur at $x = 0$, 15, or 30, that is, at (0,0), (15,225), or (30,0). Thus, the absolute maximum value of their product is 225, which happens when $x = 15$.

3. Let x = the indicated length.

Then $0 \le x \le 400$ and the width is $\dfrac{1200 - 3x}{2}$. Thus, the area $A(x)$ is given by
$$A(x) = x(\frac{1200 - 3x}{2}) = \frac{1200x - 3x^2}{2}, \quad 0 \le x \le 400.$$
Since $A'(x) = 600 - 3x = 3(200 - x)$, the absolute maximum of the continuous function $A(x)$ on the closed interval $0 \le x \le 400$ must occur at $x = 0$, $x = 200$, or $x = 400$. Since $A(0) = 0$, $A(400) = 0$, and $A(200) = 60,000$, the largest area 60,000 occurs when $x = 200$ and so the dimensions of the largest rectangle are 200 m x 300 m.

5. Let x be the number of trees exceeding 16 per acre. Then the number of grapefruit produced per tree is $200 - 10x$, $0 \le x \le 20$. Thus, the total number $g(x)$ of grapefruit produced per year per acre is

$g(x) = (x + 16)(200 - 10x) = -10x^2 + 40x + 1600, \ 0 \le x \le 20.$

Since $g'(x) = 40 - 20x$, the absolute maximum value of the continuous function $g(x)$ on the closed interval $0 \le x \le 20$ must occur at $x = 2$, $x = 0$, or $x = 20$. Since $g(2) = 3240$, $g(0) = 1600$, and $g(20) = 0$, the largest yield occurs when $x = 2$. Thus, the grower should plant $16 + 2 = 18$ trees per acre to produce the most grapefruit.

7. Let r be the radius of the circular cross section. Then the girth is $2\pi r$ and the length is $78 - 2\pi r$, $0 \le r \le 78/2\pi$. Volume $V(r)$ is
$$V(r) = \pi r^2 (78 - 2\pi r) = 78\pi r^2 - 2\pi^2 r^3, \ 0 \le r \le 78/2\pi.$$
Since $V'(r) = 156\pi r - 6\pi^2 r^2 = 6\pi^2 r \left(\frac{26}{\pi} - r\right)$, the absolute maximum value of $V(r)$ on the closed interval $0 \le r \le 78/2\pi$ must occur at $r = 0$, $r = \frac{26}{\pi}$, or $r = 78/2\pi = 39/\pi$. Since $V(0) = 0$, $V(\frac{26}{\pi}) = \frac{26^3}{\pi}$, and $V(\frac{78}{2\pi}) = 0$, the largest volume occurs when $r = \frac{26''}{\pi}$. Thus, the dimensions of the largest mailable package of this sort are radius $\frac{26''}{\pi}$ and length $78 - 2\pi(\frac{26}{\pi}) = 26''$.

11. There are 3 cases we must consider as shown below. The lengths indicated for the parts of the iguana pen follow from the figure and the fact that the sum of these lengths must be 42. Let $A(x) =$ the area of the iguana pen.

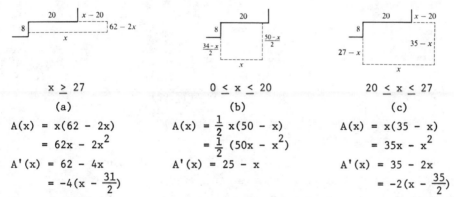

$x \ge 27$	$0 \le x \le 20$	$20 \le x \le 27$
(a)	(b)	(c)

$A(x) = x(62 - 2x)$ $A(x) = \frac{1}{2} x(50 - x)$ $A(x) = x(35 - x)$

$\quad\ = 62x - 2x^2$ $\quad\ = \frac{1}{2}(50x - x^2)$ $\quad\ = 35x - x^2$

$A'(x) = 62 - 4x$ $A'(x) = 25 - x$ $A'(x) = 35 - 2x$

$\quad\ = -4(x - \frac{31}{2})$ $\quad\ = -2(x - \frac{35}{2})$

None of the derivatives is zero at a point for which the function describes the situation. Therefore, the values of x for which $A' = 0$ need not be considered further. For case (a) $x \ge 27$, $A'(x) < 0$ so the function is decreasing and we need only consider $A(27) = 216$ for the possible absolute maximum value of $A(x)$. In cases (b) and (c) we have closed intervals and (since $A'(x) \ne 0$ for the intervals) we need only consider $A(x)$ at the endpoints of the intervals as candidates for the absolute maximum area. Thus, we find for case (b) $A(0) = \frac{1}{2}(0)(50) = 0$, $A(20) = \frac{1}{2}(20)(30) = 300$, and for case (c) $A(20) = 20(15) = 300$, $A(27) = 27(8) = 216$. Thus, the largest area occurs when $x = 20'$ so the dimensions of the largest area are $20' \times 15'$.

13. Here $x > 0$ and $C'(x) = -ax^{-2} + k = \dfrac{k(x^2 - \frac{a}{k})}{x^2} = \dfrac{k(x + \sqrt{\frac{a}{k}})(x - \sqrt{\frac{a}{k}})}{x^2}$.

Thus, $C'(x) > 0$ if $x > \sqrt{\frac{a}{k}}$ and $C'(x) < 0$ if $x < \sqrt{\frac{a}{k}}$ and so $C(x)$ is increasing

if $x > \sqrt{\frac{a}{k}}$ and $C(x)$ is decreasing if $x < \sqrt{\frac{a}{k}}$. Therefore, the smallest value of

$C(x)$ occurs at $x = \sqrt{\frac{a}{k}}$ months.

15. The revenue for x items is $x(3 - \dfrac{x}{20,000}) = 3x - \dfrac{x^2}{20,000}$. Thus, the profit $P(x)$

is $P(x) = 3x - \dfrac{x^2}{20,000} - \dfrac{x}{2} - 400 = \dfrac{5x}{2} - \dfrac{x^2}{20,000} - 400$, $x \geq 0$. Thus,

$P'(x) = \dfrac{5}{2} - \dfrac{x}{10,000} = \dfrac{-1}{10,000}(x - 25,000)$. So $P'(x) > 0$ if $x < 25,000$ and

$P'(x) < 0$ if $x > 25,000$ and therefore $P(x)$ is increasing for $x < 25,000$ and

decreasing for $x > 25,000$. Hence, the largest profit occurs when $25,000$ items

are produced per month.

21. Let n be the number of master copies made. Since there are 10 duplicating

machines, we must have $1 \leq n \leq 10$. Then, the number of copies that can be

made per hour is $5000n$ and so the number of hours the n machines must run to

make $81,000$ copies is

$$\frac{81,000}{5000n} = \frac{81}{5n}.$$

Thus, the cost $C(n)$ of making the $81,000$ copies is

$$C(n) = n(1) + \frac{81}{5n}(5 + .03n)$$

$$= n + 81n^{-1} + \frac{81(.03)}{5}, \quad 1 \leq n \leq 10.$$

Since $C'(n) = 1 - 81n^{-2} = \dfrac{n^2 - 81}{n^2} = \dfrac{(n - 9)(n + 9)}{n^2}$

the absolute minimum value of $C(n)$ for $1 \leq n \leq 10$ must occur for $n = 1$, $n = 9$,

or $n = 10$. Substituting in $C(n)$ we get $C(1) = 82.486$, $C(9) = 18.486$, and

$C(10) = 18.586$. Thus, they should make 9 master copies to minimize the cost.

23. The situation under consideration may be pictured as shown.

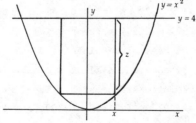

With the dimensions as indicated the perimeter of the rectangle is $P = 4x + 2z$.

Note that $z = 4 - x^2$. Thus, $P = 4x + 8 - 2x^2$ and $\dfrac{dP}{dx} = 4 - 4x$. $\dfrac{dP}{dx} = 0$ when

$x = 1$ and $\dfrac{d^2P}{dx^2} = -2 < 0$. Thus, the maximum value of P occurs when $x = 1$. Con-

sequently the dimensions of the desired rectangle are 2 units by 3 units.

25. Let r be the radius and h be the height of the can. Then, the volume V is
$V = \pi r^2 h = (1.8)(12) = 21.6 \text{ in}^3$.

So $\qquad\qquad\qquad\qquad\qquad h = \dfrac{21.6}{\pi r^2}$.

The cost of the can is $C(r) = (.0002)\pi r^2 + 2\pi rh\,(.0001) + (.0001)\pi r^2$

$$= (.0003)\pi r^2 + \frac{2\pi r(.0001)(21.6)}{\pi r^2}$$

$$= (.0003)\pi r^2 + (.0002)(21.6)r^{-1}.$$

Then $C'(r) = (.0006)\pi r - (.0002)(21.6)r^{-2} = \dfrac{.0006\pi r^3 - .00432}{r^2}$.

So $C'(r) = 0$ when $r = \sqrt[3]{\dfrac{7.2}{\pi}}$ and $C'(r) < 0$ if $r < \sqrt[3]{\dfrac{7.2}{\pi}}$ and $C'(r) > 0$ if

$r > \sqrt[3]{\dfrac{7.2}{\pi}}$. Thus, the minimum cost of the can occurs when $r = \sqrt[3]{\dfrac{7.2}{\pi}} \approx 1.32"$

and $h = \dfrac{21.6}{\pi r^2} \approx 3.95"$.

Exercises 3.8: Implicit Differentiation (page 144)

1. $3x^2 + 5y^4 y' = 0$ so $y' = -\dfrac{3x^2}{5y^4}$

3. $2x^2 y\, y' + 2xy^2 + 9xy^2 y' + 3y^3 = 0$

$(2x^2 y + 9xy^2)y' = -2xy^2 - 3y^3$

$$y = \frac{-2xy^2 - 3y^3}{2x^2 y + 9xy^2} = \frac{-2xy - 3y^2}{2x^2 + 9xy}$$

5. $\dfrac{1}{2}(xy)^{-1/2}(xy' + y) = 3x^2 y^2 y' + 2xy^3$.

Multiplying by $2(xy)^{1/2} = 2x^{1/2}y^{1/2}$, we get

$$xy' + y = 6x^{5/2}y^{5/2}y' + 4x^{3/2}y^{7/2}$$

$$(x - 6x^{5/2}y^{5/2})y' = 4x^{3/2}y^{7/2} - y$$

$$y' = \frac{4x^{3/2}y^{7/2} - y}{x - 6x^{5/2}y^{5/2}}$$

7. Differentiation produces $x^2(-3y^{-4}y') + 2xy^{-3} = y'$

$$2xy^{-3} = (1 + 3x^2 y^{-4})y'$$

$$y' = \frac{2xy^{-3}}{1 + 3x^2 y^{-4}} = \frac{2xy}{y^4 + 3x^2}$$

11. $(x^4 - y^5)^7 = (x^3 + y)^5$

$7(x^4 - y^5)^6(4x^3 - 5y^4 y') = 5(x^3 + y)^4(3x^2 + y')$

$28x^3(x^4 - y^5)^6 - 35y^4(x^4 - y^5)^6 y' = 15x^2(x^3 + y)^4 + 5(x^3 + y)^4 y'$

$28x^3(x^4 - y^5)^6 - 15x^2(x^3 + y)^4 = [35y^4(x^4 - y^5)^6 + 5(x^3 + y)^4]y'$

$$y' = \frac{28x^3(x^4 - y^5)^6 - 15x^2(x^3 + y)^4}{35y^4(x^4 - y^5)^6 + 5(x^3 + y)^4}$$

15. $(x^3y^{-2} + xy^2)^3(xy + 2)^2 = 2$

$3(x^3y^{-2} + xy^2)^2(3x^2y^{-2} - 2x^3y^{-3}y' + y^2 + 2xy\,y')(xy + 2)^2$

$\qquad + 2(x^3y^{-2} + xy^2)^3(xy + 2)(xy' + y) = 0$

Divide by $(x^3y^{-2} + xy^2)^2(x + 2y)$ to simplify giving

$3(xy + 2)(3x^2y^{-2} - 2x^3y^{-3}y' + y^2 + 2xy\,y') + 2(x^3y^{-2} + xy^2)(xy' + y) = 0$

or $[6x^3y^{-3}(xy + 2) + 6xy(xy + 2) - 2x(x^3y^{-2} + xy^2)]y'$

$\qquad = 3(xy + 2)(3x^2y^{-2} + y^2) + 2y(x^3y^{-2} + xy^2)$

so $y' = \dfrac{3(xy + 2)(3x^2y^{-2} + y^2) + 2y(x^3y^{-2} + xy^2)}{6x^3y^{-3}(xy + 2) + 6xy(xy + 2) - 2x(x^3y^{-2} + xy^2)}$

17. (a) $6x - 2y\,y' = 0$ so $y' = \dfrac{6x}{2y} = \dfrac{3x}{y}$.

   (b) $y'' = \dfrac{y(3) - 3xy'}{y^2} = \dfrac{3y - 3xy'}{y^2}$.

   (c) Replacing y' by $\dfrac{3x}{y}$, we get

   $y'' = \dfrac{3y - 3x\left(\frac{3x}{y}\right)}{y^2} = \dfrac{3y^2 - 9x^2}{y^3} = \dfrac{-3(3x^2 - y^2)}{y^3}$.

   (d) Using $3x^2 - y^2 = 7$, we obtain $y'' = \dfrac{-3(7)}{y^3} = \dfrac{-21}{y^3}$.

19. Differentiating we get $2x + 10y\,y' = 0$ so $y' = \dfrac{-2x}{10y} = \dfrac{-x}{5y}$.

   Therefore, $y'' = \dfrac{5y(-1) - (-x)5y'}{25y^2} = \dfrac{-y + xy'}{5y^2}$.

   Replacing y' by $\dfrac{-x}{5y}$ and using $x^2 + 5y^2 = 7$, we have

   $y'' = \dfrac{-y - \frac{x^2}{5y}}{5y^2} = \dfrac{-5y^2 - x^2}{25y^3} = \dfrac{-7}{25y^3}$.

25. First find y'. $2y\,y' = 3x^2y^2y' + 2xy^3 - 7$

$\qquad\qquad (2y - 3x^2y^2)y' = 2xy^3 - 7$

$\qquad\qquad\qquad\qquad y' = \dfrac{2xy^3 - 7}{2y - 3x^2y^2}$.

So the slope of the tangent at $(1,-2)$ is

$\dfrac{2(1)(-2)^3 - 7}{2(-2) - 3(1)^2(-2)^2} = \dfrac{-23}{-16} = \dfrac{23}{16}$.

Thus, an equation of the tangent at $(1,-2)$ is

$y + 2 = \dfrac{23}{16}(x - 1)$ or $16y - 23x + 55 = 0$.

27. First find y'. $3x^2 + 3y^2 y' = 0$, $y' = -\dfrac{x^2}{y^2}$.

So the tangent at $(2,0)$ is vertical. Thus, a line perpendicular to the tangent at $(2,0)$ is horizontal and has equation $y = 0$.

29. $f'(x) = \dfrac{4}{5} x^{-1/5} + \dfrac{4}{5} x^{-9/5} = \dfrac{4}{5} (\dfrac{1}{x^{1/5}} + \dfrac{1}{x^{9/5}}) = \dfrac{4}{5} (\dfrac{x^{8/5} + 1}{x^{9/5}})$.

Since the function is not continuous for the closed interval $-1 \leq x \leq 1$, because it is not defined for $x = 0$, an absolute maximum or minimum may not exist. Since $x^{8/5} + 1 \geq 1$ for all x, we have $f'(x) < 0$ for $-1 \leq x < 0$ and $f'(x) > 0$ for $0 < x \leq 1$. Thus, $f(x)$ is decreasing for $-1 \leq x < 0$ and increasing for all $0 < x \leq 1$ and so $f(x)$ is never positive. Also, for positive or negative values of x near zero, the function takes on negative values of arbitrarily large absolute value and so there is no absolute minimum value. A sketch of the graph is

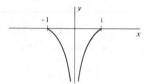

Thus, the absolute maximum occurs at $(-1,0)$ and $(1,0)$ and there is no absolute minimum value.

35. The area enclosed is given by $A = xy/2$. Moreover, $x^2 + y^2 = L^2$ so $y = \sqrt{L^2 - x^2}$. Thus,
$$A = \frac{1}{2} x \sqrt{L^2 - x^2} \text{ and,}$$
$$A' = \frac{1}{2} (L^2 - x^2)^{1/2} - \frac{1}{2} x^2 (L^2 - x^2)^{-1/2}.$$

If we set $A' = 0$ and solve for x, we find $x = \pm L/\sqrt{2}$. Since $x > 0$, we must have $x = L/\sqrt{2}$, and $y = L/\sqrt{2}$. These values of x and y maximize the enclosed area.

Exercises 3.9: Differentials (page 148)

1. $\dfrac{dy}{dx} = 14x - 3$. Thus, $dy = (14x - 3)\, dx$

5. $\dfrac{dy}{dx} = \dfrac{(x^2 + 2)(3x^2 - 3) - (x^3 - 3x)(2x)}{(x^2 + 2)^2} = \dfrac{x^4 + 9x^2 - 6}{(x^2 + 2)^2}$

Thus, $dy = \dfrac{x^4 + 9x^2 - 6}{(x^2 + 2)^2}\, dx$.

13. $g'(x) = 3 + 15x^{-4}$. Then using
$$g(x + dx) \approx g(x) + g'(x)\, dx$$
we get
$$g(x + dx) \approx 3x - 5x^{-3} + (3 + 15x^{-4})\, dx.$$
Thus, $g(1 + .06) \approx -2 + (18)(.06)$ or,
$$g(1.06) \approx -.92.$$

15. $g'(x) = x(3 + x^2)^{-1/2}$. Thus, $g'(1) = 1/2$. Use of

$$g(1.01) \approx g(1) + g'(1)(.01)$$

gives

$$g(1.01) \approx \sqrt{4} + \frac{1}{2}(.01)$$

or

$$g(1.01) \approx 2.005.$$

19. $f'(x) = \frac{1}{3}(x + 3)^{-2/3}$. Thus, $dy = \frac{1}{3}(x + 3)^{-2/3}$ dx. Since dy approximates the change in $f(x)$ as x changes by dx, the approximate change in $f(x)$ as x changes from 24 to 23.95 is

$$\frac{1}{3}(24 + 3)^{-2/3}(.05) \approx -.00185.$$

21. The surface area of a sphere is $S = 4\pi r^2$. Thus, $dS = 8\pi r$ dr. The change in the surface area as the radius increases from 5 cm to 5.03 cm is approximated by

$$dS = 8\pi(5)(.03) = .048\pi \text{ cm}^2.$$

Exercises 3.10: Differentiation; The Derivatives of the Trigonometric Functions (page 153)

7. $f'(x) = 5 \cdot 3 (\sin^2 5x)(\cos 5x)(\cos^4 6x) + (\sin^3 5x)(4 \cos^3 6x)(-\sin 6x)\,6$

$= 15 \sin^2 5x \cos 5x \cos^4 6x - 24 \sin^3 5x \cos^3 6x \sin 6x$

11. $f'(x) = \dfrac{\cos^2 u + \sin^2 u}{\cos^2 u} \dfrac{du}{dx} = \dfrac{1}{\cos^2 u} \dfrac{du}{dx} = \sec^2 u \dfrac{du}{dx}$

15. $f'(x) = \dfrac{\sin u}{\cos^2 u} \dfrac{du}{dx} = \dfrac{1}{\cos u} \dfrac{\sin u}{\cos u} \dfrac{du}{dx} = \sec u \tan u \dfrac{du}{dx}$

23. $f'(x) = 2 \sec(5x) \sec(5x) \tan(5x)(5 \tan^3 2x) + (\sec^2 5x)\,3 \tan^2 2x \sec^2 2x\,(2)$

$= 10 \sec^2 5x \tan 5x \tan^3 2x + 6 \sec^2 5x \tan^2 2x \sec^2 2x$

27. $(xy' + y) \cos xy - \sin x = 7 - y' \sec^2 y$

$y'(x \cos xy + \sec^2 y) = -y \cos xy + \sin x + 7$

Thus, $y' = \dfrac{7 - y \cos xy + \sin x}{x \cos xy + \sec^2 y}.$

43. A cross section of the trough is shown below

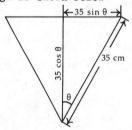

As indicated the cross sectional area is

$$A = \frac{1}{2}(35)^2 \cos\theta \sin\theta.$$

The volume of the trough is a maximum when the cross sectional area A is a maximum

$$\frac{dA}{d\theta} = \frac{1}{2}(35)^2 [\cos^2\theta - \sin^2\theta].$$

Note $\frac{dA}{d\theta}$ = 0 if and only if $\cos^2 \theta = \sin^2 \theta$. That is, if and only if $\tan^2 \theta = \pm 1$. Since $0 \leq \theta \leq \pi/2$, we must have $\theta = \pi/4$. Of course, since this is half the desired angle, the sheet metal must be bent in an angle of $\pi/2$ radions.

45.

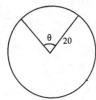

The figure illustrates how the circular sheet of paper is cut to form a cone. The volume of the cone is $V = \frac{1}{3} \pi r^2 h$, but since $h = 20 \cos \alpha$ and $r = 20 \sin \alpha$, we have

$$V = \frac{20^3 \pi}{3} \sin^2 \alpha \cos \alpha, \quad 0 \leq \alpha \leq \pi/2.$$

Consequently,

$$\frac{dV}{d\alpha} = \frac{20^3 \pi}{3} (2 \sin \alpha \cos^2 \alpha - \sin^3 \alpha).$$

Note $\frac{dV}{d\alpha}$ = 0 if and only if

$$\sin \alpha (2 \cos^2 \alpha - \sin^2 \alpha) = 0.$$

Thus, $\frac{dV}{d\alpha} = 0$ when $\alpha = 0$ or when $2 \cos^2 \alpha = \sin^2 \alpha$. If $2 \cos^2 \alpha = \sin^2 \alpha$, we have $\tan^2 \alpha = 2$ or $\tan \alpha = \sqrt{2}$. This value of α must determine the maximum volume since the volume is zero when $\alpha = 0$ or $\alpha = \pi/2$. We must now determine the value of θ associated with the value of α at which $\tan \alpha = \sqrt{2}$, $0 \leq \alpha \leq \pi/2$.

Note that the circumference of the base of the cone is the arc length subtended by the angle $2\pi - \theta$ on a circle of radius 20 cm. Thus, we have $2\pi r = 20(2\pi - \theta)$. Thus, since $r = 20 \sin \alpha$, we have $40\pi \sin \alpha = 20(2\pi - \theta)$ or,

$$\theta = 2\pi(1 - \sin \alpha).$$

Since $\tan \alpha = \sqrt{2}$, $0 \leq \alpha \leq \pi/2$, we may use the following diagram to find $\sin \alpha$.

Note $\sin \alpha = \sqrt{\frac{2}{3}}$. Thus, we have $\theta = 2\pi(1 - \sqrt{\frac{2}{3}})$ radions.

Exercises 3.11: Velocity, Acceleration, and Related Rates (page 158)

7. $s = 100 + 160t - 16t^2$, $v(t) = 160 - 32t = 32(5 - t)$

   (a) when $t = 0$, $v(t) = 160$ ft/sec.

   (b) The highest point occurs when $v(t) = 0$, that is when $t = 5$ sec.

   (c) Since the highest point occurs when $t = 5$ sec, its height above the ground is $s(5) = 100 + 160(5) - 16(5)^2 = 500'$.

9. Let the area be A and the side length s. Then $A = s^2$ and $\frac{dA}{dt} = -5$ cm^2/min. But $\frac{dA}{dt} = 2s \frac{ds}{dt}$. Thus, when $s = 15$ cm, $-5 = 2(15) \frac{ds}{dt}$ and so $\frac{ds}{dt} = -1/6$ cm/min.

11. Let A be the area and let r be the radius of the sheet. Then $\frac{dr}{dt} = 2$ in/min and $A = \pi r^2$, so $\frac{dA}{dt} = 2\pi r \frac{dr}{dt} = 4\pi r$. Thus, when $r = 12''$, $\frac{dA}{dt} = 4\pi(12) = 48\pi$ in^2/min.

13. Let A be the area and let s be the side length. Then $A = \frac{s^2\sqrt{3}}{4}$ and $\frac{ds}{dt} = .2$ cm/hr. So, $\frac{dA}{dt} = \frac{s\sqrt{3}}{2} \frac{ds}{dt}$. Thus, when $s = 4$ cm, $\frac{dA}{dt} = \frac{4\sqrt{3}}{2} (.2) = 2\sqrt{3}/5$ cm^2/hr.

15. Let V be the volume of wine in the tank when its depth is h and the radius of the surface of the wine is r as shown. Then, since $\frac{dV}{dt} = -12\pi$ ft^3/min and $V = \frac{1}{3}\pi r^2 h$.

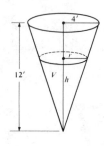

But $\frac{r}{4} = \frac{h}{12}$ so $r = \frac{h}{3}$ and $V = \frac{1}{3}\pi(\frac{h}{3})^2 h = \frac{\pi}{27}h^3$.

Thus, $-12\pi = \frac{dV}{dt} = \frac{\pi}{9}h^2 \frac{dh}{dt}$.

So, when $h = 9'$, $-12\pi = \frac{\pi}{9}(9)^2 \frac{dh}{dt}$ and hence,

$\frac{dh}{dt} = -\frac{12\pi}{9\pi} = -\frac{4}{3}$ ft/min. That is, the depth of the wine is decreasing at the rate of 4/3 ft/min.

17. Let x be the distance of the base of the ladder from the wall and let y be the height of the top of the ladder. Then, $\frac{dx}{dt} = 2$ m/sec and $x^2 + y^2 = 13^2$.

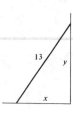

Thus, $2x \frac{dx}{dt} + 2y \frac{dy}{dt} = 0$. So when $y = 5$ m, $x^2 + 5^2 = 13^2$, $x^2 = 169 - 25 = 144$, $x = 12$ and $2(12)(2) + 2(5) \frac{dy}{dt} = 0$. Thus, $\frac{dy}{dt} = \frac{-48}{10} = \frac{-24}{5}$ m/sec and the top of the ladder is moving down the wall at 4 4/5 m/sec.

19. Let r be the radius and h be the height of the trunk. Then,

$\frac{dr}{dt} = \frac{1}{2}(\frac{1}{4}) = \frac{1}{8}$ in/yr $= \frac{1}{96}$ ft/yr,

$\frac{dh}{dt} = 1$ ft/yr and the volume V is

$V = \pi r^2 h$. Since r and h are functions

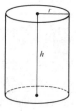

of t, $\frac{dV}{dt} = 2\pi r h \frac{dr}{dt} + \pi r^2 \frac{dh}{dt}$. Thus, when $r = \frac{3}{2}$ and $h = 50'$,

$\frac{dV}{dt} = 2\pi (\frac{3}{2})(50)(\frac{1}{96}) + \pi (\frac{3}{2})^2(1) = \frac{25\pi}{16} + \frac{9\pi}{4} = \frac{61\pi}{16}$ ft^3/yr.

21. Let y denote the height of the balloon and s the distance from the observer. Then $s^2 = y^2 + (100)^2$ so,

$$2s \frac{ds}{dt} = 2y \frac{dy}{dt}.$$

Since y = 15 ft at the end of one second, we must have $s = \sqrt{(15)^2 + (100)^2}$ at that time. Then since $\frac{dy}{dt} = 15$ ft/sec, we have

$$\frac{ds}{dt} = \frac{15}{\sqrt{(15)^2 + (100)^2}} \cdot 15 = \frac{45}{\sqrt{409}} \quad 2.23 \text{ ft/sec.}$$

25. The area of a circular plate is given by

$$A = \pi r^2.$$

Then

$$\frac{dA}{dt} = 2\pi r \frac{dr}{dt}.$$

If $\frac{dA}{dt} = \frac{dr}{dt}$, we must have $1 = 2\pi r$ or, $r = 1/2\pi$.

Technique Review Exercises, Chapter 3 (page 162)

2. $f'(x) = \dfrac{(x^2 - 3x + 2)(1) - x(2x - 3)}{(x^2 - 3x + 2)^2} = \dfrac{2 - x^2}{(x^2 - 3x + 2)^2}$

3. $f'(x) = 4x^5(x^2 + 3x)^3(2x + 3) + 5x^4(x^2 + 3x)^4$

   $= x^4(x^2 + 3x)^3(13x^2 + 17x) = x^8(x + 3)^3(13x + 17)$

4. $f'(x) = 2[3 + (x^2 - 5x)^3](3)(x^2 - 5x)^2(2x - 5) = 6(x^2 - 5x)^2(2x - 5)[3 + (x^2 - 5x)^3]$

5. Since $f'(x) = 3 - 3x^2 = 3(1 - x^2) = 3(1 - x)(1 + x)$, $f'(x) > 0$ if $-1 < x < 1$ and $f'(x) < 0$ if $x < -1$ or $x > 1$. Thus, f is increasing if $-1 < x < 1$ and decreasing if $x < -1$ or $x > 1$, and $(-1,1)$ is a local minimum point and $(1,5)$ is a local maximum point. Since $f''(x) = -6x$, $f''(x) > 0$ if $x < 0$ and $f''(x) < 0$ if $x > 0$. Thus, $(0,3)$ is an inflection point and f is concave upward for $x < 0$ and concave downward for $x > 0$.

6. Since $f'(x) = 3x^2 - 6x = 3x(x - 2)$, $f'(x) > 0$ if $x < 0$ or $x > 2$ and $f'(x) < 0$ if $0 < x < 2$, and $f'(x) = 0$ for $x = 0$ and $x = 2$. Thus, $f(x)$ is increasing for $-2 < x < 0$ or $2 < x < 3$ and $f(x)$ is decreasing for $0 < x < 2$. Hence, $(-2,-18)$ and $(2,-2)$ are local minimum points while $(0,2)$ and $(3,2)$ are local maximum points.

7. Since $f'(x) = 6x^2 + 6x - 12 = 6(x^2 + x - 2) = 6(x + 2)(x - 1)$, $f'(x) = 0$ when $x = -2$ or $x = -1$. Since $f(x)$ is continuous on a closed interval, it must have an absolute maximum and an absolute minimum value and these can occur only at points where $f'(x) = 0$ or $f'(x)$ does not exist or at an endpoint of the interval. Thus, the absolute extrema can only occur when x has the value $-3,-3,1$ or 10. Since $f(-3) = 14$, $f(-2) = 25$, $f(1) = -2$, and $f(10) = 2185$, the absolute maximum value of $f(x)$ is 2185 and the absolute minimum value of $f(x)$ is -2.

8. Let the dimensions be x and y as pictured.

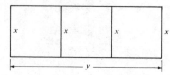

Then, $xy = 1800$ so $y = 1800x^{-1}$. Thus, the length L of the fence is
$L = 4x + 2y = 4x + 3600x^{-1}$, $x > 0$. Since

$$L' = 4 - 3600x^{-2} = \frac{4x^2 - 3600}{x^2} = \frac{4(x - 30)(x + 30)}{x^2}, \quad L' < 0 \text{ if } 0 < x < 30 \text{ and}$$

$L' > 0$ if $x > 30$, while $L' = 0$ if $x = 30$. Thus, L is decreasing for $0 < x < 30$
and increasing for $x > 30$ and so L has an absolute minimum value when $x = 30$.
Therefore, the dimensions requiring the least amount of fencing are $x = 30$ yds
and $y = 60$ yds.

9. Let A be the area and s be the side length. Then $\frac{dA}{dt} = -5$ in^2/min and $A = s^2$.
Thus, $\frac{dA}{dt} = 2s \frac{ds}{dt}$. Hence, when $s = 12''$, $-5 = 2(12) \frac{ds}{dt}$, so $\frac{ds}{dt} = \frac{-5}{24}$. Thus, the
side length is decreasing at $\frac{5}{24}$ in/min.

10. $2x^4 y\, y' + 4x^3 y^2 = 3x^2 + 4y^3 y'$

$(2x^4 y - 4y^3)y' = 3x^2 - 4x^3 y^2$

$y' = \dfrac{3x^2 - 4x^3 y^2}{2x^4 y - 4y^3}$

11. The volume of a cone with base radius r and height h is given by $V = \frac{1}{3}\pi r^2 h$.
Thus, if s denotes the diameter of the cone we have

$$V = \frac{1}{12}\pi s^2 h.$$

Consequently, $\qquad\qquad dV = \frac{1}{12}\pi\, 2sh\, ds.$

Since $ds = \frac{1}{50}$ m, $h = 3$ m and $s = 5$ m, we have

$$dV = \frac{1}{12}\pi\, 2(5)(3)\,\frac{1}{50} = \frac{\pi}{20}\, m^3.$$

Additional Exercises, Chapter 3 (page 163)

3. $(x - 7)(x + 2) = x^2 - 5x - 14$ so the derivative is $2x - 5$.

9. Use of the product rule gives $(x^2 + 5x)(3x^2 - 3) + (2x + 5)(x^3 - 3x)$. This
equals $3x^4 + 15x^3 - 3x^2 - 15x + 2x^4 + 5x^3 - 6x^2 - 15x = 5x^4 + 20x^3 - 9x^2 - 30x$.

11. Use of the quotient rule gives $\dfrac{(x + 1)(1) - (x - 7)(1)}{(x + 1)^2} = \dfrac{8}{(x + 1)^2}$

15. We get $(x^2 + 1)\frac{d}{dx}[(x^3 + 1)(x^4 + 1)] + (x^3 + 1)(x^4 + 1)\frac{d}{dx}$
$(x^2 + 1)$, or $(x^2 + 1)[(x^3 + 1)(4x^3) + (x^4 + 1)(3x^2)] + (x^3 + 1)$
$(x^4 + 1)(2x) = (x^2 + 1)(4x^6 + 4x^3 + 3x^6 + 3x^2) + 2x(x^3 + 1)(x^4 + 1)$
$\qquad = (x^2 + 1)(7x^6 + 4x^3 + 3x^2) + 2x(x^3 + 1)(x^4 + 1).$

17. $\dfrac{dy}{dx} = \dfrac{dy}{du}\dfrac{du}{dx} = (18u^2 + 6u)(-x^{-2} - 2x^{-3})$

$\qquad = (18(x^{-1} + x^{-2})^2 + 6(x^{-1} + x^{-2}))(-x^{-2} - 2x^{-3})$

21. $\dfrac{dy}{dx} = \dfrac{dy/dt}{dx/dt} = \dfrac{3t^2 - 2}{7t^6 + 3t^{-2}}$

25. $\dfrac{dy}{dx} = \dfrac{(2x - 1)^5\, 2(x + 3) - (x + 3)^2\, 5(2x - 1)^4\, 2}{(2x - 1)^{10}}$

$\qquad = \dfrac{2(2x - 1)(x + 3) - 10(x + 3)^2}{(2x - 1)^6}$

27. Since $y' = 2x - 6 = 2(x - 3)$, $y' > 0$ if $x > 3$ and $y' < 0$ if $x < 3$. Thus, we get (a) $x > 3$, (b) $x < 3$. Since $y'' = 2 > 0$ for all x we get (c) all x, (d) no x. The only critical point is at $(3,-8)$. A sketch of the graph is in the answer section of the text.

31. $y = (x^2 - 1)^3$

$y' = 3(x^2 - 1)^2(2x) = 6x(x - 1)^2(x + 1)^2$

A sketch of the graph of y' is

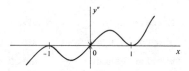

Thus, $y' > 0$ if $0 < x < 1$ or $x > 1$ and $y' < 0$ if $x < -1$ or $-1 < x < 0$ we get (a) $0 < x < 1$ or $x > 1$, (b) $x < -1$ or $-1 < x < 0$.

Now, $y'' = 6x(x^2 - 1)(2x) + (x^2 - 1)^2(6) = (x^2 - 1)[12x^2 + 6(x^2 - 1)]$

$\qquad = (x^2 - 1)(18x^2 - 6) = 18(x - 1)(x + 1)(x - \dfrac{1}{\sqrt{3}})(x + \dfrac{1}{\sqrt{3}}).$

A sketch of the graph of y'' is

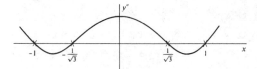

So, $y'' > 0$ if $x < -1$ or $-\dfrac{1}{\sqrt{3}} < x < \dfrac{1}{\sqrt{3}}$ or $x > 1$ and $y'' < 0$ if $-1 < x < -\dfrac{1}{\sqrt{3}}$ or $\dfrac{1}{\sqrt{3}} < x < 1$. Thus, we get

(c) $x < -1$ or $-\dfrac{1}{\sqrt{3}} < x < \dfrac{1}{\sqrt{3}}$ or $x > 1$, (d) $-1 < x < -\dfrac{1}{\sqrt{3}}$ or $\dfrac{1}{\sqrt{3}} < x < 1$. Critical points are at $(-1,0)$, $(0,-1)$, and $(1,0)$. A sketch is

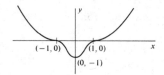

33. Since $f'(x) = -2 - 2x = -2(x + 1)$ and there are no endpoints, $(-1,8)$ is the only point where local maxima or minima can occur. Now, $f''(x) = -2 < 0$ so $(-1,8)$ is a local maximum point and there are no local minima.

37. Since $f'(x) = 2x + 4 = 2(x + 2)$, local extrema can occur only at the endpoints $(-5,-2)$ and $(-1,-10)$ or at $(-2,-11)$ where $f'(x) = 0$. Also, $f'(x) < 0$ for $x < -2$ and $f'(x) > 0$ for $x > -2$. Thus, $(-2,-11)$ is a local minimum point while $(-5,-2)$ and $(-1,-10)$ are local maximum points.

39. $x^3 - 2$. Since $f'(x) = 3x^2$ and there are no endpoints. $(0,-2)$ is the only point where local maxima or minima can occur. However, $f'(x) > 0$ if $x < 0$ and also $f'(x) > 0$ if $x > 0$. Thus, $f(x)$ is increasing for $x < 0$ and also for $x > 0$. Therefore, $(0,-2)$ is not a local maximum or minimum point and hence there are none.

45. Since x^{-1} does not exist at $x = 0$, it is not continuous on the closed interval $-2 \le x \le 1$ and may not have absolute extrema. Neither an absolute maximum point nor an absolute minimum point exists because for values of x near zero the function assumes both positive and negative values whose absolute values are arbitrarily large.

47. Since the function is not restricted to a closed interval, absolute extrema may not exist. Neither an absolute maximum point nor an absolute minimum point occurs since the function assumes both positive and negative values whose absolute values are arbitrarily large.

51. Let h be the height of the cylinder and let r be the radius and height of the cone as shown. Then, the volume V of the container is given by

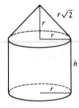

(1) $V = \pi r^2 h + \frac{1}{3} \pi r^3$.

The total surface area S is given by

$S = \pi r^2 + 2\pi rh + \pi r(r\sqrt{2}) = \pi r^2(1 + \sqrt{2})$

$+ 2\pi rh$. Since V is a constant, h is a function of r. Using this fact, we find $\frac{dS}{dr} = 2\pi r(1 + \sqrt{2}) + 2\pi r \frac{dh}{dr} + 2\pi h$. To find $\frac{dh}{dr}$, we differentiate (1) remembering that V is a constant. We get

$0 = \pi r^2 \frac{dh}{dr} + 2\pi rh + \pi r^2$. Thus,

$$\frac{dh}{dr} = \frac{-2\pi rh - \pi r^2}{\pi r^2} = \frac{-2h - r}{r}.$$

So, $\frac{dS}{dr} = 2\pi r(1 + \sqrt{2}) + 2\pi(-2h - r) + 2\pi h = 2\pi r\sqrt{2} - 2\pi h = 2\pi\sqrt{2}(r - \frac{h}{\sqrt{2}})$.

Therefore, $\frac{dS}{dr} < 0$ and S is a decreasing function of r, if $r < \frac{h}{\sqrt{2}}$. Also, $\frac{dS}{dr} > 0$ and S is an increasing function of r if $r > \frac{h}{\sqrt{2}}$. Consequently, the absolute minimum value of S occurs when $r = \frac{h}{\sqrt{2}}$ or $h = r\sqrt{2}$.

53. $x^3y^4 - 2xy^2 = 3$

$3x^2y^4 + 4x^3y^3y' - 4xy\,y' - 2y^2 = 0$ so

$(4x^3y^3 - 4xy)y' = 2y^2 - 3x^2y^4$

$y' = \dfrac{2y^2 - 3x^2y^4}{4x^3y^3 - 4xy} = \dfrac{2y - 3x^2y^3}{4x^3y^2 - 4x}$

63. Let $f(x) = \sqrt{x}$. Then $f'(x) = \dfrac{1}{2\sqrt{x}}$ so $f(x + dx) \approx f(x) + \dfrac{dx}{2\sqrt{x}}$. Consequently,

taking $x = 4$ and $dx = .03$ we get

$$\sqrt{4.03} \approx \sqrt{4} + \frac{.03}{2\sqrt{4}} = 2 + .0075 = 2.0075.$$

71. Let V be the volume, let r be the base radius, and let h be the height of the cone. Then,

$V = \dfrac{\pi}{3} r^2 h$ so $\dfrac{dV}{dt} = \dfrac{\pi}{3} r^2 \dfrac{dh}{dt} + \dfrac{2\pi}{3} rh \dfrac{dr}{dt}$. Thus, using $\dfrac{dr}{dt} = 3$, $\dfrac{dh}{dt} = -2$, when $r = 12$

$h = 8$, $\dfrac{dV}{dt} = \dfrac{\pi}{3} (12)^2(-2) + \dfrac{2\pi}{3} (12)(8)(3) = -96\pi + 192\pi = 96\pi$ cm/min.

CHAPTER 4: INTEGRATION

Exercises 4.1: The Area Problem (page 176)

1. $\displaystyle\sum_{i=1}^{5} f(x_i{}^*)h = \sum_{i=1}^{5} 4x_i{}^*h = \sum_{i=1}^{5} 4ih^2 = 4h^2 \sum_{i=1}^{5} i$

$\qquad = 4h^2(1 + 2 + 3 + 4 + 5) = 4h^2(15) = 60h^2$

3. $\displaystyle\sum_{i=1}^{7} f(x_i{}^*)h = \sum_{i=1}^{7} 4x_i{}^*h = \sum_{i=1}^{7} 4(i - 1)h^2 = 4h^2 \sum_{i=1}^{7} (i - 1)$

$\qquad = 4h^2(0 + 1 + 2 + 3 + 4 + 5 + 6) = 4h^2(21) = 84h^2$

5. $\displaystyle\sum_{i=1}^{n} f(x_i{}^*)h = \sum_{i=1}^{n} 4x_i{}^*h = \sum_{i=1}^{n} 4ih^2 = 4h^2 \sum_{i=1}^{n} i$

$\qquad = 4h^2(1 + 2 + 3 + \ldots + n)$

$\qquad = 4h^2 \dfrac{n(n + 1)}{2} = 2h^2n(n + 1)$

7. $\displaystyle\sum_{i=1}^{n} f(x_i{}^*)h = \sum_{i=1}^{n} 4x_i{}^*h = \sum_{i=1}^{n} 4(i - 1)h^2 = 4h^2 \sum_{i=1}^{n} (i - 1)$

$\qquad = 4h^2[0 + 1 + 2 + \ldots + (n - 1)]$

$\qquad = 4h^2 \dfrac{(n - 1)n}{2} = 2h^2n(n - 1)$

9. $\displaystyle\sum_{i=1}^{4} f(x_i{}^*)h = \sum_{i=1}^{4} (x_i{}^*)^2 h = \sum_{i=1}^{4} (ih)^2 h = h^3 \sum_{i=1}^{4} i^2$

$\qquad = h^3(1^2 + 2^2 + 3^2 + 4^2) = 30h^3$

11. Here, $h = (7 - 2)/n = 5/n$. Thus, $n = 5/h$. Since $f(x) = 9$ for all x, $f(x_i^*) = 9$ for any choice of x_i^*. Therefore,

$$\sum_{i=1}^{n} f(x_i^*)h = \sum_{i=1}^{n} 9h = 9h + 9h + \ldots + 9h \text{ (n terms)}$$

$$= 9nh = 9\left(\frac{5}{h}\right)h = 45.$$

Hence, $\displaystyle\int_{2}^{7} 9\,dx = \lim_{h\to 0} \sum_{i=1}^{n} f(x_i^*)h = \lim_{h\to 0} 45 = 45.$

15. Here, $h = (7 - 3)/n = 4/n$. Thus, $n = 4/h$. Take $x_1^*, x_2^*, \ldots, x_n^*$ at the right-hand endpoints. Then, $x_1^* = 2 + h$, $x_2^* = 2 + 2h$, $\ldots$, $x_i^* = 2 + ih$, $\ldots$, $x_n^* = 2 + nh = 7$. Then,

$$\sum_{i=1}^{n} f(x_i^*)h = \sum_{i=1}^{n} (2 + ih - 1)h = \sum_{i=1}^{n} (h + ih^2)$$

$$= (h + h + \ldots + h) + h^2(1 + 2 + \ldots + n)$$

$$= hn + \frac{n(n + 1)}{2} h^2, \text{ since } 1 + 2 + \ldots + n = \frac{n(n + 1)}{2}.$$

Replacing n by $4/h$, we get

$$\sum_{i=1}^{n} f(x_i^*)h = h\left(\frac{4}{h}\right) + \frac{h^2}{h}\left(\frac{4}{h}\right)\left(\frac{4}{h} + 1\right) = 4 + 2(4 + h).$$

Therefore, $\displaystyle\int_{3}^{7} (x - 1)\,dx = \lim_{h\to 0} [4 + 2(4 + h)] = 12.$

19. $\displaystyle\int_{3}^{8} f(x)\,dx = \lim_{h\to 0} \sum_{i=1}^{n} f(x_i^*)h$ where $h = \dfrac{8 - 3}{n} = \dfrac{5}{n}$ and x_i^* is an arbitrary point

in the i-th subinterval. Since each subinterval contains both rational and irrational points, we may select a rational point or an irrational point in each subinterval. If we select x_i^* to be rational, we have $f(x_i^*) = 3$ for every i. Then,

$$\sum_{i=1}^{n} f(x_i^*)h = \sum_{i=1}^{n} 3h = 3nh = 3n\left(\frac{5}{n}\right) = 15. \text{ But if we select each } x_i^* \text{ to be ir-}$$

rational, we get $f(x_i^*) = 5$ for every i and so

$$\sum_{i=1}^{n} f(x_i^*)h = \sum_{i=1}^{n} 5h = 5nh = 5n\left(\frac{5}{n}\right) = 25. \text{ Thus,}$$

$\displaystyle\lim_{h\to 0} \sum_{i=1}^{n} f(x_i^*)h$ is not independent of the choices of the x_i^*'s and so the

integral does not exist.

1. The net area is $\int_{-3}^{5} 2dx = \lim\limits_{h \to 0} \sum\limits_{i=1}^{n} f(x_i{}^*)h$ where $f(x) = 2$. Here

$h = \dfrac{5 - (-3)}{n} = \dfrac{8}{n}$ and since $f(x) = 2$ for all x, $f(x_i{}^*) = 2$ for any choice of

$x_i{}^*$. Therefore,

$$\sum\limits_{i=1}^{n} f(x_i{}^*)h = \sum\limits_{i=1}^{n} 2h = 2nh = 2(8) = 16. \quad \text{Therefore,}$$

$\lim\limits_{h \to 0} \sum\limits_{i=1}^{n} f(x_i{}^*)h = \lim\limits_{h \to 0} 16 = 16$. So the net area is 16 square units.

5. The net area is $\int_{2}^{6} x\,dx = \lim\limits_{h \to 0} \sum\limits_{i=1}^{n} f(x_i{}^*)h$ where $f(x) = x$.

Here $h = \dfrac{6 - (2)}{n} = \dfrac{4}{n}$ and if we take $x_i{}^*$ to be the right-hand endpoint of the

i-th subinterval, we have $x_i{}^* = 2 + ih$. Thus,

$$\sum\limits_{i=1}^{n} f(x_i{}^*)h = f(2 + h)h + f(2 + 2h)h + \ldots + f(2 + nh)h$$

$$= h[(2 + h) + (2 + 2h) + \ldots + (2 + nh)]$$

$$= h[2n + (1 + 2 + \ldots + n)h]$$

$$= h[2n + \dfrac{n(n + 1)}{2} h]. \quad \text{But } h = \dfrac{4}{n} \text{ so } n = \dfrac{4}{h} \text{ and this becomes}$$

$h[\dfrac{8}{h} + \dfrac{2}{h} (\dfrac{4}{h} + 1)h] = 8 + 8 + 4h = 16 + 2h$. Thus,

$\lim\limits_{h \to 0} \sum\limits_{i=1}^{n} f(x_i{}^*)h = \lim\limits_{h \to 0} (16 + 2h) = 16$. So the net area is 16 square units.

11. The net area is $\int_{0}^{4} 3x^2\,dx = \lim\limits_{h \to 0} \sum\limits_{i=1}^{n} f(x_i{}^*)h$ where $f(x) = 3x^2$.

Here, $h = \dfrac{4 - 0}{n} = \dfrac{4}{n}$. Take $x_i{}^*$ to be the right-hand endpoint of the i-th sub-

interval, so that $x_i{}^* = ih$. Then,

$$\sum\limits_{i=1}^{n} f(x_i{}^*)h = f(h)h + f(2h)h + \ldots + f(nh)h$$

$$= 3(1^2h^2)h + 3(2^2h^2)h + \ldots + 3(n^2h^2)h$$

$$= 3h^3(1^2 + 2^2 + \ldots + n^2) = 3h^3 \dfrac{n(n + 1)(2n + 1)}{6}$$

$$= \dfrac{1}{2} h^3 (\dfrac{4}{h})(\dfrac{4}{h} + 1)(\dfrac{8}{h} + 1) = 2(4 + h)(8 + h).$$

Thus, $\int_{0}^{4} 3x^2\,dx = \lim\limits_{h \to 0} [2(4 + h)(8 + h)] = 64$ and the net area is 64 square

units.

17. Let $A(x)$ be the area from 1 to x. Then, $A(1) = 0$ and the area we seek is $A(4)$. Since $A'(x) = 2x$, $A(x) = x^2 + C$. But $A(1) = 0 = 1 + C$ so $C = -1$ and $A(x) = x^2 - 1$. Thus, the area we want is
$$A(4) = 4^2 - 1 = 15 \text{ sq units.}$$

19. Let $A(x)$ be the area from 1 to x. Then $A'(x) = 3$ so $A(x) = 3x + C$. Since $A(1) = 0$, $3(1) + C = 0$ so $C = -3$ and $A(x) = 3x - 3$. Thus, the area from 1 to 5 is $A(5) = 3(5) - 3 = 12$ sq units.

21. Let $A(x)$ be the area from 2 to x. Then $A'(x) = x^2 + 2$ so $A(x) = \frac{x^3}{3} + 2x + C$. Since $A(2) = 0$, $\frac{8}{3} + 4 + C = 0$, $C = -\frac{20}{3}$ and we have $A(x) = \frac{x^3}{3} + 2x - \frac{20}{3}$. Hence, the area from 2 to 5 is $A(5) = \frac{5^3}{3} + 2(5) - \frac{20}{3} = 45$ sq units.

23. Let $A(x)$ be the area from 1 to x. Then $A'(x) = x^3$ so $A(x) = \frac{x^4}{4} + C$. Since $A(1) = 0 = \frac{1}{4} + C$, $C = -\frac{1}{4}$ and thus $A(x) = \frac{x^4}{4} - \frac{1}{4}$. Hence, the area from 1 to 2 is
$$A(2) = \frac{16}{4} - \frac{1}{4} = \frac{15}{4} \text{ sq units.}$$

25. Let $A(x)$ be the area from 2 to x. Then $A'(x) = 10x^4 - 6x^2$ so $A(x) = 2x^5 - 2x^3 + C$. Since $A(2) = 0$, $2(2)^5 - 2(2)^3 + C = 0$ and $C = -48$. Thus, $A(x) = 2x^5 - 2x^3 - 48$ and so the area from 2 to 4 is $A(4) = 2(4)^5 - 2(4)^3 - 48 = 1872$ sq units.

27. Let $A(x)$ be the net area from -3 to x. Then $A'(x) = 2x$ so $A(x) = x^2 + C$. Since $A(-3) = 0$, $(-3)^2 + C = 0$ and hence $C = -9$. Thus, $A(x) = x^2 - 9$. The net area is $A(2) = 4 - 9 = -5$ sq units.

29. Let $A(x)$ be the net area from -7 to x. Then $A'(x) = x - 1$ so $A(x) = \frac{x^2}{2} - x + C$. Since $A(-7) = 0$, $C = \frac{-63}{2}$ and $A(x) = \frac{x^2}{2} - x - \frac{63}{2}$. Thus, the net area is $A(10) = \frac{100}{2} - 10 - \frac{63}{2} = \frac{17}{2}$ sq units.

33. Since $\frac{dv}{dt} = -32$, $v = -32t + C_1$ where C_1 is a constant. But, $v = 8$ when $t = 0$, so $8 = -32(0) + C_1$ and $C_1 = 8$. Thus, $v = \frac{ds}{dt} = -32t + 8$. Hence, $s = -16t^2 + 8t + C_2$, where C_2 is a constant. Since $s = 24$ when $t = 0$, $24 = C_2$ and the equation of motion is $s = -16t^2 + 8t + 24$. When the stone strikes the ground $s = 0$, and we solve $-16t^2 + 8t + 24 = 0$. We get $t = -1$ or $t = 3/2$. Since $t \geq 0$, we reject $t = -1$ and hence the time required for the stone to strike the ground is $3/2$ second. At that time, its velocity is $-32(3/2) + 8 = -40$ ft/sec.

39. Since $\frac{dv}{dt} = -32$, $v = -32t + C_1$ where C_1 is a constant. But $v = 20$ when $t = 0$, so $20 = -32(0) + C_1$ and $C_1 = 20$. Thus, $v = \frac{ds}{dt} = -32t + 20$. Hence, $s = -16t^2 + 20t + C_2$, where C_2 is a constant. Since $s = C_2$ when $t = 0$, C_2 is the initial height we are to find. Since $s = 0$ when $t = 5$, $0 = -16(5)^2 + 20(5) + C_2$ and so $C_2 = 300$. Thus, the stone was thrown from a height of 300 ft.

1. Since $(\frac{x^2}{2})' = x$, $\int_2^5 x \, dx = \frac{x^2}{2} \Big|_2^5 = \frac{25}{2} - \frac{4}{2} = \frac{21}{2}$.

3. Since $(2x^3)' = 6x^2$, $\int_{-5}^{-3} 6x^2 \, dx = 2x^3 \Big|_{-5}^{-3} = -2(27) + 2(125) = 196$.

5. Since $(5x)' = \int_{-7}^{-3} 5dx = 5x \Big|_{-7}^{-3} = (-15) - (-35) = 20$.

7. Since $(\frac{x^4}{4})' = x^3$, $\int_2^6 x^3 \, dx = \frac{x^4}{4} \Big|_2^6 = \frac{6^4}{4} - \frac{2^4}{4} = 320$.

9. Since $(\frac{x^2}{2} + 2x)' = x + 2$, $\int_{-3}^2 (x + 2) \, dx = (\frac{x^2}{2} + 2x) \Big|_{-3}^2$

$$= \frac{4}{2} + 4 - (\frac{9}{2} - 6) = \frac{15}{2}.$$

13. Since $(x)' = 1$, $\int_3^7 dx = \int_3^7 1 \, dx = x \Big|_3^7 = 7 - 3 = 4$.

21. Since $(\frac{x^8}{8})' = x^7$, $\int_{-3}^3 x^7 \, dx = \frac{x^8}{8} \Big|_{-3}^3 \frac{3^8}{8} - \frac{(-3)^8}{8} = 0$.

23. Since $(4x^9)' = 36x^8$, $\int_{-1}^1 36x^8 \, dx = (4x^9) \Big|_{-1}^1 4(1)^9 - 4(-1)^9 = 4 + 4 = 8$.

25. $\int_a^x (t^2 - 3t + 2) \, dt = (\frac{t^3}{3} - \frac{3t^2}{2} + 2t) \Big|_a^x = \frac{x^3}{3} - \frac{3x^2}{2} + 2x - \frac{a^3}{3} + \frac{3a^2}{2} - 2a$

So $\frac{d}{dx} \int_a^x (t^2 - 3t + 2) \, dt = \frac{d}{dx} (\frac{x^3}{3} - \frac{3x^2}{2} + 2x - \frac{a^3}{3} + \frac{3a^2}{2} - 2a) = x^2 - 3x + 2$.

1. $\int_2^5 -x \, dx = -\int_2^5 x \, dx = -(\frac{x^2}{2}) \Big|_2^5 = -[\frac{25}{2} - \frac{4}{2}] = -\frac{21}{2}$.

3. Since the upper and lower limits of integration are equal, $\int_3^3 x^9 \, dx = 0$.

7. Since $(\frac{x^4}{4})' = x^3$, $\int_3^1 x^3 \, dx = \frac{x^4}{4} \Big|_3^1 = \frac{1}{4} - \frac{81}{4} = -20$.

11. $\int_{-2}^5 0 \, dx = 0 \int_0^5 dx = 0$.

13. Since $|x| = -x$ if $x \le 0$, $\int_{-3}^7 |x| \, dx = \int_{-3}^0 |x| \, dx + \int_0^7 |x| \, dx$

$$= \int_{-3}^0 -x \, dx + \int_0^7 x \, dx = \frac{-x^2}{2} \Big|_{-3}^0 + \frac{x^2}{2} \Big|_0^7$$

$$= 0 - (\frac{-9}{2}) + \frac{7^2}{2} - 0 = \frac{58}{2} = 29.$$

15. Since $|x| = -x$ if $x \leq 0$, $x|x| = -x^2$ if $x \leq 0$. Thus,

$$\int_{-2}^{3} x|x|\,dx = \int_{-2}^{0} x|x|\,dx + \int_{0}^{3} x|x|\,dx = \int_{-2}^{0} -x^2\,dx + \int_{0}^{3} x^2\,dx$$

$$= \frac{-x^3}{3}\Big|_{-2}^{0} + \frac{x^3}{3}\Big|_{0}^{3} = 0 - \frac{8}{3} + \frac{27}{3} - \frac{0}{3} = \frac{19}{3}.$$

19. $\int_{-4}^{0} |x + 3|\,dx$. Since $x + 3 < 0$ if $x < -3$ and $x + 3 > 0$ if

$x > -3$, $|x + 3| = \begin{cases} -(x + 3) = -x - 3 \text{ if } x < -3 \\ x + 3 \text{ if } x > -3 \end{cases}$

and so $\int_{-4}^{0} |x + 3|\,dx = \int_{-4}^{-3} |x + 3|\,dx + \int_{-3}^{0} |x + 3|\,dx$

$$= \int_{-4}^{-3} (-x - 3)\,dx + \int_{-3}^{0} (x + 3)\,dx$$

$$= (-\frac{x^2}{2} - 3x)\Big|_{-4}^{-3} + (\frac{x^2}{2} + 3x)\Big|_{-3}^{0}$$

$$= -\frac{9}{2} + 9 + \frac{16}{2} - 12 + 0 + 0 - \frac{9}{2} + 9 = 5.$$

23. $\int_{-2}^{2} x|x + 1|\,dx$. Since $|x + 1| = \begin{cases} x + 1 \text{ if } x \geq -1 \\ -(x + 1) \text{ if } x \leq -1 \end{cases}$ we have

$$\int_{-2}^{2} x|x + 1|\,dx = \int_{-2}^{-1} (-x^2 - x)\,dx + \int_{-1}^{2} (x^2 + x)\,dx$$

$$= (\frac{-x^3}{3} - \frac{x^2}{2})\Big|_{-2}^{-1} + (\frac{x^3}{3} + \frac{x^2}{2})\Big|_{-1}^{2}$$

$$= \frac{1}{3} - \frac{1}{2} - \frac{8}{3} + \frac{4}{2} + \frac{8}{3} + 2 + \frac{1}{3} - \frac{1}{2} = 3\frac{2}{3}.$$

25. $\int_{-2}^{3} 6|x + 2 - x^2|\,dx$. Since $x + 2 - x^2 = (2 - x)(x + 1)$ is positive for

$-1 < x < 2$ and negative for $x < -1$ or $x > 2$ we get

$$\int_{-2}^{3} 6|x + 2 - x^2|\,dx = \int_{-2}^{-1} 6(x^2 - x - 2)\,dx + \int_{-1}^{2} 6(x + 2 - x^2)\,dx$$

$$+ \int_{2}^{3} 6(x^2 - x - 2)\,dx = (2x^3 - 3x^2 - 12x)\Big|_{-2}^{-1} + (3x^2 + 12x - 2x^3)\Big|_{-1}^{2}$$

$$+ (2x^3 - 3x^2 - 12x)\Big|_{2}^{3} = 49.$$

<u>Exercises 4.6:</u> Computing Areas (page 200)

1. $A = \int_{2}^{5} x\,dx = \frac{x^2}{2}\Big|_{2}^{5} = \frac{25}{2} - \frac{4}{2} = \frac{21}{2}$ sq units

3. $A = -\int_{2}^{6} -x\,dx = \int_{2}^{6} x\,dx = \frac{x^2}{2}\Big|_{2}^{6} = \frac{36}{2} - \frac{4}{2} = 16$ sq units

5. $A = -\int_1^4 -x^2 \, dx = \int_1^4 x^2 \, dx = \frac{x^3}{3} \Big|_1^4 = \frac{64}{3} - \frac{1}{3} = 21$ sq units

7. $A = \int_{-1}^3 (3 - x) \, dx = (3x - \frac{x^2}{2}) \Big|_{-1}^3 = (9 - \frac{9}{2}) - (-3 - \frac{1}{2}) = 8$ sq units

9. $A = -\int_{-5}^{-1} (x + 1) \, dx + \int_{-1}^1 (x + 1) \, dx = -(\frac{x^2}{2} + x) \Big|_{-5}^{-1} + (\frac{x^2}{2} + x) \Big|_{-1}^1$

   $= 10$ sq units

11. $A = -\int_1^2 (3x^2 - 6x) \, dx + \int_2^4 (3x^2 - 6x) \, dx = -(x^3 - 3x^2) \Big|_1^2 + (x^3 - 3x^2) \Big|_2^4$

   $= 22$ sq units

13. $A = -\int_{-3}^0 4x^3 \, dx + \int_0^2 4x^3 \, dx = -x^4 \Big|_{-3}^0 + x^4 \Big|_0^2 = 97$ sq units

19. $A = \int_2^8 (2x - x) \, dx = \int_2^8 x \, dx = \frac{x^2}{2} \Big|_2^8 = 30$ sq units

23. $A = \int_{-1}^2 [(5 - x^2) - (x^2 - 9)] \, dx = \int_{-1}^2 14 = 2x^2 = (14x - \frac{2x^3}{3}) \Big|_{-1}^2 = 36$ units^2

25. $A = \int_0^3 (3 - x) \, dx = (3x - \frac{x^2}{2}) \Big|_0^3 = 9 - \frac{9}{2} - 0 = \frac{9}{2}$ sq units

27. $A = \int_0^4 (3x - 2x) \, dx = \int_0^4 x \, dx = \frac{x^2}{2} \Big|_0^4 = 8$ sq units

29. Intersection points of $y = x^2$ and $y = x$ occur when $x^2 = x$ or $x^2 - x = 0$,
    $x(x - 1) = 0$, so at $x = 0$, $x = 1$. So,
    $A = \int_0^1 (x - x^2) \, dx = (\frac{x^2}{2} - \frac{x^3}{3}) \Big|_0^1 = \frac{1}{2} - \frac{1}{3} = \frac{1}{6}$ sq units.

31. $A = \int_0^5 [0 - (x^2 - 5x)] \, dx = \int_0^5 (5x - x^2) \, dx$

   $= (\frac{5x^2}{2} - \frac{x^3}{3}) \Big|_0^5 = \frac{125}{2} - \frac{125}{3} = \frac{125}{6}$ sq units

Exercises 4.7: The Indefinite Integral (page 207)

5. $\int_1^2 x^{-3} \, dx = \frac{x^{-2}}{-2} \Big|_1^2 = -\frac{1}{8} - (\frac{1}{-2}) = \frac{3}{8}$

7. $\int \frac{dx}{\sqrt{x}} = \int x^{-1/2} \, dx = 2x^{1/2} + C$

11. $\int t(3 - t^2)^2 \, dt = -\frac{1}{2} \int (3 - t^2)^2 (-2t) \, dt = -\frac{1}{6} (3 - t^2)^3 + C$

13. $\int (3 - x^2)^2 \, dx = \int (9 - 6x^2 + x^4) \, dx = 9x - 2x^3 + \frac{x^5}{5} + C$

15. $\int (1 - v)^3 \, dv = (-1) \int (1 - v)^3 (-1) \, dv = - \dfrac{(1 - v)^4}{4} + C$

17. $\int x(x^2 + 4)^{1/2} \, dx = \dfrac{1}{2} \int (x^2 + 4)^{1/2} (2x) \, dx = \dfrac{1}{2} \cdot \dfrac{2}{3} (x^2 + 4)^{3/2} + C$

$$= \dfrac{1}{3} (x^2 + 4)^{3/2} + C$$

19. $\int (x^4 + 8x^2)^5 (x^3 + 4x) \, dx = \dfrac{1}{4} \int (x^4 + 8x^2)^5 (4x^3 + 16x) \, dx = \dfrac{1}{24} (x^4 + 8x^2)^6 + C$

21. $\int_2^4 \dfrac{x}{\sqrt{x^2 - 2}} \, dx = \dfrac{1}{2} \int_2^4 (x^2 - 2)^{-1/2} (2x) \, dx = (x^2 - 2)^{1/2} \Big|_2^4 = \sqrt{14} - \sqrt{2}$

23. $\int_0^1 \dfrac{x^2 \, dx}{(x^3 + 1)^5} = \dfrac{1}{3} \int_0^1 (x^3 + 1)^{-5} (3x^2) \, dx = - \dfrac{1}{12} (x^3 + 1)^{-4} \Big|_0^1 = \dfrac{5}{64}$

25. $\int (3u + 5)^8 \, du = \dfrac{1}{3} \int (3u + 5)^8 (3) \, du = \dfrac{1}{27} (3u + 5)^9 + C$

27. $\int x\sqrt{9 - x^2} \, dx = - \dfrac{1}{2} \int (9 - x^2)^{1/2} (-2x \, dx) = - \dfrac{1}{3} (9 - x^2)^{3/2} + C$

29. $\int_0^1 \dfrac{x \, dx}{(5 + 3x^2)^{2/3}} = \dfrac{1}{6} \int (5 + 3x^2)^{-2/3} (6x \, dx) = \dfrac{1}{2} (5 + 3x^2)^{1/3} \Big|_0^1 = \dfrac{1}{2} (2 - 5^{1/3})$

Exercises 4.8: The Mean Value Theorem for Integrals; Average Value of a Function (page 210)

1. $\dfrac{1}{6 - (-2)} \int_{-2}^6 3x \, dx = \dfrac{1}{8} \left(\dfrac{3x^2}{2}\right) \Big|_{-2}^6 = \dfrac{1}{8} (54 - 6) = 6$

3. $\dfrac{1}{2 - (-5)} \int_{-5}^2 x^2 \, dx = \dfrac{1}{7} \left(\dfrac{x^3}{3}\right) \Big|_{-5}^2 = \dfrac{1}{7} \left(\dfrac{8}{3} + \dfrac{125}{3}\right) = \dfrac{133}{21} = \dfrac{19}{3}$

5. $\dfrac{1}{3 - (-2)} \int_{-2}^3 (3x^2 - 4x + 5) \, dx = \dfrac{1}{5} (x^3 - 2x^2 + 5x) \Big|_{-2}^3 = 10$

7. $\dfrac{1}{2 - (-3)} \int_{-3}^2 (6x^2 - 4x^3 + 2) \, dx = \dfrac{1}{5} (2x^3 - x^4 + 2x) \Big|_{-3}^2 = 29$

9. $\dfrac{1}{2 - (-1)} \int_{-1}^2 (10x^4 - 2x + 7) \, dx = \dfrac{1}{3} (2x^5 - x^2 + 7x) \Big|_{-1}^2 = 28$

11. $f(c) = 4c = \dfrac{1}{6} \int_1^7 4x \, dx = \dfrac{1}{6} (2x^2) \Big|_1^7 = \dfrac{1}{6} (98 - 2) = 16.$ So $c = 4$.

13. $f(c) = 6c^2 = \dfrac{1}{4} \int_1^5 6x^2 \, dx = \dfrac{1}{4} (2x^3) \Big|_1^5 = \dfrac{1}{4} (250 - 2) = 62.$ So $c^2 = \dfrac{31}{3}$,

$c = \sqrt{31/3}$ since we must have $1 \le c \le 5$.

19. $\dfrac{1}{2} \int_0^2 60x^2 \, dx = \dfrac{1}{2} (20x^3) \Big|_0^2 = \dfrac{1}{2} (160) = 80° \, C$

48

21. $w = 20t - 5t^2 + 50$, $0 \leq t \leq 3$, so $w' = 20 - 10t = 10(2 - t)$. Thus, the absolute maximum and minimum values of the continuous function $w = 20t - 5t^2 + 50$, $0 \leq t \leq 3$ must occur at $t = 2$ or at one of the endpoints $t = 0$, $t = 3$. Since $w(0) = 50$, $w(2) = 70$, and $w(3) = 65$, we have (a) minimum speed of 50 words per minute, (b) maximum speed of 70 words per minute.

(c) av. speed $= \dfrac{1}{3} \int_0^3 (20t - 5t^2 + 50)\, dt = \dfrac{1}{3} (10t^2 - \dfrac{5}{3} t^3 + 50t) \Big|_0^3$

$= \dfrac{1}{3} (90 - 45 + 150) = \dfrac{195}{3} = 65$ words per minute.

(d) Total no. of words typed = (average no. words per minute) $\cdot$ (total no. of minutes) $= 65(3)(60) = 11{,}700$ words.

23. (a) $\dfrac{1}{8 - 2} \int_2^8 gt\, dt = \dfrac{1}{6} (\dfrac{gt^2}{2}) \Big|_2^8 = \dfrac{1}{6} (32g - 2g) = 5g$ ft/sec.

(b) First we need to find s for $t = 2$ and $t = 8$. Since $g = gt = \sqrt{2gs}$, $g^2 t^2 = 2gs$ and so $s = \dfrac{1}{2} gt^2$. Thus, when $t = 2$, $s = 2g$, and when $t = 8$, $s = 32g$. Hence, the average of v with respect to s for the interval $t = 2$ to $t = 8$ is

$\dfrac{1}{32g - 2g} \int_{2g}^{32g} \sqrt{2gs}\, ds = \dfrac{\sqrt{2g}}{30g} \int_{2g}^{32g} s^{1/2}\, ds = \dfrac{\sqrt{2g}}{30g} \cdot \dfrac{2}{3} s^{3/2} \Big|_{2g}^{32g}$

$= \dfrac{\sqrt{2g}}{45g} (32g\sqrt{32g} - 2g\sqrt{2g}) = \dfrac{1}{45} (256g - 4g) = \dfrac{252g}{45} = \dfrac{28g}{5}$ ft/sec.

Exercises 4.9: Integrals of Certain Trigonometric Functions (page 212)

1. $\int \sin 2x\, dx = \dfrac{1}{2} \int \sin 2x(2\, dx) = -\dfrac{1}{2} \cos 2x + C$

5. $\int \sec 2x \tan 2x\, dx = \dfrac{1}{2} \int \sec 2x \tan 2x\, (2\, dx) = \dfrac{1}{2} \sec 2x + C$

7. $\int x \sec^2 (x^2)\, dx = \dfrac{1}{2} \int \sec^2 (x^2)(2x\, dx) = \dfrac{1}{2} \tan (x^2) + C$

11. $\int \sin^3 x \cos x\, dx = \int \sin^3 x\, d(\sin x) = \dfrac{1}{4} \sin^4 x + C$

15. $\int \csc^7 x \csc x \cot x\, dx = -\int \csc^7 x\, d(\csc x) = -\dfrac{1}{8} \csc^8 x + C$

17. $\int \sec^3 t \tan t\, dt = \int \sec^2 t \sec t \tan t\, dt = \int \sec^2 t\, d(\sec t) = \dfrac{1}{3} \sec^3 t + C$

21. $\int_0^{\pi/2} \cos x\, dx = \sin x \Big|_0^{\pi/2} = 1$ unit^2

1. $A = \int_3^{10} 8\ dx = \lim_{h \to 0} \sum_{i=1}^{n} f(x_i^*)h = \lim_{h \to 0} \sum_{i=1}^{n} 8h = \lim_{h \to 0} 8nh$. But $h = \dfrac{10 - 3}{n} = \dfrac{7}{n}$

   so $n = \dfrac{7}{h}$ and we have $A = \lim_{h \to 0} 8(\dfrac{7}{h})(h) = \lim_{h \to 0} 56 = 56$ sq units.

2. Let $A(x)$ be the net area between -2 and x. Then $A'(x) = -x^3 + 5$ so

   $A(x) = \dfrac{-x^4}{4} + 5x + C$. But $A(-2) = 0$ so $0 = \dfrac{-(-2)^4}{4} + 5(-2) + C$ and hence

   $C = 14$ and so $A(x) = \dfrac{-x^4}{4} + 5x + 14$. The area above the x-axis minus the area

   below the x-axis between -2 and 4 is $A(4)$ so we get

   $A(4) = -\dfrac{4^4}{4} + 5(4) + 14 = -30$ sq units.

4. $\int_{-3}^{2} (x - 6x^2)\ dx = (\dfrac{x^2}{x} - 2x^3)\ \Big|_{-3}^{2}\ [\dfrac{2^2}{2} - 2(8)] - [\dfrac{(-3)^2}{2} - 2(-3)^3] = -\dfrac{145}{2}$

5. (a) $\int_1^4 (x^3 - 3x)\ dx = (\dfrac{x^4}{4} - \dfrac{3x^2}{2})\ \Big|_1^4 = (64 - 24) - (\dfrac{1}{4} - \dfrac{3}{2}) = 41\ 1/4$ sq units.

   (b) $-\int_1^{\sqrt{3}} (x^3 - 3x)\ dx + \int_{\sqrt{3}}^{4} (x^3 - 3x)\ dx = -(\dfrac{x^4}{4} - \dfrac{3x^2}{2})\ \Big|_1^{\sqrt{3}}$

   $+ (\dfrac{x^4}{4} - \dfrac{3x^2}{2})\ \Big|_{\sqrt{3}}^{4} = -(\dfrac{9}{4} - \dfrac{9}{2}) + (\dfrac{1}{4} - \dfrac{3}{2}) + (64 - 24) - (\dfrac{9}{4} - \dfrac{9}{2})$

   $= 43\ \dfrac{1}{4}$ sq units.

6. Intersection points occur when $x^2 - 2 = x$, so when $x = -2$, $x = 1$.

   $A = \int_{-2}^{1} (-x - x^2 + 2)\ dx = (\dfrac{-x^2}{2} - \dfrac{x^3}{3} + 2x)\ \Big|_{-2}^{1} = 4\ \dfrac{1}{2}$ sq units.

7. $\int x^2(x^3 - 7)^{10}\ dx = \dfrac{1}{3} \int (x^3 - 7)^{10}(3x^2\ dx) = \dfrac{1}{33} (x^3 - 7)^{11} + C$

8. $\int \dfrac{x^3\ dx}{(2x^4 + 1)^3} = \dfrac{1}{8} \int (2x^4 + 1)^{-3}(8x^3\ dx) = -\dfrac{1}{16} (2x^4 + 1)^{-2} + C$

9. $\int (x^2 + 3x)^2\ dx = \int (x^4 + 6x^3 + 9x^2)\ dx = \dfrac{x^5}{5} + \dfrac{3}{2} x^4 + 3x^3 + C$

10. $\int_0^1 \dfrac{x^3\ dx}{(x^4 + 10)^3} = \dfrac{1}{4} (x^4 + 10)^{-3}(4x^3\ dx) = -\dfrac{1}{8} (x^4 + 10)^{-2}\ \Big|_0^1 = \dfrac{21}{96,800}$

11. Since $-x^2 - 2x = -x(x + 2)$ is ≤ 0 if $x \le -2$ and is ≥ 0 if $-2 \le x \le 0$,

   $|-x^2 - 2x| = \begin{cases} x^2 + 2x & \text{if } x \le -2 \\ -x^2 - 2x & \text{if } -2 \le x \le 0 \end{cases}$

   $\int_{-3}^{0} |-x^2 - 2x|\ dx = \int_{-3}^{-2} |-x^2 - 2x|\ dx + \int_{-2}^{0} |-x^2 - 2x|\ dx$

$$= \int_{-3}^{-2} (x^2 + 2x)\, dx + \int_{-2}^{0} (-x^2 - 2x)\, dx = \left(\frac{x^3}{3} + x^2\right)\Big|_{-3}^{-2} + \left(\frac{-x^3}{3} - x^2\right)\Big|_{-2}^{0} = \frac{8}{3}$$

12. $\dfrac{1}{4} \int_{-3}^{1} (7 - 8x^3 - 9x^2)\, dx = \dfrac{1}{4}(7x - 2x^4 - 3x^3)\Big|_{-3}^{1} = 26$

Additional Exercises, Chapter 4 (page 215)

1. $\int_{-4}^{2} 17\,dx = \lim_{h \to 0} \sum_{i=1}^{n} f(x_i{}^*)h = \lim_{h \to 0} \sum_{i=1}^{n} 4h = \lim_{h \to 0} 4nh.$ But $= \dfrac{2 - (-4)}{n} = \dfrac{6}{n}$ so

$nh = 6.$ Therefore, $\int_{-4}^{2} 17\,dx = \lim_{h \to 0} \sum_{i=1}^{n} 17nh = \lim_{h \to 0} 17(6) = 102.$

3. $\int_{-3}^{1} x\,dx = \lim_{h \to 0} \sum_{i=1}^{n} f(x_i{}^*)h = \lim_{h \to 0} \sum_{i=1}^{n} x_i{}^*h.$ Taking $x_i{}^*$ to be the right end-

point of the i-th subinterval, we have $x_i{}^* = -3 + ih.$ Thus,

$\displaystyle\sum_{i=1}^{n} x_i{}^*h = h[(-3 + h) + (-3 + 2h) + \ldots + (-3 + nh)]$

$\qquad = h[-3n + (1 + 2 + \ldots + n)h] = h[-3n + \dfrac{n(n-1)}{2}h].$

But $h = \dfrac{1 - (-3)}{n} = \dfrac{4}{n}$ so $n = \dfrac{4}{h}$ and so

$\displaystyle\sum_{i=1}^{n} x_i{}^*h = h[\dfrac{-12}{h} + \dfrac{2}{h}(\dfrac{4}{h} - 1)h] = -12 + 8 - 2h = -4 - 2h.$ Thus,

$\int_{-3}^{1} x\,dx = \lim_{h \to 0}(-4 - 2h) = -4.$

9. Let $A(x)$ be the area from -2 to x. Then $A'(x) = 14x$ so $A(x) = 7x^2 + C.$
Since $A(-2) = 0$, $7(-2)^2 + C = 0$ so $C = -28$ and $A(x) = 7x^2 - 28$. Thus, the area from -2 to 5 is
$$A(5) = 7(5)^2 - 28 = 147 \text{ sq units.}$$

13. Let $A(x)$ be the area from 2 to x. Then $A'(x) = 8x^3$ so $A(x) = 2x^4 + C.$
Since $A(2) = 0$, $2(2)^4 + C = 0$ so $C = -32$ and $A(x) = 2x^4 - 32$. Thus, the area from 2 to 4 is
$$A(4) = 2(4)^4 - 32 = 480 \text{ sq units.}$$

15. Since $\left(\dfrac{x^2}{2}\right)' = x$, $\int_{-3}^{1} x\,dx = \dfrac{x^2}{2}\Big|_{-3}^{1} = \dfrac{1}{2} - \dfrac{9}{2} = -4.$

19. Since $(3x^4)' = 12x^3$, $\int_{-1}^{1} 12x^3\,dx = 3x^4\Big|_{-1}^{1} = 3(1 - 1) = 0.$

23. Since the upper and lower limits of integration are equal, the result is zero.

27. Since $x - 3 < 0$ if $x < 3$ and $x - 3 > 0$ if $x > 3$,
$$|x - 3| = \begin{cases} -(x - 3) = 3 - x & \text{if } x < 3 \\ x - 3 & \text{if } x > 3 \end{cases}$$

and so $\int_{-7}^{2} |x - 3| \, dx = \int_{-7}^{2} (3 - x) \, dx = (3x - \frac{x^2}{2}) \Big|_{-7}^{2} = 49 \, 1/2.$

29. Since $-5x^4 \leq 0$ for $-1 \leq x \leq 3$, the area is

$$-\int_{-1}^{3} -5x^4 \, dx = \int_{-1}^{3} 5x^4 \, dx = x^5 \Big|_{-1}^{3} = 243 + 1 = 244 \text{ sq units.}$$

33. $\int_{-7}^{1} (6x^2 + 12x) \, dx = (2x^3 + 6x^2) \Big|_{-7}^{1} = 400$ sq units

35. $\int_{0}^{1} (|x| - x^3) \, dx = \int_{0}^{1} (x - x^3) \, dx = (\frac{x^2}{2} - \frac{x^4}{4}) \Big|_{0}^{1} = \frac{1}{4}$ sq units

37. $\int_{-2}^{-1} \frac{24}{x^2} \, dx = \int_{-2}^{-1} 24x^{-2} \, dx = -24x^{-1} \Big|_{-2}^{-1} = \frac{-24}{-1} + \frac{24}{-2} = 12$

41. $\int x \sqrt{x^2 - 1} \, dx = \frac{1}{2} \int (x^2 - 1)^{1/2} (2x \, dx) = \frac{1}{2} \cdot \frac{2}{3} (x^2 - 1)^{3/2} + C$

$$= \frac{1}{3} (x^2 - 1)^{3/2} + C$$

43. $\int (2x^5 - 1)(x^6 - 3x)^8 \, dx = \frac{1}{3} \int (x^6 - 3x)^8 (6x^5 - 3) \, dx = \frac{1}{27} (x^6 - 3x)^9 + C$

47. $\frac{1}{2 + 1} \int_{-1}^{2} 8x(x^2 - 3)^3 \, dx = \frac{4}{3} \int_{-1}^{2} (x^2 - 3)^3 (2x \, dx) = \frac{4}{3} \frac{(x^2 - 3)^4}{4} \Big|_{-1}^{2} = -5.$

49. $\int_{1}^{3} 8x^3 \, dx = 2x^4 \Big|_{1}^{3} = 2(81 - 1) = 160.$ So $16C^3 = 160$, so $C = \sqrt[3]{10}.$

51. $\int \cos 7x \, dx = \frac{1}{7} \int \cos 7x \, (7 \, dx) = \frac{1}{7} \sin 7x + C$

55. $\int t^2 \sin (t^3 + 7) \, dt = \frac{1}{3} \int \sin (t^3 + 7)(3t^2 \, dt) = -\frac{1}{3} \cos (t^3 + 7) + C$

CHAPTER 5: APPLICATIONS OF THE DEFINITE INTEGRAL

Exercises 5.1: Volume (page 230)

1. $V = \int_{0}^{1} (2x + 3)^{1/2} \, dx = \frac{1}{2} \int_{0}^{1} (2x + 3)^{1/2} (2dx) = \frac{1}{3} (2x + 3)^{3/2} \Big|_{0}^{1}$

$$= \frac{1}{3} (5^{3/2} - 3^{3/2}) \text{ cubic units.}$$

3. The base of the region is as pictured.

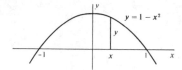

The cross section at x is a square of side $1 - x^2$ so the area function of a cross section is $A(x) = (1 - x^2)^2$. Thus, the volume is

$$V = \int_{-1}^{1} (1 - x^2)^2 \, dx = \int_{-1}^{1} (1 - 2x^2 + x^4) \, dx = 1 \tfrac{1}{15} \text{ cubic units.}$$

5. The cross section at x is an isosceles right triangle with sides $\sqrt{1 - x^2}$ so the area function of a cross section is $A(x) = \frac{1}{2} \sqrt{1 - x^2} \sqrt{1 - x^2} = \frac{1}{2} (1 - x^2)$. Thus, the volume is

$$V = \int_{-1}^{1} \frac{1}{2} (1 - x^2) \, dx = \frac{1}{2} (x - \frac{x^3}{3}) \Big|_{-1}^{1} = \frac{2}{3} \text{ cubic units.}$$

7. The volume of the cone is obtained by rotating the triangle indicated about the x-axis.

Thus, the volume is

$$V = \pi \int_{0}^{h} (\tfrac{r}{h} x)^2 \, dx = \frac{\pi r^2}{h^2} \int_{0}^{h} x^2 \, dx = \frac{\pi r^2 h}{3}.$$

9. $V = \pi \int_{0}^{5} (x^2)^2 \, dx = \pi \int_{0}^{5} x^4 \, dx = 625\pi$ cubic units

11. i) $V = 128\pi - \int_{0}^{2} \pi (x^3)^2 \, dx = 128\pi - \frac{\pi}{7} x^7 \Big|_{0}^{2} = \frac{768\pi}{7}$ units^3

ii) $V = \int_{0}^{8} \pi (y^{1/3})^2 \, dy = \frac{3\pi}{5} y^{5/3} \Big|_{0}^{8} = \frac{96\pi}{5}$ units^3

iii) $V = 128\pi - \pi \int_{0}^{8} (4 - y^{1/3})^2 \, dy = 128\pi - \pi \int_{0}^{8} (16 - 8y^{1/3} + y^{2/3}) \, dy$

$$= 128\pi - \pi (16y - 6y^{4/3} + \frac{3}{5} y^{5/3}) \Big|_{0}^{8} = \frac{384}{5} \pi \text{ units}^3$$

iv) $V = 200\pi - \pi \int_{0}^{2} (2 + x^3)^2 \, dx = 200\pi - \pi \int_{0}^{2} (4 + 4x^3 + x^6) \, dx$

$$= 200\pi - \pi (4x + x^4 + \frac{1}{7} x^7) \Big|_{0}^{2} = \frac{1104\pi}{7} \text{ units}^3$$

13. i) $V = \int_0^2 \pi(4 - x^2)^2 \, dx - \int_0^2 \pi(2 - x)^2 \, dx = \pi\int_0^2 (12 - 9x^2 + x^4 + 4x) \, dx$

$= \pi(12x - 3x^3 + \frac{1}{5}x^5 + 2x^2) \Big|_0^2 = \frac{72\pi}{5} \text{ units}^3$

ii) $V = \pi\int_2^4 (\sqrt{4 - y})^2 \, dy + \pi\int_0^2 ((\sqrt{4 - y})^2 - (2 - y)^2) \, dy$

$= \pi\int_0^4 (4 - y) \, dy - \pi\int_0^2 (4 - 4y + y^2) \, dy$

$= \pi(4y - \frac{1}{2}y^2) \Big|_0^4 - \pi(4y - 2y^2 + \frac{1}{3}y^3) \Big|_0^2 = \frac{16\pi}{3} \text{ units}^3$

iii) $V = \int_0^2 [\pi(7 - 2 + x)^2 - \pi(7 - 4 + x^2)^2] \, dx = \int_0^2 (16 + 10x - 5x^2 - x^4) \, dx$

$= \pi(16x + 5x^2 - \frac{5}{3}x^3 - \frac{1}{5}x^5) \Big|_0^2 = \frac{484\pi}{5} \text{ units}^3$

iv) $V = \pi\int_2^4 [(1 + \sqrt{4 - y})^2 - 1] \, dy + \pi\int_0^2 [(1 + \sqrt{4 - y})^2 - (1 + 2 - y)^2] \, dy$

$= \pi\int_0^4 (1 + \sqrt{4 - y})^2 \, dy - \pi\int_2^4 dy - \pi\int_0^2 (3 - y)^2 \, dy$

$= \pi\int_0^4 (1 + 2\sqrt{4 - y} + 4 - y) \, dy - \pi\int_2^4 dy - \pi\int_0^2 (9 - 6y + y^2) \, dy$

$= \pi(5y - \frac{4}{3}(4 - y)^{3/2} - \frac{1}{2}y^2) \Big|_0^4 - 2\pi - \pi(9y - 3y^2 + \frac{1}{3}y^3) \Big|_0^2$

$= \frac{52\pi}{3} \text{ units}^3$

Exercises 5.2: Work Done by a Variable Force (page 236)

1.

Let $100 + x$ be the number of feet of chain unwound as shown. The weight of this amount of chain is $5(100 + x)$ lbs. We are to find the work done by gravity as x goes from 0 to 30'. Thus, the work done is

$$W = \int_0^{30} 5(100 + x) \, dx = 5(100x + \frac{x^2}{2}) \Big|_0^{30} = 17{,}250 \text{ ft-lbs.}$$

3.

Let x be the number of feet the upper end of the cable is above the base of the building. The weight of the cable and cord being raised at this point is $x + .1(400 - x)$ pounds. Hence, the work necessary to raise the upper end of the cable to the top of the building is

$$\int_0^{400} [x + .1(400 - x)] \, dx = \int_0^{400} (x + 40 - \frac{x}{10}) dx$$
$$= 88,000 \text{ ft-lbs.}$$

5. Let k be the area of the piston in square feet and let h be the distance in feet from the top of the piston to the end of the container. If $h = h_1$ when $v = 9$ ft^3 and $h = h_2$ when $v = 4$ cu ft, the work W done in compressing the gas from 9 ft^3 to 4 ft^3 is

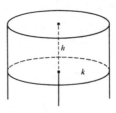

$W = -\int_{h_1}^{h_2} kp \, dh$ since the force is acting in the direction of decreasing h. But

since $v = kh$ or $h = \frac{1}{k} v$, $dh = \frac{1}{k} dv$. Using $p = 20,000v^{-3/2}$ to write the integral in terms of v, we get

$$W = -\int_9^4 k(20,000v^{-3/2}) \frac{1}{k} \, dv = 20,000 \int_9^4 v^{-3/2} \, dv = 6,666 \ 2/3 \text{ ft-lbs.}$$

7. Using $F(x) = kx$, where x is the displacement of the end of the spring and k is the spring constant, we have $F(x) = 2x$. Since the natural length of the spring is 5 cm, x goes from 0 to 3 when the spring is extended from 5 cm to 8 cm. Thus, the work done is

$$\int_0^3 2x \, dx = x^2 \Big|_0^3 = 9 \text{ kg-cm.}$$

9. Using $F(x) = kx$, we can find k from the given information that $F(2) = 10$. We get $k(2) = 10$ so $k = 5$ pounds per inch. Therefore, the work done in stretching the spring from its natural length to a length 4" greater is

$$\int_0^4 5x \, dx = \frac{5x^2}{2} \Big|_0^4 = 40 \text{ in-lbs.}$$

15. The work W needed to pump the solution to a height of 2' above the top of the tank is equal to the amount W_1 of work needed to move the solution to the top of the tank plus the work W_2 needed to move the solution from the top of the tank to a height of 2' above the top. Since the amount of work is independent of the way it is done, we will consider that W_1 is done by a piston pushing the liquid from the bottom of the tank to the top. The weight of the solution above the piston when the piston is x feet above the bottom of the tank is $(3)(5)(4 - x)(60) = 900(4 - x)$ pounds. Thus,

$$W_1 = \int_0^4 900(4 - x) \, dx = 900(4x - \frac{x^2}{2}) \Big|_0^4$$
$$= 900(16 - 8) = 7200 \text{ ft-lbs.}$$

Also, since the weight of the solution is $(3)(5)(4)(60) = 3600$ lbs
$$W_2 = 3600(2) = 7200 \text{ ft-lbs.}$$
Thus, $W = W_1 + W_2 = 14,400$ ft-lbs.

Exercises 5.3: Arc Length (page 241)

1. $y' = x^{1/2}$ so $L = \int_0^{15} (1 + x)^{1/2}\, dx = \frac{2}{3} (1 + x)^{3/2} \Big|_0^{15} = 42$ units.

3. $y' = 9x^{1/2}$ so $L = \int_1^4 (1 + 81x)^{1/2}\, dx = \frac{2}{243} [325^{3/2} - 82^{3/2}]$ units.

5. $y' = (x - 5)^{1/2}$ so $L = \int_6^8 (1 + x - 5)^{1/2}\, dx = \int_6^8 (x - 4)^{1/2}\, dx$

   $= \frac{2}{3} (8 - 2\sqrt{2})$ units.

7. $y' = x(x^2 + 2)^{1/2}$ so $[1 + (y')^2]^{1/2} = [x^4 + 2x^2 + 1]^{1/2} = x^2 + 1$

   so $L = \int_3^6 (x^2 + 1)\, dx = 66$ units.

9. $y' = \frac{x^2}{4} - x^{-2}$ so $[1 + (y')^2]^{1/2} = [1 + \frac{x^4}{16} - \frac{1}{2} + x^{-4}]^{1/2}$

   $= [\frac{x^4}{16} + \frac{1}{2} + x^{-4}]^{1/2} = \frac{x^2}{4} + x^{-2}$. Thus, $L = \int_1^{12} \frac{x^2}{4} + x^{-2}\, dx$

   $= (\frac{x^3}{12} - x^{-1}) \Big|_1^{12} = 144\ 5/6$ units.

13. We use $L = \int_0^4 [1 + (\frac{dx}{dy})^2]^{1/2}\, dy$. Since $\frac{dx}{dy} = \frac{3}{2} y^{1/2}$, we get

   $L = \int_0^4 [1 + \frac{9}{4} y]^{1/2}\, dy = \frac{4}{9} \int_0^4 [1 + \frac{9}{4} y]^{1/2}(\frac{9}{4}\, dy) = \frac{8}{27} [10^{3/2} - 1]$ units.

15. $[1 + (y')^2]^{1/2} = [1 + \frac{1}{36} (81x - 18 + x^{-1})]^{1/2} = \frac{1}{6} (9x^{1/2} + x^{-1/2})$.

   Therefore, $L = \frac{1}{6} \int_1^4 (9x^{1/2} + x^{-1/2})\, dx = \frac{1}{6} (6x^{3/2} + 2x^{1/2}) \Big|_1^4 = 7\ 1/3$ units.

17. Since $\frac{dx}{dt} = e^t \cos t - e^t \sin t$ and $\frac{dy}{dt} = e^t \sin t + e^t \cos t$ use of the formula

   $L = \int_c^d \sqrt{(\frac{dx}{dt})^2 + (\frac{dy}{dt})^2}\, dt$ gives

   $L = \int_0^\pi \sqrt{e^{2t} \cos^2 t + e^{2t} \sin^2 t + e^{2t} \sin^2 t + e^{2t} \cos^2 t}\, dt$.

   Thus, $L = \int_0^\pi \sqrt{2}\, e^t\, dt = \sqrt{2}\, (e^\pi - 1)$ units.

19. Since $\frac{dx}{d\theta} = \theta$ and $\frac{dy}{d\theta} = (2\theta + 1)^{1/2}$, use of the arc length formula gives

   $L = \int_0^9 \sqrt{\theta^2 + 2\theta + 1}\, d\theta = \int_0^9 (\theta + 1)\, d\theta = \frac{99}{2}$ units.

1. The number x of units sold when $p = 26$ satisfies $26 = 90 - x^3$,

so $x^3 = 64$, $x = 4$. Thus, $C = \int_0^4 (90 - x^3)\, dx - 26(4) = 192$.

3. The price p when 100 units are sold is $p = 20 - \dfrac{100}{50} = 18$. Thus,

$$C = \int_0^{100} (20 - \tfrac{x}{50})\, dx - 18(100) = 100.$$

9. The price p when 4 units are sold is $p = 200 + \dfrac{100}{5^2} - 4^2 = 188$.

Thus, $C = \int_0^4 [200 + 100(x + 1)^{-2} + x^2]\, dx - 188(4) = 106\,\tfrac{2}{3}$.

13. When $x = 4$, $p = 100 - 4\sqrt{16 + 9} = 80$. Thus,

$$C = \int_0^4 [100 - x\sqrt{x^2 + 9}]\, dx - 4(80) = 47\,\tfrac{1}{3}.$$

Exercises 5.7:   The Method of Cylindrical Shells (page 253)

1. $V = 2\pi \int_0^3 x(2x)\, dx = 4\pi\, (\tfrac{x^3}{3})\,\Big|_0^3 = 36\pi$ cubic units

3. $V = 2\pi \int_0^6 y(3 - \tfrac{y}{2})\, dy = 2\pi \int_0^6 (3y - \tfrac{y^2}{2})\, dy = 36\pi$ cubic units

5. This is like Exercise 1, except the element 2x moves around a circle of radius
$x + 2$ instead of x. We get

$$V = 2\pi \int_0^3 (x + 2)(2x)\, dx = 2\pi \int_0^3 (2x^2 + 4x)\, dx = 72\pi \text{ cubic units.}$$

7. This is like Exercise 3 except the element $3 - \tfrac{y}{2}$ moves around a circle of
radius $y + 3$ instead of y. We get.

$$V = 2\pi \int_0^6 (y + 3)(3 - \tfrac{y}{2})\, dy = 2\pi \int_0^6 (\tfrac{3y}{2} - \tfrac{y^2}{2} + 9)\, dy = 90\pi \text{ cubic units.}$$

9. This is like Exercise 1but the element 2x moves around a circle of radius
$5 - x$. We get

$$V = 2\pi \int_0^3 (5 - x)(2x)\, dx = 2\pi \int_0^3 (10x - 2x^2)\, dx = 54\pi \text{ cubic units.}$$

11. This is like Exercise 3 except the element $3 - \tfrac{y}{2}$ moves around a circle of
radius $8 - y$. We get

$$V = 2\pi \int_0^6 (8 - y)(3 - \tfrac{y}{2})\, dy = 2\pi \int_0^6 (24 - 7y + \tfrac{y^2}{2})\, dy = 108\pi \text{ cubic units.}$$

13.

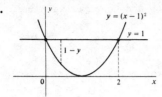

Intersection:
$$(x - 1)^2 = 1$$
$$x^2 - 2x + 1 = 1$$
$$x^2 - 2x = 0$$
$$x(x - 2) = 0$$
$$x = 0, \ x = 2$$

$$1 - y = 1 - (x - 1)^2 = -x^2 + 2x$$

$$V = 2\pi \int_0^2 x(-x^2 + 2x) \, dx = 2\pi \int_0^2 (-x^3 + 2x^2) \, dx = \frac{8\pi}{3} \text{ cubic units.}$$

15. $V = 2\pi \int_0^1 y(y^{1/3}) \, dy = 2\pi \int_0^1 y^{4/3} \, dy = \frac{6\pi}{7} \text{ cubic units}$

21.

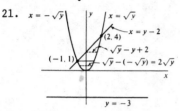

We need to add two integrals because the horizontal element goes between $x = y - 2$ and $x = \sqrt{y}$ above the line $y = 1$ and between $x = -\sqrt{y}$ and $x = \sqrt{y}$ below the line $y = 1$.

$$V = 2\pi \int_0^1 (y + 3)(2\sqrt{y}) \, dy + 2\pi \int_1^4 (y + 3)(\sqrt{y} - y + 2) \, dy = \frac{207}{5} \pi \text{ cubic units.}$$

Exercises 5.8: Area of a Surface of Revolution (page 258)

1. $A = 2\pi \int_1^4 (2x - 1)\sqrt{1 + 2^2} \, dx = 2\pi\sqrt{5} \int_1^4 (2x - 1) \, dx = 24\pi\sqrt{5} \text{ square units.}$

3. In terms of y, we have $x = \frac{1}{2}(y + 1)$, $1 \le y \le 7$, so

$$A = 2\pi \int_1^7 \frac{1}{2}(y + 1)\sqrt{1 + (\tfrac{1}{2})^2} \, dy = 15\pi\sqrt{5} \text{ square units.}$$

5. $A = 2\pi \int_1^2 y^3 \sqrt{1 + 9y^4} \, dy = \frac{\pi}{27} [(145)^{3/2} - 10^{3/2}] \text{ square units.}$

7. $f'(x) = \frac{x^2}{4} - x^{-2}$. So $\sqrt{1 + [f'(x)]^2} = \sqrt{\frac{x^4}{16} + \frac{1}{2} + x^{-4}} = \frac{x^2}{4} + x^{-2}$.

Thus, $A = 2\pi \int_1^2 (\frac{x^3}{12} + x^{-1})(\frac{x^2}{4} + x^{-2}) \, dx = 2\pi \int_1^2 (\frac{x^5}{48} + \frac{x}{3} + x^{-3}) \, dx$

$$= 2\pi (\frac{x^6}{288} + \frac{x^2}{6} - \frac{x^{-2}}{2}) \Big|_1^2 = \frac{35\pi}{16} \text{ sq in.}$$

9. $f'(x) = \dfrac{1}{2\sqrt{x}}$ so $\sqrt{1 + [f'(x)]^2} = \sqrt{1 + \dfrac{1}{4x}}$. Therefore,

$$|f(x)|\sqrt{1 + [f'(x)]^2} = \sqrt{x}\sqrt{1 + \dfrac{1}{4x}} = \sqrt{x + \dfrac{1}{4}}, \text{ so}$$

$$A = 2\pi \int_1^2 \left(x + \dfrac{1}{4}\right)^{1/2} dx = \dfrac{4\pi}{3}\left(x + \dfrac{1}{4}\right)^{3/2}\Big|_1^2 = \dfrac{\pi}{6}(27 - 5^{3/2}) \text{ square units.}$$

11. Since $x = \sqrt{y}$, we have $\dfrac{dx}{dy} = \dfrac{1}{2}y^{-1/2}$. Thus,

$$A = 2\pi \int_0^2 \sqrt{y}\sqrt{1 + \dfrac{1}{4y}}\, dy = 2\pi \int_0^2 \sqrt{y + \dfrac{1}{4}}\, dy = \dfrac{4}{3}\pi\left(y + \dfrac{1}{4}\right)^{3/2}\Big|_0^2 = \dfrac{13\pi}{3} \text{ units}^2.$$

13. In this case, $x = 4 - y^2$ or $y = \sqrt{4 - x}$. Thus, $y' = -\dfrac{1}{2}(4 - x)^{-1/2}$ and

$$A = 2\pi \int_0^4 \sqrt{4 - x}\sqrt{1 + \dfrac{1}{4(4 - x)}}\, dx = 2\pi \int_0^4 \sqrt{\dfrac{17}{4} - x}\, dx = -\dfrac{4}{3}\pi\left(\dfrac{17}{4} - x\right)^{3/2}\Big|_0^4$$

$$= \dfrac{\pi}{6}(17^{3/2} - 1) \text{ units}^2.$$

15. Since $y = 2x - 1$, $y' = 2$. The radius through which an element of arc is rotated is $|4 - (2x - 1)|$. Thus,

$$A = 2\pi \int_0^3 |4 - (2x - 1)|\sqrt{1 + 4}\, dx = 2\sqrt{5}\pi \int_0^3 |5 - 2x|\, dx$$

$$= 2\sqrt{5}\pi \left[\int_0^{5/2} (5 - 2x)\, dx + \int_{5/2}^3 (2x - 5)\, dx\right]$$

$$= 2\sqrt{5}\pi \left[(5x - x^2)\Big|_0^{5/2} + (x^2 - 5x)\Big|_{5/2}^3\right] = 13\sqrt{5}\pi \text{ units}^2.$$

Exercises 5.9: Moments and Centroids (page 263)

1. The area in question is a triangle with area $A = \dfrac{1}{2} \cdot 4 \cdot 3 = 6$. Since $y = \dfrac{1}{4}(12 - 3x)$,

$$\bar{x} = \dfrac{1}{6}\int_0^4 x \cdot \dfrac{1}{4}(12 - 3x)\, dx = \dfrac{1}{24}\int_0^4 (12x - 3x^2)\, dx$$

$$= \dfrac{1}{24}(6x^2 - x^3)\Big|_0^4 = \dfrac{4}{3}.$$

Since $x = \dfrac{1}{3}(12 - 4y)$,

$$\bar{y} = \dfrac{1}{6}\int_0^3 y\,\dfrac{1}{3}(12 - 4y)\, dy = \dfrac{1}{18}\int_0^3 (12y - 4y^2)\, dy$$

$$= \dfrac{1}{18}(6y^2 - \dfrac{4}{3}y^3)\Big|_0^3 = 1.$$

Consequently, the centroid is $(4/3, 1)$.

3.

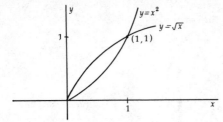

The area of the region in question is given by

$$A = \int_0^1 (\sqrt{x} - x^2)\ dx = \frac{1}{3}.$$

Then since the height of the region above x is $\sqrt{x} - x^2$, we have

$$\overline{x} = 3 \int_0^1 x(\sqrt{x} - x^2)\ dx = 3 \int_0^1 (x^{3/2} - x^3)\ dx$$

$$= 3(\frac{2}{5} x^{5/2} - \frac{1}{4} x^4)\ \Big|_0^1 = \frac{9}{20}.$$

By symmetry we also have $\overline{y} = \frac{9}{20}$. Thus, the centroid is $(9/20, 9/20)$.

7.

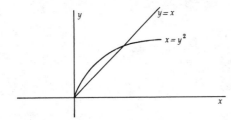

The area of the region in question is given by

$$A = \int_0^1 (y - y^2)\ dy = 1/6.$$

Since the height of the region above x is $\sqrt{x} - x$, we have

$$\overline{x} = 6 \int_0^1 x(\sqrt{x} - x)\ dx = 6 \int_0^1 (x^{3/2} - x^2)\ dx$$

$$= 6(\frac{2}{5} x^{5/2} - \frac{1}{3} x^3)\ \Big|_0^1 = 2/5.$$

Since the width of the region opposite y is $y - y^2$, we obtain

$$\overline{y} = 6 \int_0^1 y(y - y^2)\ dy = 6(\frac{1}{3} y^3 - \frac{1}{4} y^4)\ \Big|_0^1 = 1/2.$$

Consequently, the centroid is the point $(2/5, 1/2)$.

13.

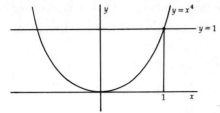

The area of the region in question is given by

$$A = 2 \int_0^1 (1 - x^4) \, dx = 8/5.$$

Since the region is symmetric about the y-axis, we know $\bar{x} = 0$. Since the width of the region opposite a point y is $2y^{1/4}$, we have

$$\bar{y} = \frac{5}{8} \int_0^1 y \, 2y^{1/4} \, dy = \frac{5}{4} \int_0^1 y^{5/4} \, dy = \frac{5}{9}.$$

Thus, the centroid is $(0, 5/9)$.

15.

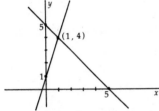

The area of the region is 21/2. Since the height of the region above the point x is given by

$$3x + 1 \text{ for } 0 \leq x \leq 1$$
$$5 - x \text{ for } 1 \leq x \leq 5,$$

we have

$$\bar{x} = \frac{2}{21} \left[\int_0^1 x(3x + 1) \, dx + \int_1^5 x(5 - x) \, dx \right]$$

$$= \frac{2}{21} \left[(x^3 + \frac{1}{2} x^2) \Big|_0^1 + (\frac{5}{2} x^2 - \frac{1}{3} x^3) \Big|_1^5 \right] = \frac{121}{63}.$$

The width of the region opposite the point y is

$$5 - y \text{ for } 0 \leq y \leq 1$$
$$5 - y - \frac{1}{3} (y - 1) \text{ for } 1 \leq y \leq 4.$$

Thus,

$$\bar{y} = \frac{2}{21} \left[\int_0^1 y(5 - y) \, dy + \int_1^4 y(5 - y - \frac{1}{3} (y - 1)) \, dy \right]$$

$$\bar{y} = \frac{2}{21} \left[(\frac{5}{2} y^2 - \frac{1}{3} y^3) \Big|_0^1 + (\frac{8}{3} y^2 - \frac{4}{9} y^3) \Big|_1^4 \right] = \frac{85}{63}.$$

Consequently, the centroid is $(\frac{121}{63}, \frac{85}{63})$.

19. By Exercise 2, the centroid of the region in question is (3/2,3/8). Consequently, by the Theorem of Pappus, the volume formed by the rotation of the region is
$$V = \frac{1}{2} \cdot \frac{9}{2} \cdot \frac{9}{8} \cdot 2\pi \; (5 - \frac{3}{2}) = \frac{567\pi}{32} \text{ units}^3.$$

21. By Exercise 9, $(\bar{x},\bar{y}) = (3/5,12/35)$. Since the area of the region is 1/12 units^2, we may use the Theorem of Pappus to obtain
$$V = \frac{1}{12} \; 2\pi \; (3 - 3/5) = 2\pi/5 \text{ units}^3.$$

<u>Exercises 5.10:</u> Force Exerted by a Fluid on a Vertical Surface
(page 266)

1. Take the positive direction of the y-axis downward. Then an equation of the right side of the triangle is $y - 5 = \frac{-10}{4} \; (x - 4)$ or $x = 6 - \frac{2y}{5}$. So $w(y) = 12 - \frac{4y}{5}$ and
$$F = \int_5^{15} (62.5)y(12 - \frac{4y}{5}) \; dy = 62.5 \int_5^{15} (12y - \frac{4y^2}{5}) \; dy = 20{,}833 \; 1/3 \text{ lbs.}$$

3. The right half of the plate is bounded by $x = \sqrt{25 - y^2} = w(y)$. To get the force on the whole plate, we multiply by 2 to get
$$F = 2 \int_0^5 (62.5)y(25 - y^2)^{1/2} \; dy = -\frac{125}{2} \int_0^5 (25 - y^2)^{1/2}(-2y \; dy)$$
$$= 5208 \; 1/3 \text{ lbs.}$$

5. An ellipse with a = 6 and b = 4 has equation $\frac{x^2}{36} + \frac{y^2}{16} = 1$. Thus, the right half of the semiellipse has an equation $x = \frac{3}{2} \; (16 - y^2)^{1/2}$. Thus,
$$F = 2 \int_0^4 (62.5)y(\frac{3}{2})(16 - y^2)^{1/2} \; dy = 4000 \text{ lbs.}$$

7.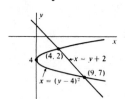

Intersection points:
$(y - 4)^2 = y + 2$
$y^2 - 9y + 14 = 0$
$(y - 7)(y - 2) = 0$
$y = 7, \; y = 2.$

Here $w(y) = y + 2 - (y - 4)^2 = y + 2 - y^2 + 8y - 16 = 9y - y^2 - 14.$

$$F = \int_2^7 62.5y(9y - y^2 - 14) \; dy = 62.5 \int_2^7 (9y^2 - y^3 - 14y) \; dy \approx 5{,}859 \text{ lbs.}$$

11. Here the y-axis is directed upward and since $y = 2x^2$, $x = \pm\sqrt{\frac{y}{2}}$ so $w(y) = \sqrt{\frac{y}{2}} - (-\sqrt{\frac{y}{2}}) = 2\sqrt{\frac{y}{2}} = \sqrt{2y}$. Since the depth of $w(y)$ is

$6 - y$, we get $F = \int_0^3 50(6 - y)\sqrt{2y} \; dy = 62.5\sqrt{2} \int_0^3 (6y^{1/2} - y^{3/2}) \; dy \approx 1029 \text{ lbs.}$

1. $2\pi \int_0^1 x(100 - x)\, dx = 2\pi(50 - \frac{1}{3}) \approx 312$

3. $2\pi \int_1^3 (100x - x^2)\, dx \approx 2{,}459$

5. $2\pi \int_0^3 x(x^2 - 100x + 1000)\, dx \approx 22{,}747$

7. $2\pi \int_3^6 (x^3 - 100x^2 + 1000x)\, dx \approx 47{,}147$

9. $2\pi \int_0^3 \dfrac{10{,}000x}{(x^2 + 2)^3}\, dx = 10{,}000\pi \int_0^3 (x^2 + 2)^{-3}(2x\, dx) \approx 3797$

11. $2\pi \int_3^8 \dfrac{10{,}000x}{(x^2 + 2)^3}\, dx = 10{,}000\pi \int_3^8 (x^2 + 2)^{-3}(2x\, dx) \approx 126$

Technique Review Exercises, Chapter 5 (page 272)

1. The side length of a cross section at x is $|x^2 - 2x|$ so the area A(x) of a cross section at x is $(x^2 - 2x)^2$.

$$V = \int_0^2 (x^2 - 2x)^2\, dx = \int_0^2 (x^4 - 4x^3 + 4x^2)\, dx = \frac{16}{15} \text{ cubic units.}$$

2. (a) The outer radius is 5 and the inner radius is $3 + \sqrt{y}$. Thus, the volume is

$$V = \pi \int_0^4 [5^2 - (3 + \sqrt{y})^2]\, dy = \pi \int_0^4 (25 - 9 - 6\sqrt{y} - y)\, dy = 24\pi \text{ cubic units.}$$

   (b) $V = 2\pi \int_0^2 (x + 3)x^2\, dx = 2\pi \int_0^2 (x^3 + 3x^2)\, dx = 24\pi$ cubic units.

3.

When the weight is x feet from its depth of 200', the total weight is $1500 + 4(200 - x) = 2300 - 4x$.

Thus, the work W is $W = \int_0^{200} (2300 - 4x)\, dx$

$= 380{,}000$ ft-lbs.

4. $y' = 3(x + 1)^{1/2}$ so $1 + (y')^2 = 1 + 9(x + 1) = 9x + 10$. Thus,

$$L = \int_6^{10} (9x + 10)^{1/2}\, dx = \frac{1}{9} \int_6^{10} (9x + 10)^{1/2}(9\, dx) = \frac{976}{27} \text{ units.}$$

5. Since $\dfrac{dx}{dt} = 2t$ and $\dfrac{dy}{dt} = 2t^2$, substitution into

$$L = \int_c^d \sqrt{\left(\frac{dx}{dt}\right)^2 + \left(\frac{dy}{dt}\right)^2}\, dt$$

63

gives
$$L = \int_1^2 \sqrt{4t^2 + 4t^4}\ dt = 2 \int_1^2 t\sqrt{1 + t^2}\ dt$$

$$= \frac{2}{3}(1 + t^2)^{3/2}\ \Big|_1^2 = \frac{2}{3}(5^{3/2} - 2^{3/2})\ \text{units.}$$

6. The number of units sold when p = 70 is obtained from $70 = 100 - \sqrt{x}$. So $\sqrt{x} = 30$, x = 900. Thus, the consumer surplus C is given by

$$C = \int_0^{900} (100 - x^{1/2})\ dx - 70(900) = 9,000.$$

7. $y' = (x + 2)^2$ so $1 + (y')^2 = 1 + (x + 2)^4$ and

$$A = 2\pi \int_{-2}^{-1} \frac{1}{3}(x + 2)^3 [1 + (x + 2)^4]^{1/2}\ dx$$

$$= \frac{\pi}{6} \int_{-2}^{-1} [1 + (x + 2)^4]^{1/2}\ 4(x + 2)^3\ dx = \frac{\pi}{6} \cdot \frac{2}{3} [1 + (x + 2)^4]^{3/2}\ \Big|_{-2}^{-1}$$

$$= \frac{\pi}{3}(2^{3/2} - 1)\ \text{sq units.}$$

8.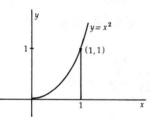

The area of the region is given by

$$A = \int_0^1 x^2\ dx = 1/3\ \text{units}^2.$$

Since the height of the region above the point x is x^2, we have

$$\bar{x} = 3 \int_0^1 x\ x^2\ dx = 3/4.$$

Since the width of the region opposite y is $1 - \sqrt{y}$, we have

$$\bar{y} = 3 \int_0^1 y(1 - \sqrt{y})\ dy = 3(\frac{1}{2} y^2 - \frac{2}{5} y^{5/2})\ \Big|_0^1 = 3/10.$$

Consequently, the centroid is (3/4, 3/10).

9. Take the origin at the center of the semicircle and the y-axis downward. Then $w(y) = 2(4 - y^2)^{1/2}$ so

$$F = \int_0^2 50y(2)(4 - y^2)^{1/2}\ dy = -50 \int_0^2 (4 - y^2)^{1/2}(-2y\ dy) = 266\ 2/3\ \text{lbs.}$$

10. $P = 2\pi \int_1^3 x(2000 - 10x^3)\ dx = 2\pi \int_1^3 (2000x - 10x^4)\ dx \approx 47,224$

1. Intersection points: $x^2 = 2x + 3$ so at $x = 3$, $x = -1$. The cross section at x is a square of side $2x + 3 - x^2$, so the area function of a cross section is $A(x) = (2x + 3 - x^2)^2$. Thus, the volume is

$$V = \int_{-1}^{3} A(x)\ dx = \int_{-1}^{3} (2x + 3 - x^2)^2\ dx = 34 \frac{2}{15} \text{ cubic units.}$$

3. $V = \int_{-\sqrt{3}}^{\sqrt{3}} [5^2 - (x^2 + 2)^2]\ dx$

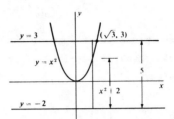

$\quad = \int_{-\sqrt{3}}^{\sqrt{3}} (25 - x^4 - 4x^2 - 4)\ dx$

$\quad = \pi \int_{-\sqrt{3}}^{\sqrt{3}} (21 - x^4 - 4x^2)\ dx$

$\quad = \dfrac{153\pi\sqrt{3}}{5}$ cubic units

5. Intersection points:

$x^2 = 6 - x$

$x^2 + x - 6 = 0$

$(x + 3)(x - 2) = 0$

$x = -3,\ x = 2$

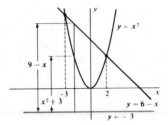

$$V = \pi \int_{-3}^{2} [(9 - x)^2 - (x^2 + 3)^2]\ dx = \frac{875\pi}{3} \text{ cubic units.}$$

7. Outer radius $10 - x^2$, inner radius $= 10 - (6 - x) = 4 + x$.

$$V = \pi \int_{-3}^{2} [(10 - x^2)^2 - (4 + x)^2]\ dx = 250\pi \text{ cubic units.}$$

9. $V = \pi \int_{0}^{5} [(x^2 + 3)^2 - 3^2]\ dx = 875\pi$ cubic units

11. 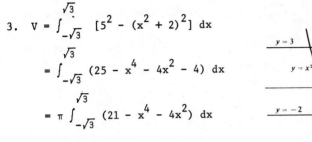 When the 500 pound weight is x feet above the ground, the total weight is $3(200 - x) + 500 = 1100 - 3x$. Thus, the work is

$$\int_{0}^{60} (1100 - 3x)\ dx = 60,600 \text{ ft-lbs.}$$

13. $F(x) = kx$. Since $F(3) = 24 = k(3)$, $k = 8$. Thus, $F(x) = 8x$. Hence,

$$W = \int_{0}^{-2} 8x\ dx = 16 \text{ in-lbs.}$$

15. Let W_1 be the work to move all of the liquid to the level of the top of the tank and let W_2 be the work in moving the liquid from the top to a level 5' above the top. To find W_1, let a variable piston of radius $r(x)$ move the liquid over the interval $0 \le x \le 12$. Then

$$W_1 = \int_0^{12} \pi [r(x)]^2 (12 - x) 60 \, dx = 60\pi \int_0^{12} \frac{x^2}{9} (12 - x) \, dx$$

$$= \frac{20\pi}{3} \int_0^{12} (12x^2 - x^3) \, dx = \frac{20\pi}{3} (4x^3 - \frac{x^4}{4}) \Big|_0^{12} = \frac{20\pi}{3} (6912 - 5184) \approx 36,191$$

ft-lbs. Also, since the volume of the liquid is $\frac{\pi}{3} (4)^2 (12)$ or 64π cubic feet, $W_2 = 5(60)(64\pi) \approx 60,319$ ft-lbs. Thus, the total work W is $W = W_1 + W_2$ $\approx 96,510$ ft-lbs.

17. $y' = x^{1/2}$ so $L = \int_3^{35} (1 + x)^{1/2} \, dx = \frac{416}{3}$ units

19. $y' = \frac{3x^2}{2} - \frac{x^{-2}}{6}$ so $[1 + (y')^2]^{1/2} = [1 + \frac{9x^4}{4} - \frac{1}{2} + \frac{x^{-4}}{36}]^{1/2}$

$= [\frac{9x^4}{4} + \frac{1}{2} + \frac{x^{-4}}{36}]^{1/2} = \frac{3x^2}{2} + \frac{x^{-2}}{6}$. Thus, $L = \int_1^2 (\frac{3x^2}{2} + \frac{x^{-2}}{6}) \, dx = \frac{43}{12}$ units.

21. $y' = x(x^2 - 2)^{1/2}$ so $[1 + (y')]^{1/2} = [x^4 - 2x^2 + 1]^{1/2} = x^2 - 1$.

Thus, $L = \int_0^6 (x^2 - 1) \, dx = (\frac{x^3}{3} - x) \Big|_0^6 = 66$ units.

25. Since $\frac{dx}{d\theta} = a \sin \theta$ and $\frac{dy}{d\theta} = a \cos \theta$,

$$L = \int_0^{2\pi} \sqrt{a^2 \sin^2 \theta + a^2 \cos^2 \theta} \, d\theta = |a| \int_0^{2\pi} d\theta = 2\pi |a| \text{ units.}$$

27. Here $30 = 155 - x^3$, $x^3 = 125$, $x = 5$. Thus, $C = \int_0^5 (155 - x^3) \, dx - 30(5)$
$= 468 \frac{3}{4}$.

29. Here $p = 300 - 10\sqrt{9} = 270$. Thus, $C = \int_0^6 [300 - 10(x + 3)^{1/2}] \, dx - 270(6)$
$= 20\sqrt{3}$.

31. $V = 2\pi \int_0^2 x(6 - 3x) \, dx = 2\pi \int_0^2 (6x - 3x^2) \, dx = 8\pi$ cubic units

33. $2\pi \int_0^2 (5 - x)(6 - 3x) \, dx = 52\pi$ cubic units

35. (a) $2\pi \int_0^6 (x + 2)(6 - 3x) \, dx = 32\pi$ cubic units

(b) $2\pi \int_0^6 (9 - y)(\frac{y}{3}) \, dy = 60\pi$ cubic units

37. Intersection: $x^2 = 2 - x$, $x = -2$, $x = 1$. We need to use two integrals. Above the line $y = 1$, $w_1 = 2 - y - (-\sqrt{y}) = 2 - y + \sqrt{y}$. Below the line $y = 1$, $w_2 = \sqrt{y} - (-\sqrt{y}) = 2\sqrt{y}$. The volume is

$$2\pi \int_1^4 (5 - y)\, w_1 \, dy + 2\pi \int_0^1 (5 - y)\, w_2 \, dy = 2\pi \int_1^4 (5 - y)(2 - y + \sqrt{y})\, dy$$

$$+ \, 2\pi \int_0^1 (5 - y)(2\sqrt{y})\, dy = \frac{153\pi}{5} \text{ cubic units.}$$

39. $V = 2\pi \int_{-2}^1 (x + 3)(2 - x - x^2)\, dx = 2\pi \int_{-2}^1 (6 - x - 4x^2 - x^3)\, dx = \frac{45\pi}{2}$ cubic units.

41. $y' = \sqrt{7}$ so $A = 2\pi \int_2^5 (\sqrt{7}x - 10)\sqrt{1 + (\sqrt{7})^2}\, dx = (42\pi\sqrt{14} - 120\pi\sqrt{2})$ sq units.

43. $y' = 2x^2$ so $A = 2\pi \int_0^{\sqrt[4]{2}} \frac{2}{3} x^3 (1 + 4x^4)^{1/2}\, dx = \frac{\pi}{12} \int_0^{\sqrt[4]{2}} (1 + 4x^4)^{1/2}(16x^3 \, dx)$

$$= \frac{13\pi}{9} \text{ sq units.}$$

45. Here $x = y^3$, $0 \le y \le 1$ and $\frac{dx}{dy} = 3y^2$ so we get

$$A = 2\pi \int_0^1 y^3 [1 + (3y^2)^2]^{1/2}\, dy = \frac{\pi}{18} \int_0^1 [1 + 9y^4]^{1/2}(36y^3 \, dy)$$

$$= \frac{\pi}{27} (10^{3/2} - 1) \text{ sq units.}$$

49.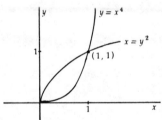

The area of the region is

$$A = \int_0^1 (\sqrt{x} - x^4)\, dx = 7/15.$$

Since the height of the region above x is $\sqrt{x} - x^4$, we have

$$\bar{x} = \frac{15}{7} \int_0^1 x(\sqrt{x} - x^4)\, dx = \frac{15}{7} \left(\frac{2}{5} x^{5/2} - \frac{1}{6} x^6\right) \Big|_0^1 = 1/2.$$

Since the width of the region opposite y is $y^{1/4} - y^2$, we have

$$\bar{y} = \frac{15}{7} \int_0^1 y(y^{1/4} - y^2)\, dy = \frac{15}{7} \left(\frac{4}{9} y^{9/4} - \frac{1}{4} y^{1/4}\right) \Big|_0^1 = 5/12.$$

Thus, the centroid is $(1/2, 5/12)$.

51. The centroid of the region in question is $(1, 4)$. Thus, by the Theorem of Pappus, the volume of revolution is

$$V = \frac{1}{2} \cdot 6 \cdot 3 \cdot 2\pi \cdot 2 = 36\pi \text{ units}^3.$$

53. $F = \int_6^{18} (62.5)\, y\, (10)\, dy = 90,000$ lbs.

55. The right half of the plate has equation $x = \sqrt{36 - y^2} = w(y)$. Thus,

$$F = 2\int_0^6 (62.5)\, y\, (36 - y^2)^{1/2}\, dy = -\frac{125}{2}\int_0^6 (36 - y^2)^{1/2}(-2y\, dy) = 9,000 \text{ lbs.}$$

57. (a) $2\pi \int_0^2 x(1000 - x^3)\, dx = 2\pi \int_0^2 (1000x - x^4)\, dx \approx 12,526.$

(b) $2\pi \int_1^3 (1000x - x^4)\, dx = 2\pi(500\, x^2 - \frac{x^5}{5})\, \Big|_1^3 \approx 24,829.$

59. (a) $2\pi \int_0^1 x[\frac{50,000}{(x^2 + 3)^2}]\, dx = 50,000\pi \int_0^1 (x^2 + 3)^{-2}(2x\, dx) \approx 13,090.$

(b) $2\pi \int_1^2 x[\frac{50,000}{(x^2 + 3)^2}]\, dx = -50,000\, \pi\, (x^2 + 3)^{-1}\, \Big|_1^2 \approx 16,830.$

CHAPTER 6: EXPONENTIAL AND LOGARITHMIC FUNCTIONS

Exercises 6.1: The Exponential Function (page 281)

1. $8^{1/3} = \sqrt[3]{8} = 2$

3. $(32)^{-3/5} = \frac{1}{(\sqrt[5]{32})^3} = \frac{1}{(2)^3} = \frac{1}{8}$

5. $(\frac{1}{9})^{-3/2} = 9^{3/2} = (\sqrt{9})^3 = 27$

7. $(\frac{27}{8})^{-2/3} = (\frac{8}{27})^{2/3} = (\sqrt[3]{\frac{8}{27}})^2 = \frac{4}{9}$

Exercises 6.2: Monotone Functions, Inverse Functions (page 286)

1. $y = 2x + 5$, so $x = g(y) = \frac{y - 5}{2}$, $1 \le y \le 19$, since $y = 1$ when $x = -2$ and $y = 19$ when $x = 7$.

3. $y = x^2 - 4$, $x^2 = y + 4$ so $x = g(y) = -\sqrt{y + 4}$, $y \ge -4$

5. $y = x^2 - 4$, $x^2 = y + 4$ so $x = g(y) = \sqrt{y + 4}$, $y \ge -4$

7. $y = \sqrt{x - 5}$, $y^2 = x - 5$ so $x = y^2 + 5$, $0 < y < 5$

9. This function is not continuous or monotone but does have an inverse.

$$x = g(y) = \begin{cases} \frac{y}{2} & \text{if } 0 \le y \le 2 \\ 5 - y & \text{if } 2 \le y \le 4 \end{cases}$$

11. For $|x|$, $x \ge 0$ the function is continuous and monotone and so has an inverse. Similarly for $|x|$, $x \le 0$.

13. For $(x - 3)^2$, $x \ge 3$ the function is continuous and monotone and so has an inverse. Similarly for $(x - 3)^2$, $x \le 3$.

15. For $|5 - x|$, $x \ge 5$ the function is continuous and monotone and so has an inverse. Similarly for $|5 - x|$, $x \le 5$.

17. For $6 - x^4$, $x \geq 0$ the function is continuous and monotone and so has an inverse. Similarly for $6 - x^4$, $x \leq 0$.

19. For $x^2 - 3$, $x \geq 0$ the function is continuous and monotone and so has an inverse. Similarly for $x^2 - 3$, $x \leq 0$.

21. $\frac{dy}{dx} = 5(x - 1)^4$ so $\frac{dx}{dy} = \dfrac{1}{5(x - 1)^4}$, $x \neq 1$. At $x = 2$ we get $1/5$.

23. $\frac{dy}{dx} = 4x^3 + 7$ so $\frac{dx}{dy} = \dfrac{1}{4x^3 + 7}$. At $x = 1$ we get $1/11$.

25. $\frac{dy}{dx} = -3x^4 + 2x$ so $\frac{dx}{dy} = \dfrac{1}{-3x^{-4} + 2x}$. At $x = 2$ we get $16/61$.

27. $\frac{dy}{dx} = 7x^6 + 2$ so $\frac{dx}{dy} = \dfrac{1}{7x^6 + 2}$. At $x = 0$ we get $1/2$.

Exercises 6.3: The Logarithmic Function (page 288)

1. $\log .01 = -2$ since $10^{-2} = .01$

3. $\log_3 (\frac{1}{27}) = -3$ since $3^{-3} = \frac{1}{27}$

5. $\log_{1/3} (\frac{1}{27}) = 3$ since $(\frac{1}{3})^3 = \frac{1}{27}$

7. $\log_3 81 = 4$ since $3^4 = 81$

9. $\log_{1/2} (1/32) = 5$ since $(1/2)^5 = 1/32$

11. $\log 10^{17} = 17$ since $10^{17} = 10^{17}$

17. $\log_a (x + h) - \log_a x = \log_a (\frac{x + h}{x}) = \log_a (1 + \frac{h}{x})$, so $\log_a (x + h) - \log_a x$
$$= \log_a (1 + \frac{h}{x})$$

19. Let $\log_b a = p$. Then $b^p = a$ so $(b^p)^{1/p} = a^{1/p}$ or $b = a^{1/p}$. Therefore, by definition of the $\log_a b$, $\log_a b = \frac{1}{p}$. Hence, $\log_a b = \dfrac{1}{\log_b a}$.

Exercises 6.4: Differentiation of Logarithmic Functions (page 293)

1. $\frac{d}{dx} [\ln (x^6 + 3x^2 + 5)] = \dfrac{\frac{d}{dx} (x^6 + 3x^2 + 5)}{x^6 + 3x^2 + 5} = \dfrac{6x^5 + 6x}{x^6 + 3x^2 + 5}$

3. $\frac{d}{dx} [x^2 \log (x - 1)] = x^2 \frac{d}{dx} \log (x - 1) + 2x \log (x - 1)$
$$= \dfrac{x^2 \log e}{x - 1} + 2x \log (x - 1)$$

7. $\frac{d}{dx} \ln |x^2 - 3x + 2| = \dfrac{\frac{d}{dx} (x^2 - 3x + 2)}{x^2 - 3x + 2} = \dfrac{2x - 3}{x^2 - 3x + 2}$

9. $\frac{d}{dx} [\ln |x^3 - 3x + 1|^3] = \frac{d}{dx} 3 \ln |x^3 - 3x + 1| = \dfrac{3(3x^2 - 3)}{x^3 - 3x + 1}$

11. $\frac{d}{dx} (\ln \ln |x^3 + 2|) = \frac{\frac{d}{dx} \ln |x^3 + 2|}{\ln |x^3 + 2|} = \frac{\frac{3x^2}{x^3 + 2}}{\ln |x^3 + 2|} = \frac{3x^2}{(x^3 + 2) \ln |x^3 + 2|}$

13. $\int_1^3 \frac{1}{x} \, dx = \ln |x| \, \Big|_1^3 = \ln 3 - \ln 1 = \ln 3 - 0 = \ln 3$

15. $\int \frac{x}{x^2 + 3} \, dx = \frac{1}{2} \int \frac{2x}{x^2 + 3} \, dx = \frac{1}{2} \ln |x^2 + 3| + C$

17. $\int \frac{x - 3}{x^2 - 6x + 7} \, dx = \frac{1}{2} \int \frac{2x - 6}{x^2 - 6x + 7} \, dx = \frac{1}{2} \ln |x^2 - 6x + 7| + C$

19. $\int \frac{x^3}{1 - x^4} \, dx = -\frac{1}{4} \int \frac{-4x^3}{1 - x^4} \, dx = -\frac{1}{4} \ln |1 - x^4| + C$

21. $\int \frac{dx}{x \ln x} = \int \frac{\frac{1}{x}}{\ln x} \, dx = \ln |\ln x| + C$

23. $\int_1^{e^2} \frac{(\ln x)^3}{x} \, dx = \int_1^{e^2} (\ln x)^3 (\frac{1}{x} \, dx) = \frac{(\ln x)^4}{4} \, \Big|_1^{e^2} = \frac{1}{4} [(\ln e^2)^4 - (\ln 1)^4]$

$= \frac{1}{4} [2^4 - 0^4] = 4$

25. $\int \frac{\ln x}{x} \, dx = \int \ln x \, (\frac{1}{x} \, dx) = \int \ln x \, d (\ln x) = \frac{1}{2} \ln^2 x + C$

29. $A = \int_2^3 \frac{x^3}{x^4 - 8} \, dx = \frac{1}{4} \int_2^3 \frac{4x^3}{x^4 - 8} \, dx = \frac{1}{4} (\ln |x^4 - 8|) \, \Big|_2^3$

$= \frac{1}{4} [\ln 73 - \ln 8] = \frac{1}{4} \ln (\frac{73}{8})$

<u>Exercises 6.5:</u> The Derivative of a^x; Logarithmic Differentiation
(page 297)

1. $\sqrt{2} \, (x^2 + 9)^{\sqrt{2}-1} (2x) = 2\sqrt{2} \, x(x^2 + 9)^{\sqrt{2}-1}$

7. $e^{5t^2-3t+2} \frac{d}{dx} (5t^2 - 3t + 2) = (10t - 3) \, e^{5t^2-3t+2}$

13. Let $y = x^x$, then $\ln y = x \ln x$, so $\frac{1}{y} \frac{dy}{dx} = x \, (\frac{1}{x}) + \ln x = 1 + \ln x$.   Therefore,

$\frac{dy}{dx} = y(1 + \ln x) = x^x (1 + \ln x)$.

15. Let $y = (x + 1)^{2x-1}$, then $\ln y = (2x - 1) \ln (x + 1)$, so

$\frac{1}{y} \frac{dy}{dx} = (2x - 1)(\frac{1}{x + 1}) + 2 \ln (x + 1)$.   Therefore,

$\frac{dy}{dx} = (x + 1)^{2x-1} [\frac{2x - 1}{x + 1} + 2 \ln (x + 1)]$.

17. Let $y = (x + 5)^4(x - 1)^3(x + 3)^2$, then $\ln |y| = 4 \ln |x + 5| + 3 \ln |x - 1|$

$+ 2 \ln |x + 3|$, so $\frac{1}{y} \frac{dy}{dx} = \frac{4}{x + 5} + \frac{3}{x - 1} + \frac{2}{x + 3}$. Therefore,

$\frac{dy}{dx} = (x + 5)^4(x - 1)^3(x + 3)^2 [\frac{4}{x + 5} + \frac{3}{x - 1} + \frac{2}{x + 3}]$.

21. Let $y = \frac{(2x + 7)^{3/2}(x^2 + 1)^{3/4}}{(x + 2)^3}$, then $\ln |y| = \frac{3}{2} \ln |2x + 7| + \frac{3}{4} \ln |x^2 + 1|$

$- 3 \ln |x + 2|$, so $\frac{1}{y} \frac{dy}{dx} = \frac{3}{2} (\frac{2}{2x + 7}) + \frac{3}{4} (\frac{2x}{x^2 + 1}) - 3 (\frac{1}{x + 2})$. Hence,

$\frac{dy}{dx} = \frac{(2x + 7)^{3/2}(x + 1)^{3/4}}{(x + 2)^3} [\frac{3}{2x + 7} + \frac{3x}{2(x^2 + 1)} - \frac{3}{x + 2}]$.

23. Let N be the number of bacteria at elapsed time t after 1 P.M. Then

$\frac{dN}{dt} = kN$, so $N = Ce^{kt}$ where C is constant. Since $N = 10,000$ when $t = 0$,

$10,000 = Ce^0 = C$ so $N = 10,000 e^{kt}$. But $N = 12,000$ when $t = 2$, so

$12,000 = 10,000 e^{2k}$ and we get $e^{2k} = 6/5$ or $e^k = (6/5)^{1/2}$. Therefore,

$N = 10,000 (6/5)^{t/2}$. So at 8 P.M., $N = 10,000 (6/5)^{7/2} \approx 18,929$.

27. Let T be the difference of the temperature of the casting and the surrounding

air at elapsed time t. Then $\frac{dT}{dt} = kT$, so $T = Ce^{kt}$. Since $T = 730$ when

$t = 0$, $730 = Ce^0 = C$, so $T = 730 e^{kt}$. But $T = 630$ when $t = 1$, so $630 = 730e^k$

and we get $e^k = 63/73$. Therefore, $T = 730 (63/73)^t$. Since $T = 30$ when the

casting cools to 100, $30 = 730 (63/73)^t$. Hence, $(63/73)^t = 30/730 = 3/73$,

and so $t \ln (63/73) = \ln (3/73)$ and we get $t = \frac{\ln (3/73)}{\ln (63/73)} \approx 21.7$ minutes.

29. Let W be the weight of the carbon 14 in the animal at elapsed time t after it

died. Then $\frac{dW}{dt} = kW$, so $W = Ce^{kt}$. Let $W = W_0$ when $t = 0$, so $W_0 = Ce^0 = C$

and $W = W_0 e^{kt}$. Since $W = W_0/2$ when $t = 5,600$, $W_0/2 = W_0 e^{5,600k}$. Thus,

$1/2 = (e^k)^{5,600}$, so $e^k = (1/2)^{1/5,600}$ and $W = W_0 (1/2)^{t/5,600}$. Hence, when

$W = W_0/5$, $W_0/5 = W_0 (1/2)^{t/5,600}$ and we get $(1/2)^{t/5,600} = 1/5$, or

$2^{t/5,600} = 5$. Thus, $(t/5,600) \ln 2 = \ln 5$, so $t = 5,600 \ln 5/\ln 2 \approx 13,000$

years ago that the animal died.

33. When $t = 0$, $N = 50,000$, when $t = 9$, $N = 6,250$, and when $t = 12$, $N = 3,125$.

$\frac{dN}{dt} = 50,000 (\frac{1}{3})(\frac{1}{2})^{t/3} \ln (\frac{1}{2})$. But, $\ln (\frac{1}{2}) = \ln 1 - \ln 2 = 0 - \ln 2 = -\ln 2$,

so $\frac{dN}{dt} = - \frac{(\ln 2) 50,000}{3} (\frac{1}{2})^{t/3}$. Thus, when $t = 0$ hours, $\frac{dN}{dt} = - \frac{(\ln 2) 50,000}{3}$

and when $t = 3$ hours, $\frac{dN}{dt} = - \frac{(\ln 2)(25,000)}{3}$.

1. $\int e^{3x} \, dx = \frac{1}{3} \int e^{3x} (3 \, dx) = \frac{1}{3} e^{3x} + C$

3. $\int e^{-t} \, dt = -\int e^{-t}(-dt) = -e^{-t} + C$

5. $\frac{1}{6} \int_0^2 e^{6x-3x^2} (6 - 6x) \, dx = \frac{1}{2} e^{6x-3x^2} \Big|_0^2 = \frac{1}{2} [e^0 - e^0] = 0$

7. $\frac{1}{3} \int e^{x^3+3x^2} (3x^2 + 6x) \, dx = \frac{1}{3} e^{x^3+3x^2} + C$

9. $-\frac{1}{2} \int e^{x^{-2}} (-2x^{-3}) \, dx = -\frac{1}{2} e^{x^{-2}} + C$

11. $\frac{1}{3} \int_0^1 5^{3x} (3 \, dx) = \frac{1}{3 \ln 5} (5^{3x}) \Big|_0^1 = \frac{124}{3 \ln 5}$

13. $-2 \int_0^2 e^{-\frac{x}{2}} (-\frac{1}{2}) \, dx = -2e^{-\frac{x}{2}} \Big|_0^2 = 2(1 - e^{-1})$

15. $\int e^{-6x} (e^{6x} + 2e^{3x} + 1) \, dx = \int (1 + 2e^{-3x} + e^{-6x}) \, dx$

$$= x - \frac{2}{3} e^{-3x} - \frac{1}{6} e^{-6x} + C$$

17. $\int \frac{x \, dx}{e^{x^2}} = \int e^{-x^2} x \, dx = -\frac{1}{2} \int e^{-x^2} (-2x \, dx) = -\frac{1}{2} e^{-x^2} + C$

21. $A = \int_0^{\ln 3} \frac{e^x - e^{-x}}{2} \, dx = \frac{1}{2} \int_0^{\ln 3} (e^x - e^{-x}) \, dx = \frac{2}{3}$ sq units.

1. Let $u = \ln |x|$ and $dv = x^7 \, dx$, then $du = \frac{dx}{x}$ and $v = \frac{x^8}{8}$, so

$$\int x^7 \ln |x| \, dx = \frac{x^8 \ln |x|}{8} - \frac{1}{8} \int x^7 \, dx = \frac{x^8 \ln |x|}{8} - \frac{x^8}{64} + C$$

3. Let $u = x^3$ and $dv = e^x \, dx$, then $du = 3x^2 \, dx$ and $v = e^x$, so

$$\int x^3 e^x \, dx = x^3 e^x - 3 \int x^2 e^x \, dx. \text{ Let } u = x^2 \text{ and } dv = e^x \, dx, \text{ then}$$

$du = 2x \, dx$ and $v = e^x$, so $\int x^3 e^x \, dx = x^3 e^x - 3x^2 e^x + 6 \int x e^x \, dx$. Let $u = x$ and $dv = e^x \, dx$, then $du = dx$ and $v = e^x$, so

$$\int x^3 e^x \, dx = x^3 e^x - 3x^2 e^x + 6xe^x - 6 \int e^x \, dx = x^3 e^x - 3x^2 e^x + 6xe^x - 6e^x + C$$

5. Let $u = x$ and $dv = e^{-2x} \, dx$, then $du = dx$ and $v = -\frac{1}{2} e^{-2x}$, so

$$\int xe^{-2x} \, dx = -\frac{x}{2} e^{-2x} + \frac{1}{2} \int e^{-2x} \, dx - \frac{x}{2} e^{-2x} - \frac{1}{4} e^{-2x} + C.$$

7. Let $u = \ln^2 |x|$ and $dv = x\,dx$, then $du = \dfrac{2 \ln |x|}{x}\,dx$ and $v = \dfrac{x^2}{2}$, so

$\int x \ln^2 |x|\,dx = \dfrac{x^2}{2} \ln^2 |x| - \int x \ln |x|\,dx$. Let $u = \ln |x|$ and $dv = x\,dx$,

then $du = \dfrac{dx}{x}$ and $v = \dfrac{x^2}{2}$, so

$\int x \ln^2 |x|\,dx = \dfrac{x^2}{2} \ln^2 |x| - \dfrac{x^2}{2} \ln |x| + \dfrac{1}{2} \int x\,dx = \dfrac{x^2}{2} \ln^2 |x|$

$- \dfrac{x^2}{2} \ln |x| + \dfrac{x^2}{4} + C.$

9. Let $u = \ln |x|$ and $dv = x^{-2}\,dx$, then $du = \dfrac{dx}{x}$ and $v = -x^{-1}$, so

$\int x^{-2} \ln |x|\,dx = -x^{-1} \ln |x| + \int x^{-2}\,dx = -x^{-1} \ln |x| - x^{-1} + C.$

11. Let $u = x^2$ and $dv = e^{3x}\,dx$, then $du = 2x\,dx$ and $v = \dfrac{1}{3} e^{3x}$, so

$\int x^2 e^{3x}\,dx = \dfrac{x^2}{3} e^{3x} - \dfrac{2}{3} \int x e^{3x}\,dx$. Let $u = x$ and $dv = e^{3x}$, then

$du = dx$ and $v = \dfrac{1}{3} e^{3x}$, so $\int x^2 e^{3x}\,dx = \dfrac{x^2}{3} e^{3x} - \dfrac{2}{9} x e^{3x} + \dfrac{2}{9} \int e^{3x}\,dx$

$= \dfrac{x^2}{3} e^{3x} - \dfrac{2}{9} x e^{3x} + \dfrac{2}{27} e^{3x} + C.$

13. Let $u = \ln x$ and $dv = dx$, then $du = \dfrac{1}{x}\,dx$ and $v = x$, so

$\int \ln x\,dx = x \ln x - \int dx = x \ln x - x + C.$

15. Use $p(x) = x^5$ and $q(x) = e^x$. Then $\int x^5 e^x\,dx = x^5 e^x - 5x^4 e^x + 20\, x^3 e^x$

$- 60\, x^2 e^x + 120\, x e^x - 120\, e^x + C.$

Exercises 6.8: The Hyperbolic Function (page 308)

11. $f'(x) = \cosh (x^2 + 5x) \dfrac{d}{dx} (x^2 + 5x) = (2x + 5) \cosh (x^2 + 5x)$

15. $f'(x) = -\operatorname{csch}^2 (4x^2 + 7x) \dfrac{d}{dx} (4x^2 + 7x) = -(8x + 7) \operatorname{csch}^2 (4x^2 + 7x)$

23. $\int \sinh (2x + 4)\,dx = \dfrac{1}{2} \int \sinh (2x + 4)(2\,dx) = \dfrac{1}{2} \cosh (2x + 4) + C$

25. $\int \dfrac{1}{x} \cosh (\ln x)\,dx = \int \cosh (\ln x)(\dfrac{1}{x}\,dx) = \sinh (\ln x) + C$

27. $\int \tanh x\, \operatorname{sech}^2 x\,dx = \int \tanh x\, d(\tanh x) = \dfrac{1}{2} \tanh^2 x + C$

29. $\displaystyle\int_0^1 x \sinh x^2\,dx = \dfrac{1}{2} \int_0^1 \sinh x^2\, d(x^2) = \dfrac{1}{2} \cosh x^2 \Big|_0^1 = \dfrac{1}{2} (\cosh 1 - 1)$

33. Let $u = x^2$ and $dv = \cosh x\,dx$ so $du = 2x\,dx$ and $v = \sinh x$. Thus, the integral

equals $x^2 \sinh x - 2 \int x \sinh x\,dx$. Now let $u = x$ and $dv = \sinh x\,dx$ so that

$du = dx$ and $v = \cosh x$. Thus, the integral equals $x^2 \sinh x - 2x \cosh x$

$- 2 \int \cosh x\,dx = x^2 \sinh x - 2x \cosh x + 2 \sinh x + C.$

1. Let $y = x^2 + 3$, then $4 \le y \le 52$ and $x^2 = y - 3$, so $x = g(y)$

    $= -\sqrt{y - 3}$, $4 \le y \le 52$.

3. $\ln y = (3x - 1) \ln x$ and $\frac{1}{y} y' = (3x - 1) \frac{1}{x} + 3 \ln x$, so

    $y' = x^{3x-1} [\frac{3x - 1}{x} + 3 \ln x]$

4. $y = 5 \ln |x^3 - 2|$, so $y' = \frac{5(3x^2)}{x^3 - 2} = \frac{15x^2}{x^3 - 2}$

5. $\int \frac{e^{\frac{2}{x}}}{x^2} dx = -\frac{1}{2} \int e^{2x^{-1}} (-2x^{-2}) dx = -\frac{1}{2} e^{2x^{-1}} + C$

6. $\int x^3 \ln x^2 dx = 2 \int x^3 \ln x \, dx$, since $\ln x^2 = 2 \ln x$. Let $u = \ln x$ and

    $dv = x^3 dx$, then $du = \frac{1}{x} dx$ and $v = \frac{x^4}{4}$, so $\int x^3 \ln x^2 dx = \frac{x^4}{2} \ln x$

    $- \frac{1}{2} \int x^3 dx = \frac{x^4}{2} \ln x - \frac{x^4}{8} + C.$

7. Let $u = x^2$ and $dv = e^{3x} dx$, then $du = 2x \, dx$ and $v = \frac{1}{3} e^{3x}$, so

    $\int x^2 e^{3x} dx = \frac{x^2}{3} e^{3x} - \frac{2}{3} \int xe^{3x} dx$. Let $u = x$ and $dv = e^{3x}$, then $du = dx$ and

    $v = \frac{1}{3} e^{3x}$, so $\int x^2 e^{3x} dx = \frac{x^2}{3} e^{3x} - \frac{2x}{9} e^{3x} + \frac{2}{9} \int e^{3x} dx$

    $= \frac{x^2}{3} e^{3x} - \frac{2x}{9} e^{3x} + \frac{2}{27} e^{3x} + C.$

8. $\int \frac{x \, dx}{3x^2 + 5} = \frac{1}{6} \int \frac{6x \, dx}{3x^2 + 5} = \frac{1}{6} \ln |3x^2 + 5| + C$

9. $y' = 3 \sinh^2 (x^2 + 5) \frac{d}{dx} \sinh [(x^2 + 5)] = 6x \sinh^2 (x^2 + 5) \cosh (x^2 + 5)$

10. $\int \sinh^3 (7x) \cosh (7x) dx = \frac{1}{7} \int \sinh^3 (7x) d[\sinh (7x)] = \frac{1}{28} \sinh^4 (7x) + C$

Additional Exercises, Chapter 6 (page 312)

1. $64^{2/3} = (\sqrt[3]{64})^2 = 4^2 = 16$

3. $(\frac{8}{64})^{-2/3} = (\frac{1}{8})^{-2/3} = 8^{2/3} = (\sqrt[3]{8})^2 = 4$

9. $y = 6x - 1$ so $x = g(y) = \frac{y + 1}{6}$, $-7 \le y \le 17$

11. $y = x^2 + 5$, $x^2 = y - 5$ so $x = g(y) = \sqrt{y - 5}$, $y \ge 5$

13. $\log .0001 = -4$ since $10^{-4} = .0001$

15. $\log_2 (\frac{1}{16}) = -4$ since $2^{-4} = \frac{1}{16}$

21. $\frac{d}{dx} [\log_3 (7 - x)] = \frac{-1}{7 - x} \log_3 e = \frac{\log_3 e}{x - 7}$

23. $\dfrac{d}{dx} [4 \ln |2 - x - x^3|] = \dfrac{4(-1 - 3x^2)}{2 - x - x^3}$

25. $-\ln |1 - x| \Big|_2^5 = -\ln |-4| + \ln |-1| = -\ln 4 + 0 = -\ln 4$

27. $\int (\ln x)^{-2}(\frac{1}{x} dx) = -(\ln x)^{-1} + C = \dfrac{-1}{\ln x} + C$

31. Let $y = x^{9x}$, then $\ln y = 9x \ln x$, so $\dfrac{y'}{y} = 9x \left(\frac{1}{x}\right) + 9 \ln x = 9 + 9 \ln x$.

Therefore, $y' = y(9 + 9 \ln x) = 9x^{9x} (1 + \ln x)$.

33. Let $y = \dfrac{(x + 17)^4}{(x - 6)^5(x - 8)^9}$, then $\ln |y| = 4 \ln |x + 17| - 5 \ln$

$|x - 6| - 9 |x - 8|$, so $\dfrac{y'}{y} = \dfrac{4}{x + 17} - \dfrac{5}{x - 6} - \dfrac{9}{x - 8}$. Therefore,

$y' = \dfrac{(x + 17)^4}{(x - 6)^5(x - 8)^9} [\dfrac{4}{x + 17} - \dfrac{5}{x - 6} - \dfrac{9}{x - 8}]$.

35. $\int e^{15x} dx = \frac{1}{15} \int e^{15x} (15 dx) = \frac{1}{15} e^{15x} + C$

37. $\int x^3 e^{x^4+5} dx = \frac{1}{4} \int e^{x^4+5} (4x^3 dx) = \frac{1}{4} e^{x^4+5} + C$

39. $\int \dfrac{e^{1/x^3}}{x^4} dx = -\frac{1}{3} \int e^{1/x^3} (-3x^{-4}) dx = -\frac{1}{3} e^{1/x^3} + C$

41. $-\frac{1}{3} \int_0^1 e^{-t^3} (-3t^2 dt) = -\frac{1}{3} e^{-t^3} \Big|_0^1 = -\frac{1}{3} (e^{-1} - 1) = \dfrac{1 - e^{-1}}{3}$

43. Let $u = x$ and $dv = e^{4x} dx$, then $du = dx$ and $v = \frac{1}{4} e^{4x}$, so

$\int xe^{4x} dx = \frac{x}{4} e^{4x} - \frac{1}{4} \int e^{4x} dx = \frac{x}{4} e^{4x} - \frac{1}{16} e^{4x} + C$

45. Let $u = \ln |x|$ and $dv = x^{10} dx$. Then $du = \frac{1}{x} dx$ and $v = \dfrac{x^{11}}{11}$, so

$\int x^{10} \ln |x| dx = \dfrac{x^{11}}{11} \ln |x| - \int \dfrac{x^{10}}{11} dx = \dfrac{x^{11}}{11} \ln |x| - \dfrac{x^{11}}{121} + C.$

47. Let $u = \ln |x^3| dx$ and $dv = x^4 dx$. Then $du = \dfrac{3x^2}{x^3} = \dfrac{3}{x}$ and $v = \dfrac{x^5}{5}$. So

$\int x^4 \ln |x^3| dx = \dfrac{x^5}{5} \ln |x^3| - \int \dfrac{3}{5} x^4 dx = \dfrac{x^5}{5} \ln |x^3| - \dfrac{3}{25} x^5 + C.$

49. Let $u = x^3$ and $dv = e^{3x} dx$. Then $du = 3x^2 dx$ and $v = \frac{1}{3} e^{3x}$, so

$\int x^3 e^{3x} dx = \dfrac{x^3}{3} e^{3x} - \int x^2 e^{3x} dx.$ Let $u = x^2$ and $dv = e^{3x} dx$. Then $du = 2x$

and $v = \frac{1}{3} e^{3x}$, so $\int x^3 e^{3x} dx = \dfrac{x^3}{3} e^{3x} - \dfrac{x^2}{3} e^{3x} + \frac{2}{3} \int xe^{3x} dx.$ Let $u = x$ and

$dv = e^{3x} dx.$ Then $du = dx$ and $v = \frac{1}{3} e^{3x}$, so

$\int x^3 e^{3x} dx = \dfrac{x^3}{3} e^{3x} - \dfrac{x^2}{3} e^{3x} + \frac{2}{9} xe^{3x} - \frac{2}{9} \int e^{3x} dx$

$$= \frac{x^3}{3} e^{3x} - \frac{x^2}{3} e^{3x} + \frac{2}{9} xe^{3x} - \frac{2}{27} e^{3x} + C.$$

53. $f'(x) = 5 \sinh^4 (x^2 - x) \frac{d}{dx} \sinh (x^2 - x) = 5(2x - 1) \sinh (x^2 - x) \cosh (x^2 - x)$

57. $\int x \sinh (5 - x^2) \, dx = -\frac{1}{2} \int \sinh (5 - x^2)(-2x \, dx) = -\frac{1}{2} \cosh (5 - x^2) + C$

CHAPTER 7: CIRCLES AND TRIGONOMETRIC FUNCTIONS

Exercises 7.1: The Derivative of the Sine Function (page 320)

1. $\lim\limits_{x \to 0} \frac{x}{\sin x} = \lim\limits_{x \to 0} \frac{1}{\frac{\sin x}{x}} = \frac{1}{1} = 1$

3. $\lim\limits_{x \to 0} \frac{\sin 9x}{2x} = \frac{9}{2} \lim\limits_{x \to 0} \frac{\sin 9x}{9x} = \frac{9}{2} (1) = \frac{9}{2}$

5. $\lim\limits_{x \to 0} \frac{1}{2x \csc 8x} = 4 \lim\limits_{x \to 0} \frac{\sin 8x}{8x} = 4(1) = 4$

7. $\lim\limits_{x \to 0} \frac{3x - 2 \sin 5x}{x} = \lim\limits_{x \to 0} (3 - \frac{2 \sin 5x}{x}) = 3 - 10 \lim\limits_{x \to 0} \frac{\sin 5x}{5x} = 3 - 10(1) = -7$

9. $\lim\limits_{x \to 0} \frac{\tan 3x}{5x} = \lim\limits_{x \to 0} (\frac{\sin 3x}{\cos 3x} \cdot \frac{1}{5x}) = \frac{3}{5} \lim\limits_{x \to 0} (\frac{\sin 3x}{3x} \cdot \frac{1}{\cos 3x})$

$$= \frac{3}{5} (\lim\limits_{x \to 0} \frac{\sin 3x}{3x})(\lim\limits_{x \to 0} \frac{1}{\cos 3x}) = \frac{3}{5} (1)(1) = \frac{3}{5}$$

11. $\lim\limits_{x \to 0} \frac{1 - \cos x}{\sin x} = \lim\limits_{x \to 0} (\frac{1 - \cos x}{\sin x} \cdot \frac{1 + \cos x}{1 + \cos x}) = \lim\limits_{x \to 0} \frac{1 - \cos^2 x}{\sin x (1 + \cos x)}$

$$= \lim\limits_{x \to 0} \frac{\sin^2 x}{\sin x (1 + \cos x)} = \lim\limits_{x \to 0} \frac{\sin x}{1 + \cos x} = \frac{0}{2} = 0$$

13. $\lim\limits_{x \to 0} \frac{x \cos x}{1 - \cos x} = \lim\limits_{x \to 0} (\frac{x \cos x}{1 - \cos x} \cdot \frac{1 + \cos x}{1 + \cos x})$

$$= \lim\limits_{x \to 0} \frac{(x \cos x)(1 + \cos x)}{1 - \cos^2 x} = \lim\limits_{x \to 0} \frac{(x \cos x)(1 + \cos x)}{\sin^2 x}$$

$$= \lim\limits_{x \to 0} [\frac{x}{\sin x} (\cos x)(1 + \cos x) \frac{1}{\sin x}] \text{ does not exist.}$$

17. $4(\sin^3 3x) 3 \cos 3x = 12 \sin^3 3x \cos 3x$

21. $\frac{d}{dx} \csc x = \frac{d}{dx} (\sin x)^{-1} = -(\sin x)^{-2} \cos x = \frac{-\cos x}{\sin^2 x}$

25. Let $y = x^{\sin x}$, then $\ln y = \sin x \ln x$, so $\frac{1}{y} \frac{dy}{dx} = \frac{\sin x}{x} + \cos x \ln x$. Hence,

$$\frac{dy}{dx} = x^{\sin x} (\frac{\sin x}{x} + \cos x \ln x).$$

27. Since $R = \dfrac{v_0^2}{g} \sin 2\theta$, $0 \le \theta \le \pi/2$, we have $\dfrac{dR}{d\theta} = \dfrac{v_0^2}{g} 2 \cos 2\theta$. Note that

$\dfrac{dR}{d\theta} = 0$ if and only if $\theta = \pi/4$. Thus, we must check $\theta = 0$, $\pi/4$, and $\pi/2$ to

determine which determines the largest range. Since $R = 0$ when $\theta = 0$ and

$\theta = \pi/2$ the angle that gives the maximum range is $\theta = \pi/4$.

29.

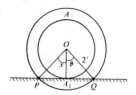

Let x be the height of the center 0 of the disc above the solution as shown.
Then $0 \le x \le 2$ and so $0 \le \theta \le \pi/2$ for the indicated angle θ. Let A be the area
of the wetted surface above the solution and let A_1 be the area of the wetted
surface of the disc below the solution. Then since $x = 2 \cos \theta$,

$A + A_1 = \pi(2^2 - x^2) = \pi(4 - 4 \cos^2 \theta) = 4\pi(1 - \cos^2 \theta) = 4\pi \sin^2 \theta$. Also,

$A_1 = $ (area of sector OPQ) $-$ (area of triangle OPQ) $= \pi(2)^2 (\dfrac{2\theta}{2\pi})$

$- (2 \cos \theta)(2 \sin \theta) = 4\theta - 4 \sin \theta \cos \theta$. Therefore $A = 4\pi \sin^2 \theta - A_1$

$= 4\pi \sin^2 \theta - 4\theta + 4 \sin \theta \cos \theta = 4\pi \sin^2 \theta - 4\theta + 2 \sin 2\theta$, $0 \le \theta \le \pi/2$

and $\dfrac{dA}{d\theta} = 8\pi \sin \theta \cos \theta - 4 + 4 \cos 2\theta = 4(2\pi \sin \theta \cos \theta - 1 + \cos 2\theta)$. Using

the identity $1 - \cos 2\theta = 2 \sin^2 \theta$ this can be expressed

$\dfrac{dA}{d\theta} = 4(2\pi \sin \theta \cos \theta - 2 \sin^2 \theta) = 8 \sin \theta \cos \theta (\pi - \tan \theta)$. So for

$0 < \theta < \pi/2$, $\dfrac{dA}{d\theta} = 0$ if $\tan \theta = \pi$, while $\dfrac{dA}{d\theta} > 0$ if $\tan \theta < \pi$ and $\dfrac{dA}{d\theta} < 0$ if

$\tan \theta > \pi$. Thus, for $0 < \theta < \pi/2$, A is increasing if $\tan \theta < \pi$ and decreasing
if $\tan \theta > \pi$. Thus, A has its largest value when $\tan \theta = \pi$. But, $x = 2 \cos \theta$

$= \dfrac{2}{\sec \theta}$. Since $1 + \tan^2 \theta = \sec^2 \theta$, $\sec \theta = \sqrt{1 + \tan^2} = \sqrt{1 + \pi^2}$,

$x = \dfrac{2}{\sqrt{1 + \pi^2}} \approx .61$ feet.

Exercises 7.2: The Derivatives of the Other Trigonometric Functions
(page 323)

1. $\dfrac{d}{dx} \cos (3 - x^2) = -\sin (3 - x^2) \dfrac{d}{dx} (3 - x^2) = 2x \sin (3 - x^2)$

3. $\dfrac{d}{dx} \cos^3 (5 - 2x) = 3 \cos^2 (5 - 2x) \dfrac{d}{dx} [\cos (5 - 2x)] = 6 \cos^2 (5 - 2x)$

$\sin (5 - 2x)$

5. $\frac{d}{dx} [\sec^2 (3x^2 - 1)] = 2 \sec (3x^2 - 1) \frac{d}{dx} [\sec (3x^2 - 1)]$

$\qquad = 2 \sec (3x^2 - 1) \sec (3x^2 - 1) \tan (3x^2 - 1) \frac{d}{dx} (3x^2 - 1)$

$\qquad = 12x \sec^2 (3x^2 - 1) \tan (3x^2 - 1)$

7. $\sec^2 \sqrt{2x - 1} \frac{d}{dx} \sqrt{2x - 1} = (\sec^2 \sqrt{2x - 1}) (\frac{1}{2}) (2x - 1)^{-1/2} (2)$

$\qquad\qquad\qquad = (2x - 1)^{-1/2} \sec^2 \sqrt{2x - 1}$

9. $2 \csc (e^x + x) \frac{d}{dx} \csc (e^x + x) = -2 \csc (e^x + x)$

$\qquad [\csc (e^x + x) \cot (e^x + x)] (e^x + 1) = -2(e^x + 1) \csc^2 (e^x + x) \cot (e^x + x)$

13. $3 \cot^2 (x^2 - 1) \frac{d}{dx} \cot (x^2 - 1) = 3 \cot^2 (x^2 - 1) [-\csc^2 (x^2 - 1)(2x)]$

$\qquad\qquad\qquad\qquad = -6x \cot^2 (x^2 - 1) \csc^2 (x^2 - 1)$

15. $(\sec^2 3x) 5 \tan^4 3x \frac{d}{dx} \tan 3x + (\tan^5 3x) 2 \sec 3x \frac{d}{dx} (\sec 3x)$

$\qquad = 5 \sec^2 3x \tan^4 3x (\sec^2 3x)(3) + (\tan^5 3x)(2 \sec 3x)(3 \sec 3x \tan 3x)$

$\qquad = 15 \sec^4 3x \tan^4 3x + 6 \sec^2 5x \tan^6 3x$

19. $d \ln |\sec 5x| = \frac{5 \sec 5x \tan 5x}{\sec 5x} dx = 5 \tan 5x \, dx$

23. $\sec 2x \, d (\tan 2x) + \tan 2x \, d (\sec 2x) = \sec 2x (2 \sec^2 2x) \, dx$

$\qquad + \tan 2x (2 \sec 2x \tan 2x) \, dx = (2 \sec^3 2x + 2 \sec 2x \tan^2 2x) \, dx$

29.

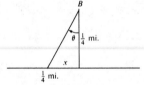

$\qquad\qquad\qquad\qquad \frac{d\theta}{dt} = 60 \text{ radians/hr}$

$x = \frac{1}{4} \tan \theta$, so $\frac{dx}{dt} = \frac{1}{4} \sec^2 \theta \frac{d\theta}{dt}$. When $x = 1/4$, $\theta = \pi/4$, so

$\frac{dx}{dt} = \frac{1}{4} (\sec^2 \pi/4) (60) = \frac{1}{4} (\sqrt{2})^2 (60) = 30 \text{ mph}$.

31.

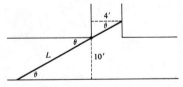

Let the length of a board touching the corner and walls indicated be $L(\theta)$.
Then $L(\theta) = 4 \sec \theta + 10 \csc \theta$, $0 < \theta < \pi/2$. Of course $L(\theta)$ varies with θ and
the smallest value of $L(\theta)$ (not the largest value) is the length of the longest
straight thin board that can be moved horizontally around the corner. Thus, we
are to find the absolute minimum value of $L(\theta)$. Now $L'(\theta) = 4 \sec \theta \tan \theta$

$- 10 \csc \theta \cot \theta = 4 \frac{\sin \theta}{\cos^2 \theta} - \frac{10 \cos \theta}{\sin^2 \theta} = \frac{4 \cos \theta}{\sin^2 \theta} [\frac{\sin^3 \theta}{\cos^3 \theta} - \frac{5}{2}]$

$$= \frac{4 \cos \theta}{\sin^2 \theta} [\tan^3 \theta - \frac{5}{2}], \quad 0 < \theta < \pi/2.$$ We note that $\sin \theta$ and $\cos \theta$ are both positive for $0 < \theta < \pi/2$. Thus, $L'(\theta) < 0$ and $L(\theta)$ is decreasing if $\tan^3 \theta < \frac{5}{2}$, while $L'(\theta) > 0$ and $L(\theta)$ is increasing if $\tan^3 \theta > \frac{5}{2}$. Hence, the absolute minimum value of $L(\theta)$ occurs when $\tan \theta = \sqrt[3]{\frac{5}{2}} = \sqrt[3]{2.5}$. Using $\sec \theta = \sqrt{1 + \tan^2 \theta}$ and $\csc \theta = \sqrt{1 + \cot^2 \theta}$, we see that the longest board has length

$$L = 4\sqrt{1 + (2.5)^{2/3}} + 10\sqrt{1 + (2.5)^{-2/3}} \approx 19.16 \text{ ft.}$$

33. The maximum value of W must occur where $\frac{dW}{d\theta} = 0$ or at one of the endpoints $\theta = 0$ or $\theta = \pi/2$. Since

$$\frac{dW}{d\theta} = \frac{F}{\mu} (-\sin \theta + \mu \cos \theta)$$

$\frac{dW}{d\theta} = 0$ only where $\sin \theta = \mu \cos \theta$. That is, where $\tan \theta = \mu$ or $\theta = \arctan \mu$. Since $\frac{d^2W}{d\theta^2} = \frac{F}{\mu} (-\cos \theta - \mu \sin \theta) < 0$ for $0 \leq \theta \leq \pi/2$, $\theta = \arctan \mu$ must determine a maximum value.

35.

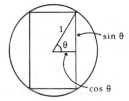

The figure shows the sphere and the inscribed right circular cylinder in cross section. If θ denotes the indicated angle, the lateral surface area of the cylinder is

$$S = 4\pi \sin \theta \cos \theta, \quad 0 \leq \theta \leq \pi/2.$$

Since

$$\frac{dS}{d\theta} = 4\pi (\cos^2 - \sin^2 \theta) = 4\pi \cos 2\theta,$$

we have

$$\frac{dS}{d\theta} = 0 \text{ if and only if } 2\theta = \pi/2, \text{ or } \theta = \pi/4.$$

Since the maximum value of S must occur at $\theta = \pi/4$, $\theta = 0$, or $\theta = \pi/2$ and since S is 0 at both 0 and $\pi/2$, $\theta = \pi/4$ must determine the maximum surface area. The radius if $\cos \pi/4 = 1/\sqrt{2}$ and the height is $2 \sin \pi/4 = 2/\sqrt{2}$.

37. On differentiating $\sin 2x = 2 \sin x \cos x$ we get

$$2 \cos 2x = 2 \cos^2 x - 2 \sin^2 x$$

or,

$$\cos 2x = \cos^2 x - \sin^2 x.$$

Exercises 7.3: The Integrals of the Trigonometric Functions (page 328)

1. $\int \sin 5x \, dx = \frac{1}{5} \int \sin 5x \, (5 \, dx) = -\frac{1}{5} \cos 5x + C$

3. $\int x \tan x^2 \, dx = \frac{1}{2} \int \tan x^2 \, (2x \, dx) = \frac{1}{2} \ln |\sec x^2| + C$

5. $\int \frac{1}{x} \sin (\ln x) \, dx = \int \sin (\ln x)(\frac{1}{x} \, dx) = -\cos (\ln x) + C$

7. $\int \sin x \cos^2 x \, dx = -\int \cos^2 x \, (-\sin x \, dx) = -\frac{1}{3} \cos^3 x + C$

11. Let u = x and dv = sin x dx, then du = dx and v = -cos x, so

$\int x \sin x \, dx = -x \cos x + \int \cos x \, dx = -x \cos x + \sin x + C$

13. $\int \csc^2 (5x - 11) \, dx = \frac{1}{5} \int \csc^2 (5x - 11)(5 \, dx) = -\frac{1}{5} \cot (5x - 11) + C$

15. $\int \sec 5x \, dx = \frac{1}{5} \int \sec 5x \, (5 \, dx) = \frac{1}{5} \ln |\sec 5x + \tan 5x| + C$

17. $\frac{1}{5} \int_0^{\pi/2} \frac{5 \cos 5x}{1 + \sin 5x} \, dx = \frac{1}{5} \ln |1 + \sin 5x| \Big|_0^{\pi/2} = \frac{\ln 2}{5}$

19. $\frac{1}{2} \int \tan^5 2x \sec^2 2x \, (2 \, dx) = \frac{1}{12} \tan^6 2x + C$

21. $\frac{1}{3} \int_0^{\pi/2} \sec 3x \tan 3x \, (3 \, dx) = \frac{1}{3} \sec 3x \Big|_0^{\pi/2} = \frac{1}{3} (\sqrt{2} - 1)$

23. $\int \sec^5 x \tan x \, dx = \int \sec^4 x \, (\sec x \tan x \, dx) = \frac{\sec^5 x}{5} + C$

25. $\int \sin x e^{\cos x} \, dx = -\int e^{\cos x} (-\sin x \, dx) = -e^{\cos x} + C$

27. Let u = x and dv = sin 2x dx. Then du = dx and v = $-\frac{1}{2} \cos 2x$. Then we get

$-\frac{x}{2} \cos 2x + \frac{1}{2} \int \cos 2x \, dx = -\frac{x}{2} \cos 2x + \frac{1}{4} \sin 2x + C.$

29. Let u = x^2 and dv = sin x dx. Then du = 2x dx and v = -cos x, so

$\int x^2 \sin x \, dx = -x^2 \cos x + 2 \int x \cos x \, dx.$ Let u = x and dv = cos x. Then

du = dx and v = sin x. $\int x^2 \sin x \, dx = -x^2 \cos x + 2x \sin x - 2 \int \sin x \, dx$

$= -x^2 \cos x + 2x \sin x + 2 \cos x + C.$

31. $\int \sin (\ln x) \cos^2 (\ln x) \frac{1}{x} \, dx = \int \cos^2 (\ln x) \sin (\ln x) \, d(\ln x)$

$= -\int \cos^2 (\ln x) \, d \cos (\ln x)$

$= -\frac{1}{3} \cos^3 (\ln x) + C$

33. $\int_0^{\pi/3} 2 \sin 3x \, dx = \frac{2}{3} \int_0^{\pi/3} \sin 3x \, (3 \, dx) = \frac{4}{3}$ sq units.

35. The volume is given by V = $2\pi \int_0^{\pi} x \sin x \, dx.$ Use of integration by parts with

u = x and dv = sin x dx gives

$$V = 2\pi(-x \cos x \Big|_0^{\pi} + \int_0^{\pi} \cos x \, dx)$$

$$= 2\pi(\pi + \sin x \Big|_0^{\pi}) = 2\pi^2 \text{ units}^3.$$

37. $W = \int_0^{\pi/2} \sin x \cos^3 x \, dx = -\int_0^{\pi/2} \cos^3 x \, d \cos x = -\frac{1}{4} \cos^4 x \Big|_0^{\pi/2} = \frac{1}{4}$ ft lbs

Exercises 7.4:   Inverse Trigonometric Functions (page 333)

1. $\arcsin \dfrac{\sqrt{3}}{2} = \pi/3$

3. $\text{arccot } \sqrt{3} = \pi/6$

5. $\arctan (-\sqrt{3}) = -\dfrac{\pi}{3}$

7. $\arccos (\dfrac{1}{2}) = \dfrac{\pi}{3}$

9. $\text{arcsec } (-\sqrt{2}) = \dfrac{3\pi}{4}$

11. $\text{arccsc } 2 = \pi/6$

13. $\sec (\text{arcsec } 5) = 5$

15. $\arctan (\tan 10\pi) = 0$

17. $\arccos (\cos \dfrac{7\pi}{6}) = \dfrac{5\pi}{6}$

19. $\text{arcsec } (\sec \dfrac{4\pi}{3}) = \dfrac{2\pi}{3}$

21. Since $\cos (\arcsin x) = \sqrt{1 - x^2}$, we have $\cos (\arcsin 1/2) = \sqrt{1 - 1/4} = \sqrt{3}/2$.

23. Since $\tan (\text{arcsec } x) = \sqrt{x^2 - 1}$ when $x \geq 1$, we have

$\tan (\text{arcsec } 5) = \sqrt{25 - 1} = 2\sqrt{6}$.

25. Since $\tan (\text{arcsec } x) = -\sqrt{x^2 - 1}$ when $x \leq -1$, we have

$\tan (\text{arcsec } (-5)) = -\sqrt{25 - 1} = -2\sqrt{6}$.

27. Since $\sec (\arctan x) = \sqrt{1 + x^2}$, we have $\sec (\arctan 10) = \sqrt{101}$.

29. Since $\sin 7\pi/8 = \sin \pi/8$, we know that $\arcsin (\sin 7\pi/8) = \arcsin (\sin \pi/8)$. Then since $-\pi/2 \leq \pi/8 \leq \pi/2$, $\arcsin (\sin \pi/8) = \pi/8$.   Thus,
$$\arcsin (\sin 7\pi/8) = \pi/8.$$

31. $y = 3 \sin 5x$, so $-3 \leq y \leq 3$; $\sin 5x = y/3$, so $5x = \arcsin y/3$.

$x = g(y) = \dfrac{1}{5} \arcsin (y/3)$, $-3 \leq y \leq 3$.

33. $y = 3 \arcsin 2x$, so $\dfrac{-3\pi}{2} \leq y \leq \dfrac{3\pi}{2}$.   Also, $\arcsin 2x = \dfrac{y}{3}$, so $2x = \sin \dfrac{y}{3}$.   Hence,

$g(y) = \dfrac{1}{2} \sin \dfrac{y}{3}$, $\dfrac{-3\pi}{2} \leq y \leq \dfrac{3\pi}{2}$.

37. $y = \sin x \cos x$, so $2y = 2 \sin x \cos x = \sin 2x$, or $y = \dfrac{1}{2} \sin 2x$.   Hence,

$-\dfrac{1}{2} \leq y \leq \dfrac{1}{2}$.   Also, $2x = \arcsin 2y$, so $x = g(y) = \dfrac{1}{2} \arcsin 2y$, $-\dfrac{1}{2} \leq y \leq \dfrac{1}{2}$.

Exercises 7.5:   Derivatives of the Inverse Trigonometric Functions
(page 336)

1. $\dfrac{1}{\sqrt{1 - 25x^2}} \dfrac{d (5x)}{dx} = \dfrac{5}{\sqrt{1 - 25x^2}}$

3. $\dfrac{1}{|x^{-2}|\sqrt{x^{-4} - 1}} \dfrac{d}{dx} (x^{-2}) = \dfrac{x^2}{\sqrt{x^{-4} - 1}} (-2x^{-3}) = \dfrac{-2}{x\sqrt{x^{-4} - 1}}$

5. $\dfrac{d}{dx} \arctan (e^x - 1) = \dfrac{1}{(e^x - 1)^2 + 1} \dfrac{d}{dx} (e^x - 1) = \dfrac{e^x}{e^{2x} - 2e^x + 2}$

15. $y' = \dfrac{1}{\sqrt{1 - x^2}} + \dfrac{-1}{\sqrt{1 - x^2}} = 0$.   Therefore, $\arcsin x + \arccos x$ is a constant.
(It is $\pi/2$.)

17. $y' = \dfrac{(-1)[\sec x \tan x + \sec^2 x]}{\sec^2 x + 2 \sec x \tan x + \tan^2 x + 1} = \dfrac{(-1)(\sec x \tan x + \sec^2 x)}{2(\sec^2 x + \sec x \tan x)} = -\dfrac{1}{2}.$

Therefore, arccot $(\sec x + \tan x) = -\dfrac{x}{2} + C$ where C is a constant. $(C = \pi/4.)$

19. Let $u = \arctan x$ and $dv = dx$, then $du = \dfrac{dx}{x^2 + 1}$ and $v = x$, so

$$\int \arctan x \, dx = x \arctan x - \int \dfrac{x \, dx}{x^2 + 1} = x \arctan x - \dfrac{1}{2} \int \dfrac{2x \, dx}{x^2 + 1}$$

$$= x \arctan x - \dfrac{1}{2} \ln (x^2 + 1) + C.$$

21. Let $u = \arccos x$ and $dv = dx$, then $du = \dfrac{-dx}{\sqrt{1 - x^2}}$ and $v = x$, so

$$\int \arccos x \, dx = x \arccos x + \int \dfrac{x \, dx}{\sqrt{1 - x^2}} = x \arccos x - (1 - x^2)^{1/2} + C.$$

23.

$\theta = \text{arccot} \dfrac{x}{20} - \text{arccot} \dfrac{x}{15}, \ x \geq 0. \quad \dfrac{d\theta}{dx} = \dfrac{-20}{x^2 + 400} + \dfrac{15}{x^2 + 225}$

$= \dfrac{1500 - 5x^2}{(x^2 + 400)(x^2 + 225)} = \dfrac{5(10\sqrt{3} - x)(10\sqrt{3} + x)}{(x^2 + 400)(x^2 + 225)}.$ Thus, $\dfrac{d\theta}{dx} > 0$ if

$0 \leq x < 10\sqrt{3}$ and $\dfrac{d\theta}{dx} < 0$ if $x > 10\sqrt{3}$, so for $x \geq 0$, $x = 10\sqrt{3}$ determines an abso-

lute maximum value of θ. Therefore, the driver can see the sign best at a distance of $10\sqrt{3}$ feet down the road.

<u>Exercises 7.6:</u> Integration Formulas Involving the Inverse Trigonometric Functions (page 338)

1. $\dfrac{1}{\sqrt{3}} \int_{-1}^{1} \dfrac{(\frac{1}{\sqrt{3}}) \, dx}{(\frac{x}{\sqrt{3}})^2 + 1} = \dfrac{1}{\sqrt{3}} \arctan (\dfrac{x}{\sqrt{3}}) \Big|_{-1}^{1} = \dfrac{1}{\sqrt{3}} [\dfrac{\pi}{6} - (-\dfrac{\pi}{6})] = \dfrac{\pi}{3\sqrt{3}}$

3. $\dfrac{1}{6} \int \dfrac{\frac{3}{2} \, dx}{(\frac{3x}{2})^2 + 1} = \dfrac{1}{6} \arctan (\dfrac{3x}{2}) + C$

5. $\dfrac{1}{2} \int \dfrac{(\frac{e^x}{2}) \, dx}{\frac{e^x}{2} \sqrt{(\frac{e^x}{2})^2 - 1}} = \dfrac{1}{2} \text{arcsec} (\dfrac{e^x}{2}) + C$

7. $\dfrac{1}{4} \int \dfrac{(\frac{3}{4}) \, dx}{|\frac{3x}{4}| \sqrt{(\frac{3x}{4})^2 - 1}} = \dfrac{1}{4} \text{arcsec} (\dfrac{3x}{4}) + C$

9. $\int \dfrac{(\frac{1}{5})\ dx}{\sqrt{1 - (\frac{x}{5})^2}} = \arcsin \left(\dfrac{x}{5}\right) + C$

11. $\dfrac{1}{6} \int \dfrac{(\frac{2x}{3})\ dx}{(\frac{x}{3})^2 \sqrt{(\frac{x}{3})^2 - 1}} = \dfrac{1}{6} \operatorname{arcsec} \left(\dfrac{x^2}{3}\right) + C$

13. $\dfrac{1}{3} \int \dfrac{(\frac{3}{2x})\ dx}{\sqrt{1 - (\frac{3\ \ln\ x}{2})^2}} = \dfrac{1}{3} \arcsin \left(\dfrac{3\ \ln\ x}{2}\right) + C$

15. $-\dfrac{1}{2} \int (16 - x^2)^{-1/2}(-2x)\ dx = -\dfrac{1}{2}(2)(16 - x^2)^{1/2} + C = -(16 - x^2)^{1/2} + C$

17. $\int \dfrac{x\ dx}{x^4 + 4} = \dfrac{1}{2} \int \dfrac{2x\ dx}{(x^2)^2 + 4} = \dfrac{1}{4} \arctan \dfrac{x^2}{2} + C$

19. $\int \dfrac{dx}{\sqrt{x}\ (1 + x)} = 2 \int \dfrac{d\sqrt{x}}{1 + (\sqrt{x})^2} = 2 \arctan \sqrt{x} + C$

21. $\int \dfrac{dx}{x + x\ (\ln\ x)^2} = \int \dfrac{\frac{1}{x}\ dx}{1 + (\ln\ x)^2} = \int \dfrac{d\ \ln\ x}{1 + (\ln\ x)^2} = \arctan\ (\ln\ x) + C$

23. $\int_{-1}^{1} \dfrac{dx}{x^2 + 1} = \arctan x \Big|_{-1}^{1} = \dfrac{\pi}{4} - \left(-\dfrac{\pi}{4}\right) = \dfrac{\pi}{2}$ sq units

25. $\int_{0}^{1} \pi \left(\sqrt{\dfrac{1}{1 + x^2}}\right)^2 dx = \pi \int_{0}^{1} \dfrac{dx}{1 + x^2} = \dfrac{\pi^2}{4}$ cubic units

Technique Review Exercises, Chapter 7 (page 340)

1. $y' = \sec\ (3x - 1)\ \tan\ (3x - 1)\ \dfrac{d}{dx}\ (3x - 1) = 3 \sec\ (3x - 1)\ \tan\ (3x - 1)$

2. $y' = 2 \cot\ (1 - x^2)[-\csc^2\ (1 - x^2)]\ \dfrac{d}{dx}\ (1 - x^2)$

   $= 4x \cot\ (1 - x^2)\ \csc^2\ (1 - x^2)$

3. $y' = 3 \cos^2\ (x^3 - 2x)^2\ [-\sin\ (x^3 - 2x)^2]\ \dfrac{d}{dx}\ (x^3 - 2x)^2$

   $= -6(x^3 - 2x)(3x^2 - 2) \cos^2\ (x^3 - 2x)^2\ \sin\ (x^3 - 2x)^2$

4. $\dfrac{1}{2} \int \sin\ (x^2 + 3)(2x\ dx) = -\dfrac{1}{2} \cos\ (x^2 + 3) + C$

5. $\dfrac{1}{2} \int \tan\ (2x - 3)(2\ dx) = \dfrac{1}{2} \ln\ |\sec\ (2x - 3)| + C$

6. $-\dfrac{1}{2} \int \csc\ (1 - 2x)(-2\ dx) = \dfrac{1}{2} \ln\ |\csc\ (1 - 2x) + \cot\ (1 - 2x)| + C$

7. $\dfrac{1}{2} \int \sec^2\ (x^2 - 2)(2x\ dx) = \dfrac{1}{2} \tan\ (x^2 - 2) + C$

8. $\dfrac{1}{2} \int \sec\ 5x\ \tan\ 5x\ (5\ dx) = \dfrac{1}{5} \sec\ 5x + C$

10. $y' = \dfrac{2x}{\sqrt{1 - (x^2 - 5)^2}} = \dfrac{2x}{\sqrt{10x^2 - x^4 - 24}}$

11. $y' = 2 \arctan (3x) \dfrac{d}{dx} [\arctan (3x)] = \dfrac{6 \arctan (3x)}{1 + 9x^2}$

12. $y' = \dfrac{2x}{|x^2 + 1| \sqrt{(x^2 + 1)^2 - 1}} = \dfrac{2x}{(x^2 + 1) \sqrt{x^4 + 2x^2}}$

13. $\displaystyle\int \dfrac{(\frac{1}{4}) \, dx}{\sqrt{1 - (\frac{x}{4})^2}} = \arcsin \left(\dfrac{x}{4}\right) + C$

14. $\dfrac{1}{3} \displaystyle\int \dfrac{3x^2 \, dx}{1 + (x^3)^2} = \dfrac{1}{3} \arctan (x^3) + C$

15. $\dfrac{1}{2} \displaystyle\int \dfrac{2x \, dx}{x^2 \sqrt{(x^2)^2 - 1}} = \dfrac{1}{2} \text{arcsec} (x^2) + C$

Additional Exercises, Chapter 7 (page 341)

1. $\displaystyle\lim_{x \to 0} \dfrac{x}{\sin 2x} = \dfrac{1}{2} \lim_{x \to 0} \dfrac{2x}{\sin 2x} = \dfrac{1}{2} \lim_{x \to 0} \dfrac{1}{\frac{\sin 2x}{2x}} = \dfrac{1}{2} \left(\dfrac{1}{1}\right) = \dfrac{1}{2}$

3. $\displaystyle\lim_{t \to 0} \dfrac{4 - \cos 2t}{3t}$ does not exist since the numerator approaches 3 and the denominator approaches zero.

5. $\displaystyle\lim_{x \to 0} \dfrac{2x}{5x - 3 \sin x} = \lim_{x \to 0} \dfrac{2}{\frac{5x - 3 \sin x}{x}} = \lim_{x \to 0} \dfrac{2}{5 - 3 \frac{\sin x}{x}} = \dfrac{2}{5 - 3} = 1$

7. $\dfrac{d}{dx} \sin 12x = 12 \cos 12x$

9. $\dfrac{d}{dx} \sin^7 4x = 7 \sin^6 4x \dfrac{d \sin 4x}{dx} = 28 \sin^6 4x \cos 4x$

11. $\dfrac{d}{dx} \ln^3 (8x) = 3 \ln^2 (8x) \dfrac{d \ln 8x}{dx} = \dfrac{3 \ln^2 (8x)}{x}$

13. $\dfrac{d}{dx} \cos (x^3 - 2) = -3x^2 \sin (x^3 - 2)$

15. $\dfrac{d}{dx} \tan^2 (x^2 - 1) = 2 \tan (x^2 - 1) \sec^2 (x^2 - 1)(2x)$

$= 4x \tan (x^2 - 1) \sec^2 (x^2 - 1)$

17. $\dfrac{d}{dx} [e^{2x} \csc 3x] = e^{2x} (-\csc 3x \cot 3x)(3) + (\csc 3x) 2e^{2x}$

$= -3^{2x} \csc 3x \cot 3x + 2e^{2x} \csc 3x$

19. $\dfrac{d}{dx} \csc^5 (3 - x^2) = -5 \csc^4 (3 - x^2) \csc (3 - x^2) \cot (3 - x^2)(-2x)$

$= 10x \csc^5 (3 - x^2) \cot (3 - x^2)$

21. $\displaystyle\int \sin 15 \, dx = \dfrac{1}{15} \int \sin 15 \, x \, (15 \, dx) = -\dfrac{1}{15} \cos 15x + C$

23. $\int x^2 \tan (2 - x^3) \, dx = -\frac{1}{3} \int \tan (2 - x^3)(-3x^2 \, dx)$

$$= -\frac{1}{3} \ln \sec (2 - x^3) + C = \frac{1}{3} \ln \cos (2 - x^3) + C$$

25. $\int x^2 \sec x^3 \tan x^3 \, dx = \frac{1}{3} \int \sec x^3 \tan x^3 (3x^2 \, dx) = \frac{1}{3} \sec x^3 + C$

27. $\int x^2 \csc x^3 \, dx = \frac{1}{3} \int \csc x^3 (3x^2 \, dx) = -\frac{1}{3} \ln |\csc x^3 + \cot x^3| + C$

29. $\int x \sin 3x \, dx$. Let $u = x$ and $dv = \sin 3x \, dx$. Then $du = dx$ and $v = -\frac{1}{3} \cos 3x$,

so $\int x \sin 3x \, dx = -\frac{x}{3} \cos 3x + \frac{1}{3} \int \cos 3x \, dx = -\frac{x}{3} \cos 3x + \frac{1}{9} \sin 3x + C$.

31. $\arcsin (-\frac{1}{2}) = -\pi/6$

33. $\arctan 0 = 0$

35. $y = 2 \cos 4x$, $\cos 4x = \frac{y}{2}$, so $4x = \arccos \frac{y}{2}$, $x = \frac{1}{4} \arccos \frac{y}{2}$, $-2 \le y \le 2$

37. $\frac{d}{dx} (\arcsin 10x) = \dfrac{10}{\sqrt{1 - 100x^2}}$

39. $\frac{d}{dx} \text{arcsec } e^{2x} = \dfrac{2e^{2x}}{e^{2x} \sqrt{e^{4x} - 1}} = \dfrac{2}{\sqrt{e^{4x} - 1}}$

41. $\frac{d}{dx} \ln (\arcsin x^2) = \dfrac{\dfrac{2x}{\sqrt{1 - x^4}}}{\arcsin x^2} = \dfrac{2x}{\sqrt{1 - x^4} \arcsin x^2}$

43. $\int \dfrac{dx}{\sqrt{49 - x^2}} = \int \dfrac{\frac{1}{7} \, dx}{\sqrt{1 - (\frac{x}{7})^2}} = \arcsin (\frac{x}{7}) + C$

45. $\int \dfrac{dx}{|x| \sqrt{36x^2 - 9}} = \frac{1}{3} \int \dfrac{2 \, dx}{|2x| \sqrt{(2x)^2 - 1}} = \frac{1}{3} \text{arcsec } (2x) + C$

47. $\int \dfrac{dx}{\sqrt{e^{2x} - 1}} = \int \dfrac{e^x \, dx}{e^x \sqrt{(e^x)^2 - 1}} = \text{arcsec } e^x + C$

49. $\int \dfrac{e^{2x} \, dx}{e^{4x} + 49} = \frac{1}{2} \int \dfrac{2e^{2x} \, dx}{(e^{2x})^2 + 7^2} = \frac{1}{7} \arctan \dfrac{e^{2x}}{7} + C$

CHAPTER 8: OTHER INTEGRATION TECHNIQUES

Exercises 8.1: Further Application of Previous Techniques (page 352)

1. Let $u = \sqrt{x + 3}$ than, $u^2 = x + 3$ so $x = u^2 - 3$ and $dx = 2u \, du$. Substitution
then gives $\int (u^2 - 2)u(2u \, du) = 2\int (u^4 - 2u^2) du = 2(\frac{1}{5} u^5 - \frac{2}{3} u^3) + C$. Then
since $u = (x + 3)^{1/2}$ we have $\int (1 + x)\sqrt{x + 3} \, dx = 2(\frac{1}{5} (x + 3)^{5/2} - \frac{2}{3} (x + 3)^{3/2})$
+ C.

3. Let $u = (x + 5)^{1/3}$. Then $u^3 = x + 5$ so $x = u^3 - 5$ and $dx = 3u^2 du$. Substitution then yields $\int ((u^3 - 5)^2 - 3) u^2 (3u^2 du) = 3\int (u^6 - 10u^3 + 22) u^4 du$

$= 3\int (u^{10} - 10u^7 + 22u^4) du = 3(\frac{1}{11} u^{11} - \frac{5}{4} u^8 + \frac{22}{5} u^5) + C$. Then since

$u = (x + 5)^{1/3}$ we have $\int (x^2 - 3)(x + 5)^{2/3} dx = 3(\frac{1}{11} (x + 5)^{11/3} - \frac{5}{4} (x + 5)^{8/3}$

$+ \frac{22}{5} (x + 5)^{5/3}) + C$.

9. We will apply integration by parts twice. For the first application let

$u = \sin x$ and $dv = e^{2x} dx$. Then, $du = \cos x\, dx$ and $v = \frac{1}{2} e^{2x}$. Use of integra-

tion by parts then yields $\int e^{2x} \sin x\, dx = \frac{1}{2} e^{2x} \sin x - \frac{1}{2} \int e^{2x} \cos x\, dx$.

Integration by parts is now applied to the second integral by letting $u = \cos x$

and $dv = e^{2x} dx$. Then, $du = -\sin x\, dx$ and $v = \frac{1}{2} e^{2x}$. Use of integration by

parts again yields $\int e^{2x} \sin x\, dx = \frac{1}{2} e^{2x} \sin x - \frac{1}{2}(\frac{1}{2} e^{2x} \cos x + \frac{1}{2} \int e^{2x} \sin x\, dx)$

so on solving for $\int e^{2x} \sin x\, dx$ we have

$\frac{5}{4} \int e^{2x} \sin x\, dx = \frac{1}{2} e^{2x} \sin x - \frac{1}{4} e^{2x} \cos x + C'$ or finally

$\int e^{2x} \sin x\, dx = \frac{4}{5} (\frac{1}{2} e^{2x} \sin x - \frac{1}{4} e^{2x} \cos x) + C$

$= \frac{1}{5} (2e^{2x} \sin x - e^{2x} \cos x) + C$

where $C = \frac{4}{5} C'$.

15. We will apply integration by parts twice. For the first application let

$u = \cos x$ and $dv = e^{3x} dx$. Then $du = -\sin x\, dx$ and $v = \frac{1}{3} e^{3x}$. Application of

integration by parts then gives $\int e^{3x} \cos x\, dx = \frac{1}{3} e^{3x} \cos x + \frac{1}{3} \int e^{3x} \sin x\, dx$.

In order to apply integration by parts again we let $u = \sin x$ and $dv = e^{3x} dx$.

Then $du = \cos x\, dx$ and $v = \frac{1}{3} e^{3x}$ so

$\int e^{3x} \cos x\, dx = \frac{1}{3} e^{3x} \cos x + \frac{1}{3} (\frac{1}{3} e^{3x} \sin x - \frac{1}{3} \int e^{3x} \cos x\, dx)$. Thus,

$\frac{10}{9} \int e^{3x} \cos x\, dx = \frac{1}{3} (e^{3x} \cos x + \frac{1}{3} e^{3x} \sin x) + C'$ or

$\int e^{3x} \cos x\, dx = \frac{3}{10} (e^{3x} \cos x + \frac{1}{3} e^{3x} \sin x) + C$ where $C = \frac{9}{10} C'$.

17. We will first concentrate on the evaluation of the corresponding indefinite

integral. For the initial application of integration by parts let $u = \cos 2x$

and $dv = \sin x\, dx$. Then $du = -2 \sin 2x\, dx$ and $v = -\cos x$. Application of

integration by parts then gives $\int \sin x \cos 2x\, dx = -\cos x \cos 2x - 2\int \cos x$

$\sin 2x\, dx$. We now apply integration by parts to the second integral. This time with u = sin 2x and dv = cos x dx. Then du = 2 cos 2x dx and v = sin x. Consequently, $\int$sin x cos 2x dx = -cos x cos 2x - 2(sin x sin 2x - 2$\int$sin x cos 2x dx). On solving for $\int$sin x cos 2x dx and adding the constant of integration we have $\int$sin x cos 2x dx = $\frac{1}{3}$ (cos x cos 2x + 2 sin x sin 2x) + C.

Thus, $\int_{-\pi}^{\pi}$ sin x cos 2x dx = $\frac{1}{3}$ (cos x cos 2x + 2 sin x sin 2x) $\Big|_{-\pi}^{\pi}$ = 0.

21. An application of integration by parts with u = sin (ln x) and dv = dx gives du = $\frac{1}{x}$ cos (ln x) and v = x. Hence, $\int$sin (ln x) dx = x sin (ln x) - $\int$cos (ln x) dx. A second application of integration by parts with u = cos (ln x) and dv = dx would yield du = - $\frac{1}{x}$ sin (ln x) and v = x. So $\int$sin (ln x) dx = x sin (ln x) - (x cos (ln x) + $\int$sin (ln x) dx). Then on solving for $\int$sin (ln x) dx and adding a constant of integration we get $\int$sin (ln x) dx = $\frac{x}{2}$ (sin (ln x) - cos (ln x)) + C.

23. If we apply integration by parts with u = x^2 and dv = sin x dx we have du = 2x dx and v = -cos x. Thus, $\int x^2$ sin x dx = $-x^2$ cos x + 2$\int$x cos x dx. Another application of integration by parts with u = x, dv = cos x dx, and hence du = dx and v = sin x, gives $\int x^2$ sin x dx = $-x^2$ cos x + 2(x sin x - $\int$sin x dx). Completion of the last integration then gives $\int x^2$ sin x dx = $-x^2$ cos x + 2x sin x + 2 cos x + C.

29. We will use integration by parts with u = sec x and dv = $\sec^2$ x dx. Then, du = sec x tan x dx and v = tan x. Consequently, $\int \sec^3$ x dx = sec x tan x - $\int$sec x $\tan^2$ x dx. Then since $\tan^2$ x = $\sec^2$ x - 1, we have $\int \sec^3$ x dx = sec x tan x - $\int \sec^3$ x dx + $\int$sec x dx or, on solving for $\int \sec^3$ x dx, $\int \sec^3$ x dx = $\frac{1}{2}$ (sec x tan x + $\int$sec x dx). Finally, since $\int$sec x dx = ln $|$sec x + tan x$|$ + C', we have $\int \sec^3$ x dx = $\frac{1}{2}$ (sec x tan x + ln $|$sec x + tan x$|$) + C.

33. Let u = x + 2, then x = u - 2 and dx = du. Thus, the integral becomes $\int \frac{(u - 2)^2 \, du}{u^3}$ = $\int \frac{u^2 - 4u + 4}{u^3}$ du = $\int$ (u^{-1} - $4u^{-2}$ + $4u^{-3}$) du

= ln $|u|$ + $4u^{-1}$ - $2u^{-2}$ + C = ln $|x + 2|$ + $4(x + 2)^{-1}$ - $2(x + 2)^{-2}$ + C

35. Since both sin 2x and sin 3x are positive for $0 \le x \le \pi/2$, y = sin 2x sin 3x is positive on this interval. Thus, in order to compute the area it is only necessary to evaluate the integral

$$A = \int_{\pi}^{\pi/2} \sin 2x \sin 3x \, dx.$$

We will apply integration by parts twice in order to accomplish this. First, let u = sin 2x and dv = sin 3x dx. Then, du = 2 cos 2x dx and dv = $-\frac{1}{3}$ cos 3x.

Thus, $\int_0^{\pi/2} \sin 2x \sin 3x \, dx = -\frac{1}{3} \sin 2x \cos 3x \Big|_0^{\pi/2} + \frac{2}{3} \int_0^{\pi/2} \cos 3x \cos 2x \, dx.$

Since the first of the two expressions on the right hand side is zero we have simply $\int_0^{\pi/2} \sin 2x \sin 3x \, dx = \frac{2}{3} \int_0^{\pi/2} \cos 3x \cos 2x \, dx.$ For the second integration by parts we will let u = cos 2x and dv = cos 3x dx. Then du = -2 sin 2x dx and v = $\frac{1}{3}$ sin 3x. Thus,

$\int_0^{\pi/2} \sin 2x \sin 3x \, dx = \frac{2}{3} [\frac{1}{3} \sin 3x \cos 2x \Big|_0^{\pi/2} + \frac{2}{3} \int_0^{\pi/2} \sin 2x \sin 3x \, dx]$ or

$\int_0^{\pi/2} \sin 2x \sin 3x \, dx = \frac{2}{9} (1 - 0) + \frac{4}{9} \int_0^{\pi/2} \sin 2x \sin 3x \, dx.$ Consequently,

$A = \frac{9}{5} \cdot \frac{2}{9} = \frac{2}{5}$ sq units.

37. Use of the method of cylindrical shells gives the volume as $V = 2\pi \int_0^{\pi} x \sin x \, dx.$ An application of integration by parts with u = x and dv = sin x dx, du = dx, and v = -cos x gives $V = 2\pi \int_0^{\pi} x \sin x \, dx = 2\pi (-x \cos x \Big|_0^{\pi} + \int_0^{\pi} \cos x \, dx).$

Thus, $V = 2\pi (-x \cos x \Big|_0^{\pi} + \sin x \Big|_0^{\pi}) = 2\pi (\pi + 0) = 2\pi^2$ cubic units.

Exercises 8.2: Integrals of Powers of Certain Trigonometric Functions
(page 356)

1. $\int \sin^4 x \cos^3 x \, dx = \int \sin^4 x \cos^2 x (\cos x \, dx) = \int \sin^4 x (1 - \sin^2 x) \, d \sin x$
$= \int (\sin^4 x - \sin^6 x) \, d \sin x = \frac{1}{5} \sin^5 x - \frac{1}{7} \sin^7 x + C$

3. $\int \cos^2 x \sin^3 x \, dx = -\int \cos^2 x \sin^2 x (-\sin x \, dx) = -\int \cos^2 x (1 - \cos^2 x) \, d \cos x$
$= -\int (\cos^2 x - \cos^4 x) \, d \cos x = -\frac{1}{3} \cos^3 x + \frac{1}{5} \cos^5 x + C$

5. $\int_0^{\pi/4} \sec^2 x \tan^2 x \, dx = \int_0^{\pi/4} \tan^2 x \, d \tan x = \frac{1}{3} \tan^3 x \Big|_0^{\pi/4} = \frac{1}{3}$

7. $\int \sec 2x \tan^3 2x \, dx = \frac{1}{2} \int \tan^2 2x (2 \sec 2x \tan 2x \, dx)$
$= \frac{1}{2} \int (\sec^2 2x - 1) \, d \sec^2 x = \frac{1}{2} [\frac{1}{3} \sec^3 2x - \sec 2x] + C$

9. $\int \tan^5 x\, dx = \int \tan^4 x \sec^{-1} x(\sec x \tan x\, dx) = \int(\sec^2 x - 1)^2 \sec^{-1} x\, d \sec x$

$= \int(\sec^4 x - 2 \sec^2 x + 1) \sec^{-1} x\, d \sec x = \int(\sec^3 x - 2 \sec x + \sec^{-1} x)$

$d \sec x = \frac{1}{4} \sec^4 x - \sec^2 x + \ln |\sec x| + C$

11. $\int \sec^6 x\, dx = \int \sec^4 x\, (\sec^2 x\, dx) = \int(1 + \tan^2 x)^2\, d \tan x$

$= \int(1 + 2 \tan^2 x + \tan^4 x)\, d \tan x = \tan x + \frac{2}{3} \tan^3 x + \frac{1}{5} \tan^5 x + C$

15. $\int_{-\pi}^{\pi} \sqrt[3]{\cos x}\, \sin^3 x\, dx = -\int_{-\pi}^{\pi} \sqrt[3]{\cos x}\, \sin^2 x\, (-\sin x\, dx)$

$= -\int_{-\pi}^{\pi} \sqrt[3]{\cos x}\, (1 - \cos^2 x)\, d \cos x = -\int_{-\pi}^{\pi} (\cos^{1/3} x - \cos^{7/3} x)\, d \cos x$

$= -[\frac{3}{4} \cos^{4/3} x - \frac{3}{10} \cos^{10/3} x]\, \Big|_{-\pi}^{\pi} = -(\frac{3}{4} - \frac{3}{10}) + (\frac{3}{4} - \frac{3}{10}) = 0$

21. $\int \csc^2 x \cot^{-5} x\, dx = -\int \cot^{-5} x\, (-\csc^2 x\, dx)$

$= -\int \cot^{-5} x\, d \cot x = \frac{1}{4} \cot^{-4} x + C$

23. Since $\cos^2 \theta = \frac{1 + \cos 2\theta}{2}$, $\int_0^{\pi/2} \cos^2 (-3x)\, dx =$

$\int_0^{\pi/2} \frac{1 + \cos (-6x)}{2}\, dx = \frac{1}{2} \int_0^{\pi/2} dx - \frac{1}{12} \int_0^{\pi/2} \cos (-6x)(-6dx)$

$= \frac{1}{2} x\, \Big|_0^{\pi/2} - \frac{1}{12} \sin (-6x)\, \Big|_0^{\pi/2} = \frac{\pi}{4}$

25. $\int \sin^2 x \cos^4 x\, dx = \int(\sin x \cos x)^2 \cos^2 x\, dx$

Then since $\sin \theta \cos \theta = \frac{\sin 2\theta}{2}$ we have $\int \sin^2 x \cos^4 x\, dx = \int \frac{\sin^2 2x}{4} \cos^2 x\, dx.$

Using the identity $\cos^2 \theta = \frac{1 + \cos 2\theta}{2}$ we obtain

$\int \sin^2 x \cos^4 x\, dx = \frac{1}{4} \int \sin^2 2x\, (\frac{1 + \cos 2x}{2})\, dx$

$= \frac{1}{8} \int(\sin^2 2x + \sin^2 2x \cos 2x)\, dx = \frac{1}{8} \int \sin^2 2x\, dx + \frac{1}{8} \int \sin^2 2x \cos 2x\, dx.$

The last integral can be evaluated in a completely straightforward way. In order to evaluate the first integral we will use the identify

$$\sin^2 \theta = \frac{1 - \cos 2\theta}{2}.$$

Then, $\int \sin^2 x \cos^4 x\, dx = \frac{1}{8} \int \frac{1 - \cos 4x}{2}\, dx + \frac{1}{16} \int \sin^2 2x\, (2 \cos 2x\, dx)$

$= \frac{1}{16} \int dx - \frac{1}{16} \cdot \frac{1}{4} \int \cos 4x\, d\, 4x + \frac{1}{16} \int \sin^2 x\, d \sin 2x$

$= \frac{x}{16} - \frac{1}{64} \sin 4x + \frac{1}{48} \sin^3 2x + C.$

29. By the disk method the volume is given by $V = \pi \int_0^{\pi/2} \sin^2 2x \, dx$. Then since

$\sin^2 \theta = \dfrac{1 - \cos 2\theta}{2}$ we have

$$V = \pi \int_0^{\pi/2} \frac{1 - \cos 4x}{2} \, dx = \frac{\pi}{2} \int_0^{\pi/2} dx - \frac{\pi}{8} \int_0^{\pi/2} \cos 4x \, (4dx)$$

$$= \frac{\pi}{2} \, x \, \Big|_0^{\pi/2} - \frac{\pi}{8} \sin 4x \, \Big|_0^{\pi/2} = \frac{\pi^2}{4} - 0 = \frac{\pi^2}{4} \text{ cubic units.}$$

Exercises 8.3: Trigonometric Substitutions (page 360)

1. Let $x = 2 \tan \theta$, $-\pi/2 < \theta < \pi/2$. Then $dx = 2 \sec^2 \theta \, d\theta$ and $\sqrt{4 + x^2} = 2 \sec \theta$. Substitution of this information into the integral yields

$$\int \frac{dx}{\sqrt{4 + x^2}} = \int \frac{2 \sec^2 \theta \, d\theta}{2 \sec \theta} = \int \sec \theta \, d\theta = \ln \left| \sec \theta + \tan \theta \right| + C.$$

Then since $\sec \theta = \frac{1}{2} \sqrt{4 + x^2}$ and $\tan \theta = \frac{x}{2}$,

$$\int \frac{dx}{\sqrt{4 + x^2}} = \ln \left| \frac{1}{2} \sqrt{4 + x^2} + \frac{x}{2} \right| + C.$$

3. Let $x = 3 \sin \theta$, $-\pi/2 \le \theta \le \pi/2$. Then $dx = 3 \cos \theta \, d\theta$ and $\sqrt{9 - x^2} = \sqrt{9 - 9 \sin^2 \theta} = 3\sqrt{1 - \sin^2 \theta} = \pm 3 \cos \theta$. Since $-\pi/2 \le \theta \le \pi/2$, $\cos \theta \ge 0$ and hence $\sqrt{9 - x^2} = 3 \cos \theta$. Substitutions then yields $\int \sqrt{9 - x^2} \, dx = \int 3 \cos \theta \, (3 \cos \theta \, d\theta) = 9 \int \cos^2 \theta \, d\theta$

$$= 9 \int \frac{1 + \cos 2\theta}{2} \, d\theta = \frac{9}{2} \left[\theta + \frac{1}{2} \sin 2\theta \right] + C = \frac{9}{2} \left[\theta + \sin \theta \cos \theta \right] + C.$$

Since $\sin \theta = \frac{x}{3}$ and $-\pi/2 \le \theta \le \pi/2$, $\theta = \arcsin \frac{x}{3}$. Then using the fact that $\cos \theta = \frac{1}{3} \sqrt{9 - x^2}$ we have $\int \sqrt{9 - x^2} \, dx = \frac{9}{2} \left[\arcsin \frac{x}{3} + \frac{1}{9} x \sqrt{9 - x^2} \right] + C.$

5. Just note that $d(9 - x^2) = -2x \, dx$ so

$$-\frac{1}{2} \int_0^1 \frac{-2x \, dx}{(9 - x^2)^{3/2}} = -\frac{1}{2} \int_0^1 (9 - x^2)^{-3/2} \, d(9 - x^2)$$

$$= -\frac{1}{2} \, (-2)(9 - x^2)^{-1/2} \, \Big|_0^1 = \frac{1}{2\sqrt{2}} - \frac{1}{3}.$$

7. $\int \dfrac{x^3}{\sqrt{9 - 2x^2}} \, dx = \int \dfrac{x^3}{\sqrt{3^2 - (\sqrt{2}x)^2}} \, dx.$ Let $\sqrt{2}x = 3 \sin \theta$, $-\pi/2 \le \theta \le \pi/2$, then

$x = \dfrac{3}{\sqrt{2}} \sin \theta$ and $dx = \dfrac{3}{\sqrt{2}} \cos \theta \, d\theta.$ Also, $\sqrt{9 - 2x^2} = 3 \cos \theta$ so

$$\int \frac{x^3}{\sqrt{9 - 2x^2}} \, dx = \int \frac{\frac{3^3}{2\sqrt{2}} \sin^3 \theta \, \frac{3}{\sqrt{2}} \cos \theta \, d\theta}{3 \cos \theta} = \frac{3^3}{4} \int \sin^3 \theta \, d\theta$$

$$= \frac{3^3}{4} \int \sin^2 \theta \sin \theta \, d\theta = -\frac{3^3}{4} \int (1 - \cos^2 \theta) \, d(\cos \theta)$$

$$= -\frac{3^3}{4} (\cos \theta - \frac{1}{3} \cos^3 \theta) + C. \quad \text{Then since } \cos \theta = \frac{1}{3} \sqrt{9 - 2x^2}, \text{ we have}$$

$$\int \frac{x^3}{\sqrt{9 - 2x^2}} \, dx = -\frac{3^3}{4} (\frac{1}{3} \sqrt{9 - 2x^2} - \frac{1}{3^4} (9 - 2x^2)^{3/2}) + C$$

$$= -\frac{9}{4} (\sqrt{9 - 2x^2} + \frac{1}{12} (9 - 2x^2)^{3/2}) + C.$$

11. Let $x = 4 \sec \theta$ with $\pi/2 < \theta \le \pi$, then $dx = 4 \sec \theta \tan \theta \, d\theta$ and

$(x^2 - 16)^{1/2} = 4 \tan \theta$. Substitution then gives

$$\int \frac{dx}{(x^2 - 16)^{3/2}} = \int \frac{4 \sec \theta \tan \theta \, d\theta}{4^3 \tan^3 \theta} = \frac{1}{16} \int \sec \theta \tan^{-2} \theta \, d\theta$$

$$= \frac{1}{16} \int \frac{\cos \theta \, d\theta}{\sin^2 \theta} = \frac{1}{16} \int \frac{d \sin \theta}{\sin^2 \theta} = -\frac{1}{16} \sin^{-1} \theta + C. \quad \text{Since}$$

$\sec \theta = \frac{x}{4}$, $\cos \theta = \frac{4}{x}$. Then using the fact that $\sin \alpha = \sqrt{1 - \cos^2 \alpha}$

if $-\pi/2 \le \alpha \le \pi/2$ we have $\sin \theta = \sqrt{1 - \frac{16}{x^2}} = \frac{\sqrt{x^2 - 16}}{x}$ or

$\sin^{-1} \theta = \frac{x}{\sqrt{x^2 - 16}}$. Thus, $\int \frac{dx}{(x^2 - 16)^{3/2}} = -\frac{x}{16\sqrt{x^2 - 16}} + C.$

13. Let $4x = \sqrt{7} \sin \theta$, where $-\pi/2 \le \theta \le \pi/2$, then $x = \frac{\sqrt{7}}{4} \sin \theta$,

$dx = \frac{\sqrt{7}}{4} \cos \theta \, d\theta$ and $\sqrt{7 - 16x^2} = \sqrt{7} \cos \theta$. Substitution then gives

$$\int \sqrt{7 - 16x^2} \, dx = \int \sqrt{7} \cos \theta \, \frac{\sqrt{7}}{4} \cos \theta \, d\theta = \frac{7}{4} \int \cos^2 \theta \, d\theta$$

$$= \frac{7}{4} \int \frac{1 + \cos 2\theta}{2} \, d\theta = \frac{7}{8} [\int d\theta + \frac{1}{2} \int \cos 2\theta \, 2 \, d\theta] = \frac{7}{8} [\theta + \frac{1}{2} \sin 2\theta] + C.$$

Since $\frac{1}{2} \sin 2\theta = \sin \theta \cos \theta$ we have $\int \sqrt{7 - 16x^2} \, dx = \frac{7}{8} [\theta + \sin \theta \cos \theta] + C.$

Then since $\sin \theta = \frac{4x}{\sqrt{7}}$ and $-\pi/2 \le \theta \le \pi/2$, $\theta = \arcsin \frac{4x}{\sqrt{7}}$. Moreover, $\cos \theta$

$= \frac{1}{\sqrt{7}} \sqrt{7 - 16x^2}$ so, $\int \sqrt{7 - 16x^2} \, dx = \frac{7}{8} [\arcsin \frac{4x}{\sqrt{7}} + \frac{4}{7} x \sqrt{7 - 16x^2}] + C$

$= \frac{7}{8} \arcsin \frac{4x}{\sqrt{7}} + \frac{x}{2} \sqrt{7 - 16x^2} + C.$

15. If we complete the square we have

$$\int_0^1 \frac{dx}{\sqrt{x^2 + 8x + 32}} = \int_0^1 \frac{dx}{\sqrt{x^2 + 8x + 16 + 16}} = \int_0^1 \frac{dx}{\sqrt{(x + 4)^2 + 16}}.$$

Then if we let $x + 4 = 4 \tan \theta$, where $-\pi/2 < \theta < \pi/2$, $dx = 4 \sec^2 \theta \, d\theta$ and

$\sqrt{(x + 4)^2 + 16} = 4 \sec \theta$. Moreover, when $x = 0$, $\tan \theta = 1$ so $\theta = \pi/4$.
When $x = 1$, $\tan \theta = 5/4$ so since $-\pi/2 < \theta < \pi/2$, $\theta = \arctan 5/4$. Substitution
then gives

$$\int_0^1 \frac{dx}{\sqrt{x^2 + 8x + 32}} = \int_{\pi/4}^{\arctan 5/4} \frac{4 \sec^2 \theta \, d\theta}{4 \sec \theta}$$

$$= \int_{\pi/4}^{\arctan 5/4} \sec \theta \, d\theta = \ln \left| \sec \theta + \tan \theta \right| \Big|_{\pi/4}^{\arctan 5/4}$$

$$= \ln \left| \sec (\arctan 5/4) + \tan (\arctan 5/4) \right| - \ln \left| \sec \pi/4 + \tan \pi/4 \right|.$$

Since $\sec \alpha = \sqrt{1 + \tan^2 \alpha}$ for $-\pi/2 < \alpha < \pi/2$, $\sec (\arctan 5/4$

$$= \sqrt{1 + \tan^2 (\arctan 5/4)} = \sqrt{1 + \frac{25}{16}} = \frac{\sqrt{41}}{4}. \quad \text{Thus,}$$

$$\int_0^1 \frac{dx}{\sqrt{x^2 + 8x + 32}} = \ln \left| \frac{\sqrt{41}}{4} + \frac{5}{4} \right| - \ln \left| \sqrt{2} + 1 \right|.$$

19. On completing the square under the radical we have

$$\int (x - 3)^3 \sqrt{x^2 - 6x + 13} \, dx = \int (x - 3)^3 \sqrt{x^2 - 6x + 9 + 4} \, dx$$

$$= \int (x - 3)^3 \sqrt{(x - 3)^2 + 4} \, dx. \quad \text{If we let } x - 3 = 2 \tan \theta \text{ where}$$

$-\pi/2 < \theta < \pi/2$, then $dx = 2 \sec^2 \theta \, d\theta$, and $\sqrt{(x - 3)^2 + 4} = 2 \sec \theta$. On
substitution of this information into the integral we get

$$\int (x - 3)^3 \sqrt{x^2 - 6x + 13} \, dx = \int 2^3 \tan^3 \theta \, (2 \sec \theta) \, 2 \sec^2 \theta \, d\theta$$

$$= 2^5 \int \tan^3 \theta \sec^3 \theta \, d\theta = 2^5 \int \tan^2 \theta \sec^2 \theta \sec \theta \tan \theta \, d\theta$$

$$= 2^5 \int (\sec^2 \theta - 1) \sec^2 \theta \, d \sec \theta = 2^5 \left[\frac{1}{5} \sec^5 \theta - \frac{1}{3} \sec^3 \theta \right] + C.$$

Then since $\sec \theta = \frac{1}{2} \sqrt{x^2 - 6x + 13}$, we have

$$\int (x - 3)^3 \sqrt{x^2 - 6x + 13} \, dx = 2^5 \left[\frac{1}{5 \cdot 2^5} (x^2 - 6x + 13)^{5/2} \right.$$

$$\left. - \frac{1}{3 \cdot 2^3} (x^2 - 6x + 13)^{3/2} \right] + C = \frac{1}{5} (x^2 - 6x + 13)^{5/2} - \frac{4}{3} (x^2 - 6x + 13)^{3/2} + C.$$

25. The area is given by $A = \int_{-2}^{2} \sqrt{4 - x^2}\, dx$. Use of the substitution $x = 2 \sin\theta$,

$-\pi/2 \le \theta \le \pi/2$ then gives $dx = 2\cos\theta\, d\theta$ and $\sqrt{4 - x^2} = 2\cos\theta$. Since

$\theta = -\pi/2$ when $x = -2$ and $\theta = \pi/2$ when $x = 2$ we have $A = 4 \int_{-\pi/2}^{\pi/2} \cos^2\theta\, d\theta$.

In order to evaluate this integral we use the identity $\cos^2\theta = \dfrac{1 + \cos 2\theta}{2}$.

Then $A = 2 \int_{-\pi/2}^{\pi/2} (1 + \cos 2\theta)\, d\theta = 2(\theta + \dfrac{1}{2}\sin 2\theta)\,\Big|_{-\pi/2}^{\pi/2}$

$= 2(\dfrac{\pi}{2} + \dfrac{\pi}{2}) = 2\pi$ square units.

29. Using $\tan x = \dfrac{\sin x}{\cos x} = \dfrac{2u}{1 - u^2}$, $\displaystyle\int \dfrac{dx}{\sin x + \tan x} = \int \dfrac{\dfrac{2\, du}{1 + u^2}}{\dfrac{2u}{1 + u^2} + \dfrac{2u}{1 - u^2}}$

$= \displaystyle\int \dfrac{(1 - u^2)\, du}{u(1 - u^2) + u(1 + u^2)} = \int \dfrac{1 - u^2}{2u}\, du = \dfrac{1}{2} \int (u^{-1} - u)\, du$

$= \dfrac{1}{2} \ln |u| - \dfrac{u^2}{4} + C = \dfrac{1}{2} \ln \left|\tan \left(\dfrac{x}{2}\right)\right| - \dfrac{1}{4} \tan^2 \left(\dfrac{x}{2}\right) + C$

Exercises 8.4: Integrals of Functions with Quadratics in the Denominator (page 364)

1. Since $\displaystyle\int \dfrac{du}{u^2 + a^2} = \dfrac{1}{a} \arctan \dfrac{u}{a} + C$ we have $\displaystyle\int \dfrac{dx}{x^2 + 4} = \dfrac{1}{2} \arctan \dfrac{x}{2} + C$.

3. Since $\displaystyle\int \dfrac{du}{u^2 - a^2} = \dfrac{1}{2a} \ln \left|\dfrac{u - a}{u + a}\right| + C$, we have $\displaystyle\int \dfrac{dx}{x^2 - 9} = \dfrac{1}{6} \ln \left|\dfrac{x - 3}{x + 3}\right| + C$.

5. $\displaystyle\int \dfrac{dx}{12 - x^2} = -\int \dfrac{dx}{x^2 - 12} = -\dfrac{1}{2\sqrt{12}} \ln \left|\dfrac{x - \sqrt{12}}{x + \sqrt{12}}\right| + C = -\dfrac{1}{4\sqrt{3}} \ln \left|\dfrac{x - 2\sqrt{3}}{x + 2\sqrt{3}}\right| + C$

7. $\displaystyle\int \dfrac{dx}{x^2 + 19} = \dfrac{1}{\sqrt{19}} \arctan \dfrac{x}{\sqrt{19}} + C$

9. $\displaystyle\int \dfrac{3x - 7}{x^2 + 16}\, dx = \dfrac{3}{2} \int \dfrac{2x}{x^2 + 16} - 7 \int \dfrac{dx}{x^2 + 16} = \dfrac{3}{2} \ln |x^2 + 16| - \dfrac{7}{4} \arctan \dfrac{x}{4} + C$

11. $\displaystyle\int \dfrac{2x + 1}{\sqrt{4 - (x + 3)^2}}\, dx = \int \dfrac{2x + 1}{\sqrt{-x^2 - 6x - 5}}\, dx = -\int \dfrac{-2x - 6}{\sqrt{-x^2 - 6x - 5}}\, dx$

$- 5 \displaystyle\int \dfrac{dx}{\sqrt{4 - (x + 3)^2}} = -\int [4 - (x + 3)^2]^{-1/2}\, (-2x - 6)\, dx - 5 \arcsin \left(\dfrac{x + 3}{2}\right) + C$

$= -2\sqrt{4 - (x - 3)^2} - 5 \arcsin \left(\dfrac{x + 3}{2}\right) + C$

13. Completion of the square gives

$\displaystyle\int \dfrac{dx}{x^2 + 4x + 8} = \int \dfrac{dx}{x^2 + 4x + 4 + 4} = \int \dfrac{dx}{(x + 2)^2 + 4}$. Use of the equation

$\int \dfrac{du}{u^2 + a^2} = \dfrac{1}{a} \arctan \dfrac{u}{a} + C$ then gives $\int \dfrac{dx}{x^2 + 4x + 8} = \dfrac{1}{2} \arctan \dfrac{x+2}{2} + C.$

15. On completing the square we get $\int \dfrac{dx}{7 - x^2 - 6x} = \int \dfrac{dx}{16 - (x^2 + 6x + 9)}$

$= \int \dfrac{dx}{16 - (x+3)^2} = -\int \dfrac{dx}{(x+3)^2 - 16}.$ Then since

$\int \dfrac{du}{u^2 - a^2} = \dfrac{1}{2a} \ln \left| \dfrac{u-a}{u+a} \right| + C$ we have

$\int \dfrac{dx}{7 - x^2 - 6x} = -\dfrac{1}{8} \ln \left| \dfrac{x+3-4}{x+3+4} \right| + C = -\dfrac{1}{8} \ln \left| \dfrac{x-1}{x+7} \right| + C.$

17. Completion of the square gives $\int \dfrac{5\ dx}{x^2 - 3x + 9} = \int \dfrac{5\ dx}{(x^2 - 3x + \frac{9}{4}) + \frac{27}{4}}$

$= \int \dfrac{5\ dx}{(x - \frac{3}{2})^2 + \frac{27}{4}}.$ Then since $\int \dfrac{du}{u^2 + a^2} = \dfrac{1}{a} \arctan \dfrac{u}{a} + C$ we have

$\int \dfrac{5\ dx}{x^2 - 3x + 9} = \dfrac{10}{\sqrt{27}} \arctan \dfrac{2x - 3}{\sqrt{27}} + C.$

19. $\int \dfrac{3\ dx}{\sqrt{x^2 + 4x}} = 3 \int \dfrac{dx}{\sqrt{x^2 + 4x + 4 - 4}} = 3 \int \dfrac{dx}{\sqrt{(x+2)^2 - 4}}$

$= 3 \ln \left| x + 2 + \sqrt{(x+2)^2 - 4} \right| + C$

23. The first goal is to express this integral as the sum of two integrals, one whose numerator is the derivative of $x^2 - 4x + 13$, and one with a constant numerator. Thus,

$\int_0^1 \dfrac{x\ dx}{x^2 - 4x + 13} = \dfrac{1}{2} \int_0^1 \dfrac{2x\ dx}{x^2 - 4x + 13} = \dfrac{1}{2} \int_0^1 \dfrac{2x - 4}{x^2 - 4x + 13}\ dx + \dfrac{1}{2} \int_0^1 \dfrac{4\ dx}{x^2 - 4x + 13}$

$= \dfrac{1}{2} \ln \left| x^2 - 4x + 13 \right| \Big|_0^1 + 2 \int_0^1 \dfrac{dx}{x^2 - 4x + 4 + 9}$

$= \dfrac{1}{2} \ln \left| x^2 - 4x + 13 \right| \Big|_0^1 + 2 \int_0^1 \dfrac{dx}{(x - 2)^2 + 9}$

$= \dfrac{1}{2} \ln \left| x^2 - 4x + 13 \right| \Big|_0^1 + \dfrac{2}{3} \arctan \dfrac{x - 2}{3} \Big|_0^1$

$= \dfrac{1}{2} \ln \left| 10 \right| - \dfrac{1}{2} \ln \left| 13 \right| + \dfrac{2}{3} \left[\arctan \left(-\dfrac{1}{3} \right) - \arctan \left(-\dfrac{2}{3} \right) \right]$

$= \ln \sqrt{10} - \ln \sqrt{13} + \frac{2}{3} [\arctan \ (\frac{2}{3}) - \arctan \ (\frac{1}{3})].$

27. $\int_0^2 \frac{3 - 2x}{7 - x + x^2} dx = - \int_0^2 \frac{2x - 3}{7 - x + x^2} dx = - \int_0^2 \frac{2x - 1}{7 - x + x^2} dx - \int_0^2 \frac{-2 \ dx}{7 - x + x^2}$

$= - \ln |7 - x + x^2| \ \Big|_0^2 + 2 \int_0^2 \frac{dx}{27/4 + x^2 - x + 1/4}$

$= - \ln 9 + \ln 7 + 2 \int_0^2 \frac{dx}{27/4 + (x - 1/2)^2}$

$= - \ln 9 + \ln 7 + \frac{2}{\sqrt{27}/2} \arctan \frac{x - 1/2}{\sqrt{27}/2} \ \Big|_0^2$

$= \ln 7/9 + \frac{4}{3\sqrt{3}} (\arctan \frac{1}{\sqrt{3}} - \arctan (-\frac{1}{3\sqrt{3}}))$. The fact that $\arctan \frac{1}{\sqrt{3}} = \pi/6$

can be used to simplify this result further.

29. $\int \frac{x + 3}{7 - x^2 - 6x} dx = - \frac{1}{2} \int \frac{-2x - 6}{7 - x^2 - 6x} dx = - \frac{1}{2} \ln |7 - x^2 - 6x| + C$

35. $\int \frac{5x - 2}{\sqrt{x^2 - 4x - 5}} dx = \int \frac{5x - 2}{\sqrt{x^2 - 4x + 4 - 9}} dx = \frac{5}{2} \int \frac{2x - \frac{4}{5} - 4 + 4}{\sqrt{(x - 2)^2 - 9}} dx$

$= \frac{5}{2} \int (x^2 - 4x - 15)^{-1/2}(2x - 4) \ dx + \frac{5}{2} \cdot \frac{16}{5} \int \frac{dx}{\sqrt{(x - 2)^2 - 9}}$

$= 5(x^2 - 4x - 5)^{1/2} + 8 \ln |x - 2 + \sqrt{x^2 - 4x - 5}| + C$

37. By the method of cylindrical shells the volume is given by

$V = 2\pi \int_0^1 \frac{x}{x^2 + 4x + 8} dx.$ Thus,

$V = \pi \int_0^1 \frac{2x + 4}{x^2 + 4x + 8} dx - \pi \int_0^1 \frac{4}{x^2 + 4x + 8} dx$

$= \pi \ln |x^2 + 4x + 8| \ \Big|_0^1 - 4\pi \int_0^1 \frac{dx}{x^2 + 4x + 4 + 4}$

$= \pi (\ln 13 - \ln 8) - 4\pi \int_0^1 \frac{dx}{(x + 2)^2 + 4}$

$= \pi (\ln \frac{13}{8}) - 4\pi \frac{1}{2} \arctan \frac{x + 2}{2} \ \Big|_0^1$

$= \pi (\ln \frac{13}{8}) - 2\pi (\arctan \frac{3}{2} - \arctan 1)$

$= \pi \ln (\frac{13}{8}) + \frac{\pi^2}{2} - 2\pi \arctan \frac{3}{2}$ cubic units.

1. Since if $b^2 > a^2$, $\int \dfrac{dx}{a + b \sin x} = \dfrac{1}{\sqrt{b^2 - a^2}} \ln \left| \dfrac{a \tan \frac{x}{2} + b - \sqrt{b^2 - a^2}}{a \tan \frac{x}{2} + b + \sqrt{b^2 - a^2}} \right|$

   we have $\int \dfrac{dx}{-2 + 7 \sin x} = \dfrac{1}{\sqrt{45}} \ln \left| \dfrac{-2 \tan \frac{x}{2} + 7 - \sqrt{45}}{-2 \tan \frac{x}{2} + 7 + \sqrt{45}} \right| + C.$

3. Since $\int \dfrac{dx}{a + b \sin x} = \dfrac{2}{\sqrt{a^2 - b^2}} \arctan \left(\dfrac{a \tan \frac{x}{2} + b}{\sqrt{a^2 - b^2}} \right)$ if $a^2 > b^2$, then

   $\displaystyle\int_0^{\pi/2} \dfrac{dx}{7 - 4 \sin x} = \dfrac{2}{\sqrt{33}} \arctan \left(\dfrac{7 \tan \frac{x}{2} - 4}{\sqrt{33}} \right) \Big|_0^{\pi/2}$

   $= \dfrac{2}{\sqrt{33}} \left(\arctan \left(\dfrac{3}{\sqrt{33}} \right) - \arctan \left(-\dfrac{4}{\sqrt{33}} \right) \right) = \dfrac{2}{\sqrt{33}} \left[\arctan \left(\dfrac{3}{\sqrt{33}} \right) + \arctan \left(\dfrac{4}{\sqrt{33}} \right) \right].$

5. Since $\int \dfrac{dx}{(x^2 + a^2)^{k+1}} = \dfrac{x}{2ka^2 (x^2 + a^2)^k} + \dfrac{2k - 1}{2ka^2} \int \dfrac{dx}{(x^2 + a^2)^k}$ if $k = 1$ and

   $a = \sqrt{3}$ we get $\int \dfrac{dx}{(x^2 + 3)^2} = \dfrac{x}{6(x^2 + 3)} + \dfrac{1}{6} \int \dfrac{dx}{x^2 + 3}.$

   Then on evaluating the last integral we get

   $\int \dfrac{dx}{(x^2 + 3)^2} = \dfrac{x}{6(x^2 + 3)} + \dfrac{1}{6\sqrt{3}} \arctan \dfrac{x}{\sqrt{3}} + C.$

7. We will apply the reduction formula

   $\int \dfrac{dx}{(x^2 + a^2)^{k+1}} = \dfrac{x}{2ka^2 (x^2 + a^2)^k} + \dfrac{2k - 1}{2ka^2} \int \dfrac{dx}{(x^2 + a^2)^k}$ twice. For the first

   application we will let $a = \sqrt{7}$ and $k = 2$. Then

   $\int \dfrac{dx}{(x^2 + 7)^3} = \dfrac{x}{28(x^2 + 7)^2} + \dfrac{3}{28} \int \dfrac{dx}{(x^2 + 7)^2}.$ We now apply the reduction

   formula to the last integral. In this case we will take $a = \sqrt{7}$ and $k = 1$.

   Then $\int \dfrac{dx}{(x^2 + 7)^3} = \dfrac{x}{28(x^2 + 7)^2} + \dfrac{3}{28} \left[\dfrac{x}{14(x^2 + 7)} + \dfrac{1}{14} \int \dfrac{dx}{x^2 + 7} \right].$

   On evaluating the last integral we obtain

   $\int \dfrac{dx}{(x^2 + 7)^3} = \dfrac{x}{28(x^2 + 7)^2} + \dfrac{3}{28} \left[\dfrac{x}{14(x^2 + 7)} + \dfrac{1}{14\sqrt{7}} \arctan \dfrac{x}{\sqrt{7}} \right] + C.$

13. Substitution of $n = 3$ into the result of Exercise 10 gives

   $\int \cos^3 x \, dx = \dfrac{1}{3} \cos^2 x \sin x + \dfrac{2}{3} \int \cos x \, dx = \dfrac{1}{3} \cos^2 x \sin x + \dfrac{2}{3} \sin x + C.$

15. Substitution of n = 4 into the result of Exercise 10 gives

$\int \cos^4 x \, dx = \frac{1}{4} \cos^3 x \sin x + \frac{3}{4} \int \cos^2 x \, dx$. Now using the same result with

n = 2 gives $\int \cos^4 x \, dx = \frac{1}{4} \cos^3 x \sin x + \frac{3}{4} [\frac{1}{2} \cos x \sin x + \frac{1}{2} \int dx]$

$= \frac{1}{4} \cos^3 x \sin x + \frac{3}{8} \cos x \sin x + \frac{1}{2} x + C.$

Exercises 8.6: Partial Fractions (page 374)

1. To evaluate this integral we first must find constants a and b such that

$\frac{5x + 11}{(x + 5)(x - 2)} = \frac{a}{x + 5} + \frac{b}{x - 2}$ for all x. Thus, we seek a and b such that

5x + 11 = a(x - 2) + b(x + 5) for all x. If x = 2 we have 21 = 7b or b = 3.

If x = -5 we have -14 = -7a or a = 2. Thus,

$\int_{-1}^{1} \frac{5x + 11}{(x + 5)(x - 2)} \, dx = \int_{-1}^{1} \frac{2}{x + 5} \, dx + \int_{-1}^{1} \frac{3}{x - 2} \, dx$

$= 2 \ln |x + 5| \, \Big|_{-1}^{1} + 3 \ln |x - 2| \, \Big|_{-1}^{1} = 2(\ln 6 - \ln 4) + 3(\ln 1 - \ln 3)$

$= 2 \ln \frac{3}{2} - 3 \ln 3 = -\ln 12.$

3. We first seek constants a, b, and c such that

$\frac{3x^2 - 8x + 12}{x(x + 3)(x - 4)} = \frac{a}{x} + \frac{b}{x + 3} + \frac{c}{x - 4}$ or $3x^2 - 8x + 12 = a(x + 3)(x - 4)$

+ bx(x - 4) + cx(x + 3). If x = 0, we have 12 = -12a or a = -1. If x = -3,

we have 63 = 21b or b = 3. If x = 4, we have 28 = 28c or c = 1. Thus,

$\int \frac{3x^2 - 8x + 12}{x(x + 3)(x - 4)} \, dx = - \int \frac{dx}{x} + 3 \int \frac{dx}{x + 3} + \int \frac{dx}{x - 4}$ or

$\int \frac{3x^2 - 8x + 12}{x(x + 3)(x - 4)} \, dx = - \ln |x| + 3 \ln |x + 3| + \ln |x - 4| + C.$

This result can also be written in the more compact form

$\int \frac{3x^2 - 8x + 12}{x(x + 3)(x - 4)} \, dx = \ln \left| \frac{(x + 3)^3 (x - 4)}{x} \right| + C.$

5. Since the degree of the numerator is the same as the degree of the denominator

we first use long division to write $\frac{x - 3}{x + 1} = 1 - \frac{4}{x + 1}$. Thus,

$\int \frac{x - 3}{x + 1} \, dx = \int dx - 4 \int \frac{dx}{x + 1} = x - 4 \ln |x + 1| + C.$

7. On factoring the denominator we have $\frac{3x^2 - 8x - 4}{x^3 - 4x} = \frac{3x^2 - 8x - 4}{x(x - 2)(x + 2)}$. Thus, we

seek constants a, b, and c such that $\frac{3x^2 - 8x - 4}{x(x - 2)(x + 2)} = \frac{a}{x} + \frac{b}{x - 2} + \frac{c}{x + 2}$ or,

$3x^2 - 8x - 4 = a(x - 2)(x + 2) + bx(x + 2) + cx(x - 2)$. If x = 0, we have

-4 = -4a or a = 1. If x = 2, we have -8 = 8b or b = -1. If x = -2, we have

$24 = 8c$ or $c = 3$. Thus, $\int \frac{3x^2 - 8x - 4}{x^3 - 4x} \, dx = \int \frac{dx}{x} - \int \frac{dx}{x - 2} + 3 \int \frac{dx}{x + 2}$

so $\int \frac{3x^2 - 8x - 4}{x^3 - 4x} \, dx = \ln |x| - \ln |x - 2| + 3 \ln |x + 2| + C$

$$= \ln \left| \frac{x(x + 2)^3}{x - 2} \right| + C.$$

9. Since the linear factor x appears to the second power we seek constants a, b, and c such that $\frac{2x^2 - 5x - 10}{x^2(x + 2)} = \frac{a}{x} + \frac{b}{x^2} + \frac{c}{x + 2}$ or $2x^2 - 5x - 10$

$= ax(x + 2) + b(x + 2) + cx^2$. Thus, if $x = 0$, we have $-10 = 2b$ or $b = -5$. If $x = -2$, we have $8 = 4c$ or $c = 2$. If $x = 1$, we have $-13 = 3a + 3b + c$. Thus, since $b = -5$ and $c = 2$, we have $-13 = 3a - 15 + 2$ or $a = 0$. Consequent-

ly, $\int_1^2 \frac{2x^2 - 5x - 10}{x^2(x + 2)} \, dx = -5 \int_1^2 \frac{dx}{x^2} + 2 \int_1^2 \frac{dx}{x + 2}$

$= 5x^{-1} \Big|_1^2 + 2 \ln |x + 2| \, \Big|_1^2 = (\frac{5}{2} - 5) + 2(\ln 4 - \ln 3) = 2 \ln \frac{4}{3} - \frac{5}{2}.$

11. Since the linear factor x + 1 appears to the second power we seek constants a and b such that $\frac{2x + 1}{(x + 1)^2} = \frac{a}{x + 1} + \frac{b}{(x + 1)^2}$ or $2x + 1 = a(x + 1) + b$. If $x = -1$, we have $-1 = b$. If $x = 0$, $1 = a + b$. Then since $b = -1$, $a = 2$. Consequently, $\int \frac{2x + 1}{(x + 1)^2} \, dx = 2 \int \frac{dx}{x + 1} - \int \frac{dx}{(x + 1)^2}$

$= 2 \ln |x + 1| + \frac{1}{x + 1} + C.$

13. Since the quadratic factor $x^2 + 1$ appears to the second power we seek constants a, b, c, and d such that $\frac{x^3 + 1}{(x^2 + 1)^2} = \frac{ax + b}{x^2 + 1} + \frac{cx + d}{(x^2 + 1)^2}$ or,

$x^3 + 1 = (ax + b)(x^2 + 1) + cx + d$. If $x = i$, then $x^2 + 1 = 0$ so $-i + 1 = ci + d$. Then since the real and imaginary parts must be equal, $c = -1$ and $d = 1$. If $x = 0$, $1 = b + d$. Thus, since $d = 1$ we must have $b = 0$. Finally, choose another value of x that will make the calculations involved easy. Say $x = 1$. In this case we have $2 = 2a + 2b + c + d$. Then since $c = -1$, $d = 1$, and $b = 0$, we have $a = 1$. Consequently,

$\int \frac{x^3 + 1}{(x^2 + 1)^2} \, dx = \int \frac{x \, dx}{x^2 + 1} + \int \frac{-x + 1}{(x^2 + 1)^2} \, dx = \frac{1}{2} \int \frac{2x \, dx}{x^2 + 1}$

$- \frac{1}{2} \int \frac{2x \, dx}{(x^2 + 1)^2} + \int \frac{dx}{(x^2 + 1)^2}$. The first two integrals are straightforward

and a reduction formula may be applied to the third. Then,

$$\int \frac{x^3 + 1}{(x^2 + 1)^2} \, dx = \frac{1}{2} \ln \left| x^2 + 1 \right| + \frac{1}{2} (x^2 + 1)^{-1} + \frac{x}{2(x^2 + 1)} + \frac{1}{2} \int \frac{dx}{x^2 + 1}.$$

Consequently, $\int \dfrac{x^3 + 1}{(x^2 + 1)^2} \, dx = \dfrac{1}{2} \ln \left| x^2 + 1 \right| + \dfrac{1}{2(x^2 + 1)} + \dfrac{x}{2(x^2 + 1)}$

$+ \dfrac{1}{2} \arctan x + C.$

15. Since the degree of the numerator exceeds the degree of the denominator, we

first use long division to write $\dfrac{x^3 - x^2 - 16x - 4}{x^2 - 2x - 15} = x + 1 + \dfrac{x + 11}{x^2 - 2x - 15}.$

Thus, $\int \dfrac{x^3 - x^2 - 16x - 4}{x^2 - 2x - 15} \, dx = \int (x + 1) \, dx + \int \dfrac{x + 11}{x^2 - 2x - 15} \, dx.$

To evaluate the last of these two integrals, we use partial fractions, and so
seek constants a and b such that

$\dfrac{x + 11}{x^2 - 2x - 15} = \dfrac{x + 11}{(x - 5)(x + 3)} = \dfrac{a}{x - 5} + \dfrac{b}{x + 3}$ or $x + 11 = a(x + 3) + b(x - 5).$

Then if x = 5, 16 = 8a, and a = 2. If x = -3, 8 = -8b so b = -1. Thus,

$\int \dfrac{x^3 - x^2 - 16x - 4}{x^2 - 2x - 15} \, dx = \int (x + 1) \, dx + \int \dfrac{2}{x - 5} \, dx + \int \dfrac{-dx}{x + 3}$

$= \dfrac{1}{2} x^2 + x + 2 \ln \left| x - 5 \right| - \ln \left| x + 3 \right| + C$

$= \dfrac{1}{2} x^2 + x + \ln \left| \dfrac{x - 5}{x + 3} \right|^2 + C.$

19. Since the quadratic in the denominator is not factorable into real linear factors, we shall find constants a, b, and c such that

$\dfrac{5x^2 + 3x + 16}{(x^2 + 2x + 5)(x - 1)} = \dfrac{ax + b}{x^2 + 2x + 5} + \dfrac{c}{x - 1}$ or

$5x^2 + 3x + 16 = (ax + b)(x - 1) + c(x^2 + 2x + 5).$ If we let x = 1, we have
24 = 8c so c = 3. Setting x = 0 yields 16 = -b + 5c. Then, since c = 3, we
have b = -1. Finally, if we let x = -1, we have 18 = 2a - 2b + 4c. Then since
c = 3 and b = -1, we have a = 2 and consequently,

$\int \dfrac{5x^2 + 3x + 16}{(x^2 + 2x + 5)(x - 1)} \, dx = \int \dfrac{2x - 1}{x^2 + 2x + 5} \, dx + \int \dfrac{3}{x - 1} \, dx$

$= \int \dfrac{2x + 2}{x^2 + 2x + 5} \, dx - \int \dfrac{3}{x^2 + 2x + 5} \, dx + \int \dfrac{3}{x - 1} \, dx$

$= \ln \left| x^2 + 2x + 5 \right| - 3 \int \dfrac{dx}{(x + 1)^2 + 4} + 3 \ln \left| x - 1 \right|$

$= \ln \left| x^2 + 2x + 5 \right| - \dfrac{3}{2} \arctan \left(\dfrac{x + 1}{2} \right) + 3 \ln \left| x - 1 \right| + C$

$= \ln \left(\left| x^2 + 2x + 5 \right| \left| x - 1 \right|^3 \right) - \dfrac{3}{2} \arctan \dfrac{x + 1}{2} + C.$

21. We seek constants a, b, and c such that

$$\frac{x^2 + 12x - 5}{(x + 1)^2(x - 7)} = \frac{a}{x + 1} + \frac{b}{(x + 1)^2} + \frac{c}{x - 7}.$$ On multiplication by $(x + 1)^2$

$(x - 7)$ we have $x^2 + 12x - 5 = a(x + 1)(x - 7) + b(x - 7) + c(x + 1)^2$. If $x = -1$, we have $-16 = -8b$ so $b = 2$. If $x = 7$, $128 = 64c$, or $c = 2$. Finally, if we let $x = 0$, we get $-5 = -7a - 7b + c$. Then since $b = 2$, and $c = 2$, we have $a = -1$. Substitution of these results gives

$$\int_0^1 \frac{x^2 + 12x - 5}{(x + 1)^2(x - 7)}\, dx = -\int_0^1 \frac{dx}{x + 1} + 2\int_0^1 \frac{dx}{(x + 1)^2} + 2\int_0^1 \frac{dx}{x - 7}$$

$$= (-\ln|x + 1| - 2(x + 1)^{-1} + 2\ln|x - 7|)\,\Big|_0^1$$

$$= \left(\frac{-2}{x + 1} + \ln\frac{|x - 7|^2}{|x + 1|}\right)\Big|_0^1 = -1 + \ln 18 + 2 - \ln 49 = 1 + \ln(18/49).$$

23. Since the quadratic $x^2 + 4$ is not factorable into real linear factors, we seek

constants a, b, c, and d such that $\dfrac{2x^2 + x + 8}{(x^2 + 4)^2} = \dfrac{ax + b}{x^2 + 4} + \dfrac{cx + d}{(x^2 + 4)^2}$ or

$2x^2 + x + 8 = (ax + b)(x^2 + 4) + cx + d$. Then if we let $x = 2i$, we have $2i = 2ci + d$. Thus, $c = 1$ and $d = 0$. If we let $x = 0$, we have $8 = 4b + d$, and since $d = 0$, we have $b = 2$. Finally, if we let $x = 1$, we have $11 = 5a + 5b + c + d$. Then since $b = 2$, $c = 1$, and $d = 0$, we have $a = 0$. Substitution then

yields $\displaystyle\int \frac{2x^2 + x + 8}{(x^2 + 4)^2}\, dx = \int \frac{2\, dx}{x^2 + 4} + \int \frac{x\, dx}{(x^2 + 4)^2} = \arctan\frac{x}{2} - \frac{1}{2(x^2 + 4)} + C.$

31. First note that $\dfrac{1}{x(a - x)\left(1 - \dfrac{m}{x}\right)} = \dfrac{1}{(a - x)(x - m)}.$ Thus, we will find constants

b and c such that $\dfrac{1}{(a - x)(x - m)} = \dfrac{b}{a - x} + \dfrac{c}{x - m}$, or $1 = b(x - m) + c(a - x)$.

If $x = m$, we have $c = 1/(a - m)$. If $x = a$, we get $b = 1/(a - m)$. Thus,

$$\frac{1}{(a - x)(x - m)} = \frac{1}{a - m}\left(\frac{1}{a - x} + \frac{1}{x - m}\right) \text{ and so}$$

$$\int \frac{dx}{kx(a - x)\left(1 - \dfrac{m}{x}\right)} = \frac{1}{k(a - m)} \int \left(\frac{1}{a - x} + \frac{1}{x - m}\right) dx$$

$$= \frac{1}{k(a - m)}\left(\ln\left|\frac{x - m}{a - x}\right|\right) + C.$$

Note that if $x < m < a$, then $kx(a - x)\left(1 - \dfrac{m}{x}\right) < 0$. Thus,

$\dfrac{dx}{dt} = kx(a - x)\left(1 - \dfrac{m}{x}\right) < 0$ and the population decreases with increasing time.

1. On multiplication and division by 2 this becomes a standard form.

    Thus, $\int \dfrac{x + 5}{x^2 + 10x}\, dx = \dfrac{1}{2} \int \dfrac{2x + 10}{x^2 + 10x}\, dx = \dfrac{1}{2} \ln \left| x^2 + 10x \right| + C.$

3. Since the degree of the numerator is the same as that of the denominator, we

    use long division to obtain $\dfrac{x^2}{x^2 - 4} = 1 + \dfrac{4}{x^2 - 4}.$ Thus, $\int \dfrac{x^2\, dx}{x^2 - 4}$

    $= \int dx + 4 \int \dfrac{dx}{x^2 - 4}.$ The second of these integrals can be evaluated by use of

    the equation $\int \dfrac{du}{u^2 - a^2} = \dfrac{1}{2a} \ln \left| \dfrac{u - a}{u + a} \right| + C.$ Thus,

    $\int \dfrac{x^2\, dx}{x^4 - 4} = x + \ln \left| \dfrac{x - 2}{x + 2} \right| + C.$

5. On factoring we find $\dfrac{1}{x^2 - 6x} = \dfrac{1}{x(x - 6)}.$ Thus, to apply partial fractions we

    seek constants a and b such that $\dfrac{1}{x^2 - 6x} = \dfrac{a}{x} + \dfrac{b}{x - 6}$ or $1 = a(x - 6) + bx.$ If

    $x = 0,$ we find $a = -\dfrac{1}{6}$ and if $x = 6,$ we have $b = \dfrac{1}{6}.$ Thus,

    $\int \dfrac{dx}{x^2 - 6x} = -\dfrac{1}{6} \int \dfrac{dx}{x} + \dfrac{1}{6} \int \dfrac{dx}{x - 6} = -\dfrac{1}{6} \ln |x| + \dfrac{1}{6} \ln |x - 6| + C$

    $\qquad = \dfrac{1}{6} \ln \left| \dfrac{x - 6}{x} \right| + C.$

7. On multiplication and division by 2 this can be recognized as a standard form.

    $\int_0^1 \dfrac{x\, dx}{x^4 + 1} = \dfrac{1}{2} \int_0^1 \dfrac{2x\, dx}{1 + (x^2)^2} = \dfrac{1}{2} \int_0^1 \dfrac{dx^2}{1 + (x^2)^2} = \dfrac{1}{2} \arctan x^2 \Big|_0^1$

    $\qquad = \dfrac{1}{2} (\arctan 1 - \arctan 0) = \pi/8.$

9. One must apply integration by parts twice to evaluate this integral. If $u = x^2$
    and $dv = \sin x\, dx,$ then $du = 2x\, dx$ and $v = -\cos x.$ Thus,

    $\int x^2 \sin x\, dx = -x^2 \cos x + 2 \int x \cos x\, dx.$ If we now let $u = x$ and $dv = \cos$

    $x\, dx,$ we have $du = dx$ and $v = \sin x.$ Thus, $\int x^2 \sin x\, dx = -x^2 \cos x$

    $+ 2(x \sin x - \int \sin x\, dx)$ and so $\int x^2 \sin x\, dx = -x^2 \cos x + 2x \sin x + 2 \cos x$

    $+ C.$

11. Since the degree of the numerator exceeds that of the denominator we first use

    long division to find $\dfrac{x^3 + 1}{x + 2} = x^2 - 2x + 4 - \dfrac{7}{x + 2}.$ Thus,

    $\int \dfrac{x^3 + 1}{x + 2}\, dx = \int (x^2 - 2x + 4)\, dx - 7 \int \dfrac{dx}{x + 2}$

$$= \frac{1}{3} x^3 - x^2 + 4x - 7 \ln |x + 2| + C.$$

13. Since d sin x = cos x dx this is a standard form

$$\int_0^{\pi/2} \frac{\cos x}{\sin x + 3} \, dx = \int_0^{\pi/2} \frac{d \sin x}{\sin x + 3} = \ln |\sin x + 3| \Big|_0^{\pi/2} = \ln 4 - \ln 3 = \ln \frac{4}{3}.$$

15. Since the denominator is in factored form we will use the method of partial fractions. Thus, we seek constants a, b, and c such that

$$\frac{x^2 + 5x + 18}{(x + 3)(x^2 + 2x + 3)} = \frac{a}{x + 3} + \frac{bx + c}{x^2 + 2x + 3} \text{ or}$$

$x^2 + 5x + 18 = a(x^2 + 2x + 3) + (bx + c)(x + 3)$. If x = -3, we have 12 = 6a so a = 2. If x = 0, we have 18 = 3a + 3c so since a = 2, c = 4. Finally, if we let x = 1, then 24 = 6a + 4b + 4c. Use of a = 2 and c = 4 then yields b = -1. Thus,

$$\int \frac{x^2 + 5x + 18}{(x + 3)(x^2 + 2x + 3)} \, dx = \int \frac{2}{x + 3} \, dx + \int \frac{-x + 4}{x^2 + 2x + 3} \, dx. \text{ The first of these}$$

two integrals is easily evaluated so we will concentrate on the second.

$$\int \frac{-x + 4}{x^2 + 2x + 3} \, dx = -\frac{1}{2} \int \frac{2x - 8}{x^2 + 2x + 3} \, dx$$

$$= -\frac{1}{2} \int \frac{2x + 2}{x^2 + 2x + 3} \, dx - \frac{1}{2} \int \frac{-10 \, dx}{x^2 + 2x + 3} = -\frac{1}{2} \ln |x^2 + 2x + 3|$$

$$+ 5 \int \frac{dx}{x^2 + 2x + 3}. \text{ We now complete the square to get}$$

$$\int \frac{-x + 4}{x^2 + 2x + 3} \, dx = -\frac{1}{2} \ln |x^2 + 2x + 3| + 5 \int \frac{dx}{(x + 1)^2 + 2}.$$

Then, use of $\int \frac{du}{u^2 + a^2} = \frac{1}{a} \arctan \frac{u}{a} + C$ yields

$$\int \frac{-x + 4}{x^2 + 2x + 3} \, dx = -\frac{1}{2} \ln |x^2 + 2x + 3| + \frac{5}{\sqrt{2}} \arctan \frac{x + 1}{\sqrt{2}} + C.$$

The complete integral is then

$$\int \frac{x^2 + 5x + 18}{(x + 3)(x^2 + 2x + 3)} \, dx = 2 \ln |x + 3| - \frac{1}{2} \ln |x^2 + 2x + 3|$$

$$+ \frac{5}{\sqrt{2}} \arctan \frac{x + 1}{\sqrt{2}} + C.$$

17. The technique is to use integration by parts twice. First we will let u = sin 7x and dv = sin x dx. Then du = 7 cos 7x dx and v = -cos x so

$$\int_{-\pi}^{\pi} \sin x \sin 7x \, dx = -\sin 7x \cos x \Big|_{-\pi}^{\pi} + 7 \int_{-\pi}^{\pi} \cos x \cos 7x \, dx.$$

Since sin 7π = sin (-7π) = 0, we have $\int_{-\pi}^{\pi} \sin x \sin 7x \, dx = 7 \int_{-\pi}^{\pi} \cos x \cos 7x$ dx.

We now apply integration by parts again with u = cos 7x and dv = cos x dx.
Then, du = -7 sin 7x dx and v = sin x so

$$\int_{-\pi}^{\pi} \sin x \sin 7x \, dx = 7 \left[\sin x \cos 7x \, \Big|_{-\pi}^{\pi} + 7 \int_{-\pi}^{\pi} \sin x \sin 7x \, dx \right].$$

Since sin π = sin (-π) = 0, we then have

$$\int_{-\pi}^{\pi} \sin x \sin 7x \, dx = 49 \int_{-\pi}^{\pi} \sin x \sin 7x \, dx. \quad \text{Consequently,}$$

$$\int_{-\pi}^{\pi} \sin x \sin 7x \, dx = 0.$$

19. An application of the reduction formula

$$\int \frac{dx}{(x^2 + a^2)^{k+1}} = \frac{x}{2ka^2(x^2 + a^2)^k} + \frac{2k - 1}{2ka^2} \int \frac{dx}{(x^2 + a^2)^k} \text{ with a = 3 and k = 1}$$

yields $\int \dfrac{dx}{(x^2 + 9)^2} = \dfrac{x}{18(x^2 + 9)} + \dfrac{1}{18} \int \dfrac{dx}{x^2 + 9}$. The last integral is the

standard form $\int \dfrac{du}{u^2 + a^2} = \dfrac{1}{a} \arctan \dfrac{u}{a} + C.$

Thus, $\int \dfrac{dx}{(x^2 + 9)^2} = \dfrac{x}{18(x^2 + 9)} + \dfrac{1}{54} \arctan \dfrac{x}{3} + C.$

21. Since d sin x = cos x dx this is a standard form. Thus,

$$\int \cos x e^{\sin x} \, dx = \int e^{\sin x} \, d \sin x = e^{\sin x} + C.$$

23. On multiplication and division by 4 this is recognized as a standard form.

Thus, $\int \dfrac{e^{4x}}{4 + e^{4x}} \, dx = \dfrac{1}{4} \int \dfrac{4e^{4x} \, dx}{4 + e^{4x}} = \dfrac{1}{4} \int \dfrac{de^{4x}}{4 + e^{4x}} = \dfrac{1}{4} \ln |4 + e^{4x}| + C.$

25. The technique is integration by parts. If we let u = ln 2x and dv = x^6 dx,

then du = $\dfrac{dx}{x}$ and v = $\dfrac{1}{7} x^7$. Consequently,

$$\int x^6 \ln 2x \, dx = \frac{1}{7} x^7 \ln 2x - \frac{1}{7} \int x^7 \frac{dx}{x} = \frac{1}{7} x^7 \ln 2x - \frac{1}{49} x^7 + C.$$

27. The substitution u = $\sqrt{x - 3}$ will enable us to evaluate this integral. If

u = $\sqrt{x - 3}$, then u^2 = x - 3 and dx = 2u du. Moreover, u = 0 when x = 3 and
u = 1 when x = 4. Thus,

$$\int_3^4 x\sqrt{x - 3} \, dx = \int_0^1 (u^2 + 3) \, u \, 2u \, du$$

$$= 2 \int_0^1 (u^4 + 3u^2) \, du = 2(\frac{1}{5} u^5 + u^3) \, \Big|_0^1 = 2(\frac{1}{5} + 1) = \frac{12}{5}.$$

31. Since d tan x = $\sec^2$ x dx this is a standard form.

$$\int \frac{\sec^2 x}{\tan x} \, dx = \int \frac{d \tan x}{\tan x} = \ln |\tan x| + C.$$

33. The trigonometric identity $\sin^2 \theta = \dfrac{1 - \cos 2\theta}{2}$ is used. Then,

$$\int \sin^2 5x \, dx = \int \frac{1 - \cos 10x}{2} \, dx = \frac{1}{2} x - \frac{1}{20} \sin 10x + C.$$

35. The trigonometric substitution $x = \sqrt{2} \sin \theta$ will yield $dx = \sqrt{2} \cos \theta \, d\theta$ and

$(2 - x^2)^{3/2} = (2 - 2 \sin^2 \theta)^{3/2} = 2^{3/2} \cos^3 \theta.$

Thus, $\displaystyle\int \frac{dx}{(2 - x^2)^{3/2}} = \int \frac{\sqrt{2} \cos \theta \, d\theta}{2^{3/2} \cos^3 \theta} = \frac{1}{2} \int \frac{d\theta}{\cos^2 \theta} = \frac{1}{2} \int \sec^2 \theta \, d\theta = \frac{1}{2} \tan \theta + C.$

It is now necessary to put this result in terms of x. To do this note that

$\sin \theta = \dfrac{x}{\sqrt{2}}$ so $\cos \theta = \sqrt{1 - \dfrac{x^2}{2}}$. Thus, $\tan \theta = \dfrac{\sin \theta}{\cos \theta} = \dfrac{\frac{x}{\sqrt{2}}}{\sqrt{1 - \frac{x^2}{2}}} = \dfrac{x}{\sqrt{2 - x^2}}.$

On substitution then we have $\displaystyle\int \frac{dx}{(2 - x^2)^{3/2}} = \frac{x}{2\sqrt{2 - x^2}} + C.$

37. The trigonometric substitution $\cos^2 \theta = \dfrac{1 + \cos 2\theta}{2}$ will be used. Thus,

$$\int_0^{\pi/3} \cos^2 3x \, dx = \int_0^{\pi/3} \frac{1 + \cos 6\theta}{2} \, d\theta = \frac{1}{2} \theta + \frac{1}{12} \sin 6\theta \, \Big|_0^{\pi/3} = \frac{\pi}{6}.$$

39. Use integration by parts with $u = x$ and $dv = \sec x \tan x \, dx$. Then $du = dx$ and

$v = \sec x$ so $\int x \sec x \tan x \, dx = x \sec x - \int \sec x \, dx$

$= x \sec x - \ln |\sec x + \tan x| + C.$

41. Multiplication and division by 2 gives

$$\int \frac{x \, dx}{1 + x^4} = \frac{1}{2} \int \frac{2x \, dx}{1 + (x^2)^2} = \frac{1}{2} \int \frac{dx^2}{1 + (x^2)^2} = \frac{1}{2} \arctan x^2 + C.$$

43. $\displaystyle\int \cos^3 7x \, dx = \frac{1}{7} \int \cos^2 7x \cos 7x \, (7 \, dx) = \frac{1}{7} \int (1 - \sin^2 7x) \, d \sin 7x$

$\qquad = \dfrac{1}{7} \sin 7x - \dfrac{1}{21} \sin^3 7x + C.$

45. Since $d(3x^2 + 1) = 6x \, dx$ this can be made into a standard form by multiplication and division by 6. Thus,

$\displaystyle\int x(3x^2 + 1)^{1/2} \, dx = \frac{1}{6} \int (3x^2 + 1)^{1/2}(6x \, dx) = \frac{1}{6} \cdot \frac{2}{3} (3x^2 + 1)^{3/2} + C$

$\qquad = \dfrac{1}{9} (3x^2 + 1)^{3/2} + C.$

47. Let $t = \sqrt{x}$, then $x = t^2$ and $dx = 2t \, dt$. We get $2 \int t \sin t \, dt$. Now let $u = t$ and $dv = \sin t \, dt$. Then $du = dt$ and $v = -\cos t$. Thus, the integral $= -2t \cos t + 2 \int \cos t \, dt = -2t \cos t + 2 \sin t + C = -2 \sqrt{x} \cos \sqrt{x} + 2 \sin \sqrt{x} + C.$

51. Let $t = \sqrt{x + 3}$, then $x = t^2 - 3$ and $dx = 2t \, dt$. We get $2 \int t \ln t \, dt$. Now let $u = \ln t$ and $dv = t \, dt$. Then $du = \dfrac{1}{t} \, dt$ and $v = \dfrac{t^2}{2}$. So

$2 \int t \ln t \, dt = t^2 \ln t - \int t \, dt = t^2 \ln t - \dfrac{t^2}{2} + C = (x + 3) \ln \sqrt{x + 3}$

$- \dfrac{x + 3}{2} + C.$

Technique Review Exercises, Chapter 8 (page 381)

1. Since the degree of the numerator exceeds the degree of the denominator we

   first apply long division to obtain $\dfrac{3x^3 - 2x^2 - 10x + 9}{x^2 - x - 6} = 3x + 1 + \dfrac{9x + 15}{x^2 - x - 6}.$

   Thus, $\int \dfrac{3x^3 - 2x^2 - 10x + 9}{x^2 - x - 6} \, dx = \dfrac{3}{2} x^2 + x + \int \dfrac{9x + 15}{x^2 - x - 6}.$

   This last integral can be evaluated by the method of partial fractions as

   follows. $\dfrac{9x + 15}{x^2 - x - 6} = \dfrac{9x + 15}{(x - 3)(x + 2)}.$ Thus, we seek constants a and b such

   that $\dfrac{9x + 15}{(x - 3)(x + 2)} = \dfrac{a}{x - 3} + \dfrac{b}{x + 2}$ or $9x + 15 = a(x + 2) + b(x - 3).$ Then if

   $x = 3$, we have $a = \dfrac{42}{5}.$ If $x = -2$, we get $b = 3/5.$ Thus,

   $\int \dfrac{9x + 15}{(x - 3)(x + 2)} \, dx = \dfrac{42}{5} \ln |x - 3| + \dfrac{3}{5} \ln |x + 2| + C.$ Consequently,

   $\int \dfrac{3x^3 - 2x^2 - 10x + 9}{x^2 - x - 6} \, dx = \dfrac{3}{2} x^2 + x + \dfrac{42}{5} \ln |x - 3| + \dfrac{3}{5} \ln |x + 2| + C.$

2. We use substitution $u = \sqrt{2x + 1}.$ Then $u^2 = 2x + 1$ so $x = (u^2 - 1)/2$ and

   $dx = u \, du.$ Substitution then gives

   $\int \dfrac{x \, dx}{(2x + 1)^{3/2}} = \int \dfrac{(u^2 - 1) u \, du}{2u^3} = \dfrac{1}{2} \int (1 - \dfrac{1}{u^2}) \, du = \dfrac{1}{2} (u + \dfrac{1}{u}) + C.$

   Then since $u = \sqrt{2x + 1}$, we have $\int \dfrac{x \, dx}{(2x + 1)^{3/2}} = \dfrac{1}{2} (\sqrt{2x + 1} + \dfrac{1}{\sqrt{2x + 1}}) + C.$

3. $\int \sin^2 x \cos^5 x \, dx = \int \sin^2 x \cos^4 x \cos x \, dx$

   $= \int \sin^2 x (1 - \sin^2 x)^2 \, d \sin x = \int (\sin^2 x - 2 \sin^4 x + \sin^6 x) \, d \sin x$

   $= \dfrac{1}{3} \sin^3 x - \dfrac{2}{5} \sin^5 x + \dfrac{1}{7} \sin^7 x + C.$

4. Use of the identity $\sin \theta \cos \theta = \dfrac{\sin 2\theta}{2}$ gives

   $\int \sin^2 \pi t \cos^2 \pi t \, dt = \dfrac{1}{4} \int \sin^2 2\pi t \, dt.$ Then since $\sin^2 \theta = \dfrac{1 - \cos 2\theta}{2},$

   we have $\int \sin^2 \pi t \cos^2 \pi t \, dt = \dfrac{1}{8} \int (1 - \cos 4\pi t) \, dt = \dfrac{1}{8} (t - \dfrac{1}{4\pi} \sin 4\pi t) + C.$

6. If we let $x = 3 \sin \theta,$ $-\pi/2 \leq \theta \leq \pi/2,$ then $dx = 3 \cos \theta \, d\theta$ and

   $\sqrt{9 - x^2} = 3 \cos \theta.$ Since $\theta = 0$ when $x = 0$ and $\theta = \pi/6$ when $x = 3/2$ we have

105

$$\int_0^{3/2} \frac{x^2 \, dx}{\sqrt{9 - x^2}} = \int_0^{\pi/6} \frac{9 \sin^2 \theta \ (3 \cos \theta) \, d\theta}{3 \cos \theta} = 9 \int_0^{\pi/6} \sin^2 \theta \, d\theta$$

$$= 9 \int_0^{\pi/6} \frac{1 - \cos 2\theta}{2} \, d\theta = \frac{9}{2} \left(\theta - \frac{1}{2} \sin 2\theta\right) \Big|_0^{\pi/6} = \frac{9}{2} \left(\frac{\pi}{6} - \frac{\sqrt{3}}{4}\right).$$

7. $\int \sec^4 2x \tan^3 2x \, dx = \int \sec^3 2x \tan^2 2x \sec 2x \tan 2x \, dx$

$= \frac{1}{2} \int \sec^3 2x \ (\sec^2 2x - 1) \ d \ (\sec 2x) = \frac{1}{2} \left(\frac{1}{6} \sec^6 2x - \frac{1}{4} \sec^4 2x\right) + C$

$= \frac{\sec^4 2x}{4} \left(\frac{\sec^2 2x}{3} - \frac{1}{2}\right) + C.$

8. On completion of the square we get

$\int \dfrac{dx}{\sqrt{5 + 4x - x^2}} = \int \dfrac{dx}{\sqrt{9 - (x - 2)^2}}.$ Then if we let $x - 2 = 3 \sin \theta$,

$-\pi/2 \leq \theta \leq \pi/2$, we have $dx = 3 \cos \theta \, d\theta$ and $\sqrt{9 - (x - 2)^2} = 3 \cos \theta$. Substitution then gives $\int \dfrac{dx}{\sqrt{5 + 4x - x^2}} = \int \dfrac{3 \cos \theta \, d\theta}{3 \cos \theta} = \theta + C.$ Since $\theta = \arcsin \dfrac{x - 2}{3}$,

we have $\int \dfrac{dx}{\sqrt{5 + 4x - x^2}} = \arcsin \dfrac{x - 2}{3} + C.$

9. We will apply integration by parts twice. First let $u = \sin 3x$ and $dv = \cos x \, dx$. Then $du = 3 \cos 3x \, dx$ and $v = \sin x$, so, $\int \sin 3x \cos x \, dx$ $= \sin 3x \sin x - 3 \int \sin x \cos 3x \, dx$. We now apply integration by parts with $u = \cos 3x$ and $dv = \sin x \, dx$. Then $du = -3 \sin 3x \, dx$ and $v = -\cos x$. Consequently, $\int \sin 3x \cos x \, dx = \sin 3x \sin x - 3 \ [-\cos x \cos 3x - 3 \int \sin 3x \cos x \, dx]$, or, $\int \sin 3x \cos x \, dx = \sin 3x \sin x + 3 \cos x \cos 3x + 9 \int \sin 3x \cos x \, dx$. Then on solving for the integral and adding a constant of integration we get $\int \sin 3x \cos x \, dx = -\frac{1}{8} \ (\sin 3x \sin x + 3 \cos x \cos 3x) + C.$

10. The first goal is to express this integral as the sum of two integrals, one whose numerator is the derivative of $x^2 - 4x - 5$, and one with constant numerator. Thus, $\int \dfrac{x \, dx}{x^2 - 4x - 5} = \dfrac{1}{2} \int \dfrac{2x \, dx}{x^2 - 4x - 5} = \dfrac{1}{2} \int \dfrac{2x - 4}{x^2 - 4x - 5} \, dx$

$+ \dfrac{1}{2} \int \dfrac{4 \, dx}{x^2 - 4x - 5} = \dfrac{1}{2} \ln \left| x^2 - 4x - 5 \right| + \dfrac{1}{2} \int \dfrac{4 \, dx}{x^2 - 4x + 4 - 9}$

$= \dfrac{1}{2} \ln \left| x^2 - 4x - 5 \right| + 2 \int \dfrac{dx}{(x - 2)^2 - 9}.$ Then since

$\int \dfrac{du}{u^2 - a^2} = \dfrac{1}{2a} \ln \left| \dfrac{u - a}{u + a} \right| + C$, we have

$\int \dfrac{x \, dx}{x^2 - 4x - 5} = \dfrac{1}{2} \ln \left| x^2 - 4x - 5 \right| + \dfrac{1}{3} \ln \left| \dfrac{x - 5}{x + 1} \right| + C.$

11. If we let $u = \sqrt{x + 1}$, then $x = u^2 - 1$ and $dx = 2u\,du$. Substitution gives

$$\int \frac{e^{\sqrt{x+1}}}{\sqrt{x + 1}}\, dx = \int \frac{e^u\, 2u\, du}{u} = 2 \int e^u\, du = 2e^u + C.$$ Since $u = \sqrt{x + 1}$, we have

$$\int \frac{e^{\sqrt{x+1}}}{\sqrt{x + 1}}\, dx = 2e^{\sqrt{x+1}} + C.$$

Additional Exercises, Chapter 8 (page 381)

1. Let $u = (x - 3)^{1/2}$, then $x = u^2 + 3$ so $dx = 2u\,du$. Since $u = 1$ when $x = 4$ and $u = \sqrt{2}$ when $x = 5$, we have

$$\int_4^5 (x - 1)\sqrt{x - 3}\, dx = \int_1^{\sqrt{2}} (u^2 + 2)\, u\, (2u)\, du$$

$$= 2 \int_1^{\sqrt{2}} (u^4 + 2u^2)\, du = 2\left(\frac{u^5}{5} + \frac{2}{3} u^3\right)\Big|_1^{\sqrt{2}}$$

$$= 2\left(\frac{4}{5}\sqrt{2} + \frac{4}{3}\sqrt{2} - \frac{1}{5} - \frac{2}{3}\right) = \frac{2}{15}(32\sqrt{2} - 13).$$

3. We will apply integration by parts twice. First, let $u = \cos 3x$ and $dv = e^x\, dx$. Then $du = -3 \sin 3x\, dx$ and $v = e^x$, so $\int e^x \cos 3x\, dx = e^x \cos 3x + 3 \int e^x \sin 3x\, dx$. Next let $u = \sin 3x$ and $v = e^x$. Then, $du = 3 \cos 3x\, dx$ and $v = e^x$, so $\int e^x \cos 3x\, dx = e^x \cos 3x + 3(e^x \sin 3x - 3 \int e^x \cos 3x\, dx)$. Solving for $\int e^x \cos 3x\, dx$ and adding the constant of integration gives $\int e^x \cos 3x\, dx = \frac{1}{10}(e^x \cos 3x + 3e^x \sin 3x) + C$.

5. We use integration by parts twice. First, let $u = \cos 3x$ and $dv = \sin x\, dx$. Then, $du = -3 \sin 3x\, dx$ and $v = -\cos x$. Thus,

$$\int_0^{\pi/2} \sin x \cos 3x\, dx = -\cos 3x \cos x \Big|_0^{\pi/2} - 3 \int_0^{\pi/2} \sin 3x \cos x\, dx$$

$$= 1 - 3 \int_0^{\pi/2} \sin 3x \cos x\, dx.$$ Next let $u = \sin 3x$ and $dv = \cos x\, dx$. Then, $du = 3 \cos 3x\, dx$ and $v = \sin x$ and so

$$\int_0^{\pi/2} \sin x \cos 3x\, dx = 1 - 3\left(\sin x \sin 3x \Big|_0^{\pi/2} - 3 \int_0^{\pi/2} \sin x \cos 3x\, dx\right)$$

$$= 4 + 9 \int_0^{\pi/2} \sin x \cos 3x\, dx.$$ Solution for $\int_0^{\pi/2} \sin x \cos 3x\, dx$ gives

$$\int_0^{\pi/2} \sin x \cos 3x\, dx = -1/2.$$

7. Let $u = x$ and $dv = \sin 2x\, dx$. Then, $du = dx$ and $v = -\frac{1}{2} \cos 2x$ and so

$\int x \sin (2x)\, dx = -\frac{1}{2} x \cos 2x + \frac{1}{2} \int \cos 2x\, dx = -\frac{1}{2} x \cos 2x + \frac{1}{4} \sin 2x + C.$

9. Since $d \tan (x^2) = 2x \sec^2 (x^2)$, we have

$$\int x \sec^2 (x^2) \tan^5 (x^2)\, dx = \frac{1}{2} \int \tan^5 (x^2)\, d \tan (x^2) = \frac{1}{12} \tan^6 (x^2) + C.$$

11. $\int_0^{\pi/2} \sin^6 x \cos^3 x\, dx = \int_0^{\pi/2} \sin^6 x \cos^2 x\, d \sin x$

$= \int_0^{\pi/2} \sin^6 x (1 - \sin^2 x)\, d \sin x = \frac{1}{7} \sin^7 x - \frac{1}{9} \sin^9 x \Big|_0^{\pi/2} = \frac{1}{7} - \frac{1}{9} = 2/63.$

13. $\int \csc^{1/4} x \cot^5 x\, dx = \int (\csc^{-3/4} x \cot^4 x)(\csc x \cot x)\, dx$

$= -\int \csc^{-3/4} x\, (\csc^2 x - 1)^2\, d \csc x$

$= -\int \csc^{-3/4} x\, (\csc^4 x - 2 \csc^2 x + 1)\, d \csc x$

$= -\frac{4}{17} \csc^{17/4} x + \frac{8}{9} \csc^{9/4} x - 4 \csc^{1/4} x + C.$

15. $\int \sec^4 x \tan^{5/2} x\, dx = \int \sec^2 x \tan^{5/2} x \sec^2 x\, dx$

$= \int (1 + \tan^2 x) \tan^{5/2} x\, d \tan x = \frac{2}{7} \tan^{7/2} x + \frac{2}{11} \tan^{11/2} x + C.$

17. We will use the substitution $x = 3 \tan \theta$, $-\pi/2 < \theta < \pi/2$. Then, $dx = 3 \sec^2 \theta\, d\theta$ and $\sqrt{9 + x^2} = 3 \sec \theta$. Thus,

$$\int \frac{dx}{\sqrt{9 + x^2}} = \int \frac{3 \sec^2 \theta\, d\theta}{3 \sec \theta} = \int \sec \theta\, d\theta = \ln |\sec \theta + \tan \theta| + C'.$$

Then using $\sec \theta = \frac{1}{3} \sqrt{9 + x^2}$, and $\tan \theta = x/3$, we have

$$\int \frac{dx}{\sqrt{9 + x^2}} = \ln \left| \frac{1}{3} \sqrt{9 + x^2} + \frac{x}{3} \right| + C' = \ln \left| \sqrt{9 + x^2} + x \right| - \ln 3 + C'.$$

Thus, if we let $C = C' - \ln 3$, we have $\int \frac{dx}{\sqrt{9 + x^2}} = \ln \left| \sqrt{9 + x^2} + x \right| + C.$

19. $\int x \sqrt{7 - x^2}\, dx = -\frac{1}{2} \int \sqrt{7 - x^2}\, (-2x)\, dx = -\frac{1}{3} (7 - x^2)^{3/2} + C.$

21. We will first evaluate the indefinite integral. On completing the square, we

have $\int \frac{dx}{\sqrt{x^2 + 4x + 16}} = \int \frac{dx}{\sqrt{x^2 + 4x + 4 + 12}}$ or,

$\int \frac{dx}{\sqrt{x^2 + 4x + 16}} = \int \frac{dx}{\sqrt{(x + 2)^2 + 12}}$. Thus, we make the substitution

$(x + 2) = \sqrt{12} \tan \theta$. Then, $dx = \sqrt{12} \sec^2 \theta\, d\theta$ and $\sqrt{(x + 2)^2 + 12} = \sqrt{12} \sec \theta$,

$-\pi/2 < \theta < \pi/2$. Consequently, $\int \frac{dx}{\sqrt{x^2 + 4x + 16}} = \int \frac{\sqrt{12} \sec^2 \theta\, d\theta}{\sqrt{12} \sec \theta}$

$= \ln |\sec \theta + \tan \theta| + C.$ Then since $\tan \theta = (x + 2)/\sqrt{12}$ and

$\sec \theta = \sqrt{(x + 2)^2/12 + 1}$, we have

$$\int \frac{dx}{\sqrt{x^2 + 4x + 16}} = \ln \left| \sqrt{\frac{(x + 2)^2}{12} + 1} + \frac{x + 2}{\sqrt{12}} \right| + C.$$

Thus, $\int_0^1 \frac{dx}{\sqrt{x^2 + 4x + 16}} = \ln \left| \sqrt{\frac{(x + 2)^2}{12} + 1} + \frac{x + 2}{\sqrt{12}} \right| \Big|_0^1$

$= \ln (\sqrt{7}/2 + \sqrt{3}/2) - \ln (2/\sqrt{3} + 1/\sqrt{3}) = \ln (\frac{\sqrt{21} + 3}{6}).$

23. Completion of the square yields

$$\int (x - 4)^3 \sqrt{10 - x^2 + 8x} \, dx = \int (x - 4)^3 \sqrt{26 - (x^2 - 8x + 16)} \, dx$$

$= \int (x - 4)^3 \sqrt{26 - (x - 4)^2} \, dx.$ Thus, we use the substitution $x - 4 = \sqrt{26} \sin \theta$,

$-\pi/2 \le \theta \le \pi/2.$ Then, $dx = \sqrt{26} \cos \theta \, d\theta$ and $\sqrt{26 - (x - 4)^2} = \sqrt{26} \cos \theta.$ On

substitution we obtain $\int (x - 4)^3 \sqrt{10 - x^2 + 8x} \, dx = \int 26^{3/2} \sin^3 \theta \, 26^{1/2}$

$\cos \theta \, 26^{1/2} \cos \theta \, d\theta = 26^{5/2} \int \sin^3 \theta \cos^2 \theta \, d\theta = 26^{5/2} \int (1 - \cos^2 \theta)$

$\cos^2 \theta \sin \theta \, d\theta = 26^{5/2} \cos^3 \theta \, (\frac{1}{3} - \frac{1}{5} \cos^2 \theta) + C.$ Then since $\cos \theta$

$= \sqrt{1 - \frac{1}{26} (x - 4)^2}$ we have $\int (x - 4)^3 \sqrt{10 - x^2 + 8x} \, dx$

$= 26^{5/2} (1 - \frac{1}{26} (x - 4)^2)^{3/2} (\frac{1}{3} - \frac{1}{5} (1 - \frac{1}{26} (x - 4)^2)) + C = (26 - (x - 4)^2)^{3/2}$

$(26/3 - \frac{1}{5} (26 - (x - 4)^2)) + C.$

25. On completing the square we have

$$\int \frac{dx}{x^2 - 6x + 5} = \int \frac{dx}{(x^2 - 6x + 9) - 4} = \int \frac{dx}{(x - 3)^2 - 4}.$$ Then since

$\int \frac{du}{u^2 - a^2} = \frac{1}{2a} \ln \left| \frac{u - a}{u + a} \right| + C,$ we have

$\int \frac{dx}{x^2 - 6x + 5} = \frac{1}{4} \ln \left| \frac{x - 3 - 2}{x - 3 + 2} \right| + C = \frac{1}{4} \ln \left| \frac{x - 5}{x - 1} \right| + C.$

27. $\int \frac{x \, dx}{x^2 - 6x + 5} = \frac{1}{2} \int \frac{2x \, dx}{x^2 - 6x + 5} = \frac{1}{2} \int \frac{2x - 6}{x^2 - 6x + 5} \, dx + \frac{1}{2} \int \frac{6 \, dx}{x^2 - 6x + 5}.$

Then as in Exercise 25, $\int \frac{x \, dx}{x^2 - 6x + 5} = \frac{1}{2} \ln |x^2 - 6x + 5| + \frac{3}{4} \ln \left| \frac{x - 5}{x - 1} \right| + C.$

37. Use of the formula

$$\int \frac{dx}{(x^2 + a^2)^{k+1}} = \frac{x}{2ka^2(x^2 + a^2)^k} + \frac{2k - 1}{2ka^2} \int \frac{dx}{(x^2 + a^2)^k} \text{ with } k = 1 \text{ and } a = 2$$

gives $\int_1^2 \dfrac{dx}{(x^2 + 4)^2} = \dfrac{x}{8(x^2 + 4)} \Big|_1^2 + \dfrac{1}{8} \int_1^2 \dfrac{dx}{x^2 + 4} = \dfrac{1}{32} - \dfrac{1}{40} + \dfrac{1}{16} \arctan \dfrac{x}{2} \Big|_1^2$

$= \dfrac{1}{160} + \dfrac{1}{16} (\dfrac{\pi}{4} - \arctan \dfrac{1}{2})$.

41. In order to apply partial fractions we will find constants a and b such that

$\dfrac{1}{(x - 2)(x + 1)} = \dfrac{a}{x - 2} + \dfrac{b}{x + 1}$ or, $1 = a(x + 1) + b(x - 2)$. If x = 2, we have

1 = 3a or a = 1/3. If x = -1, we have 1 = -3b or b = -1/3. Thus,

$\int \dfrac{dx}{(x - 2)(x + 1)} = \dfrac{1}{3} \int \dfrac{dx}{x - 2} - \dfrac{1}{3} \int$ or,

$\int \dfrac{dx}{(x - 2)(x + 1)} = \dfrac{1}{3} \ln |x - 2| - \dfrac{1}{3} \ln |x + 1| + C = \dfrac{1}{3} \ln \left|\dfrac{x - 2}{x + 1}\right| + C.$

43. Since the degree of the numerator exceeds that of the denominator we first

apply division to obtain $\dfrac{x^3}{(x - 3)(x + 1)} = x + 2 + \dfrac{7x + 6}{(x - 3)(x + 1)}$. Thus,

$\int \dfrac{x^3}{(x - 3)(x + 1)} \, dx = \int (x + 2) \, dx + \int \dfrac{7x + 6}{(x - 3)(x + 1)} \, dx$. To apply partial

fractions in the evaluation of the last integral we find constants a and b such

that $\dfrac{7x + 6}{(x - 3)(x + 1)} = \dfrac{a}{x - 3} + \dfrac{b}{x + 1}$ or, $7x + 6 = a(x + 1) + b(x - 3)$. Then if

x = 3, we have 27 = 4a or a = 27/4. If x = -1, we have -1 = -4b or b = 1/4.

Thus, $\int \dfrac{x^3}{(x - 3)(x + 1)} \, dx = \int (x + 2) \, dx + \dfrac{27}{4} \int \dfrac{dx}{x - 3} + \dfrac{1}{4} \int \dfrac{dx}{x + 1}$

$= \dfrac{1}{2} x^2 + 2x + \dfrac{27}{4} \ln |x - 3| + \dfrac{1}{4} \ln |x + 1| + C.$

45. We seek constants a, b, and c such that

$\dfrac{6x^2 + x + 24}{x(x^2 + 6)} = \dfrac{a}{x} + \dfrac{bx + c}{x^2 + 6}$ or, $6x^2 + x + 24 = a(x^2 + 6) + (bx + c) x$. If x = 0,

we have 24 = 6a or a = 4. Letting x = 1 gives 31 = 7a + b + c. If x = -1, we

have 29 = 7a + b - c. Then since a = 4 these equations become 3 = b + c and

1 = b - c. Solution of these equations yields b = 2, c = 1. Hence,

$\int_1^2 \dfrac{6x^2 + x + 24}{x(x^2 + 6)} \, dx = 4 \int_1^2 \dfrac{dx}{x} + \int_1^2 \dfrac{2x + 1}{x^2 + 6} \, dx$

$= [4 \ln |x| + \ln |x^2 + 6| + \dfrac{1}{\sqrt{6}} \arctan \dfrac{x}{\sqrt{6}}] \Big|_1^2$

$= \ln (\dfrac{160}{7}) + \dfrac{1}{\sqrt{6}} (\arctan \dfrac{2}{\sqrt{6}} - \arctan \dfrac{1}{\sqrt{6}}).$

47. We seek constants a, b, and c such that

$\dfrac{x^2 - x + 1}{x(x - 2)^2} = \dfrac{a}{x} + \dfrac{b}{x - 2} + \dfrac{c}{(x - 2)^2}$ or, $x^2 - x + 1 = a(x - 2)^2 + bx(x - 2) + cx.$

If we let x = 2 we obtain 3 = 2c or c = 3/2. When x = 0, we have 1 = 4a or

$a = 1/4$. Finally, if $x = 1$, we have $1 = a - b + c$. Then using $a = 1/4$ and $c = 3/2$, we obtain $b = 3/4$. Thus,

$$\int \frac{x^2 - x + 1}{x(x-2)^2}\, dx = \frac{1}{4} \int \frac{dx}{x} + \frac{3}{4} \int \frac{dx}{x-2} + \frac{3}{2} \int \frac{dx}{(x-2)^2}$$

$$= \frac{1}{4} \ln |x| + \frac{3}{4} \ln |x-2| - \frac{3}{2} (x-2)^{-1} + C.$$

49. $\displaystyle \int_0^{\pi/4} \frac{\sin^3 x}{\cos x}\, dx = \int_0^{\pi/4} \frac{\sin^2 x}{\cos x} \sin x\, dx = - \int_0^{\pi/4} \frac{(1 - \cos^2 x)}{\cos x}\, d\cos x$

$\displaystyle = - \ln |\cos x| + \frac{1}{2} \cos^2 x \Big|_0^{\pi/4} = - \ln \frac{1}{\sqrt{2}} + \ln 1 + \frac{1}{2} \cdot \frac{1}{2} - \frac{1}{2} = \ln \sqrt{2} - \frac{1}{4}.$

51. To apply partial fractions we seek constants a, b, c, and d such that

$\displaystyle \frac{x^2}{(x^2 + 5)^2} = \frac{ax + b}{x^2 + 5} + \frac{cx + d}{(x^2 + 5)^2}$ or, $x^2 = (ax + b)(x^2 + 5) + (cx + d)$. If we let

$x = 5i$, we obtain $-5 = 5ci + d$. Thus, $d = -5$ and $c = 0$. If we set $x = 0$, we

have $5b + d = 0$. Then since $d = -5$, we have $b = 1$. Finally, if $x = 1$, we ob-

tain $1 = 6a + 6b + c + d$. Substitution of $b = 1$, $c = 0$, and $d = -5$ gives $a = 0$.

Thus, $\displaystyle \int \frac{x^2}{(x^2 + 5)^2}\, dx = \int \frac{dx}{x^2 + 5} - 5 \int \frac{dx}{(x^2 + 5)^2}$, so on evaluating the first

integral and using a reduction formula on the second, we have

$$\int \frac{x^2}{(x^2 + 5)^2}\, dx = \frac{1}{\sqrt{5}} \arctan \frac{1}{\sqrt{5}} - 5 \left[\frac{x}{10(x^2 + 5)} + \frac{1}{10} \int \frac{dx}{x^2 + 5} \right]$$

$$= \frac{1}{2\sqrt{5}} \arctan \frac{x}{\sqrt{5}} - \frac{x}{2(x^2 + 5)} + C.$$

53. $\displaystyle \int \frac{x}{x^2 + 4x + 5}\, dx = \frac{1}{2} \int \frac{2x}{x^2 + 4x + 5}\, dx = \frac{1}{2} \int \frac{2x + 4}{x^2 + 4x + 5}\, dx - 2 \int \frac{dx}{x^2 + 4x + 5}.$

The first integral can be evaluated directly. In the second we must complete

the square. Then, $\displaystyle \int \frac{x}{x^2 + 4x + 5}\, dx = \frac{1}{2} \ln |x^2 + 4x + 5| - 2 \int \frac{dx}{x^2 + 4x + 4 + 1}$

$$= \frac{1}{2} \ln |x^2 + 4x + 5| - 2 \int \frac{dx}{(x+2)^2 + 1}$$

$$= \frac{1}{2} \ln |x^2 + 4x + 5| - 2 \arctan (x + 2) + C.$$

CHAPTER 9: NUMERICAL METHODS

Exercises 9.2: Newton's Method (page 391)

1. In this case $f(x) = x^2 - 5$ so $f'(x) = 2x$. Then substitution into

$$c_{i+1} = c_i - \frac{f(c_i)}{f'(c_i)} \text{ gives } c_{i+1} = c_i - \frac{c_i^2 - 5}{2c_i} = \frac{c_i^2 + 5}{2c_i}.$$

Since $c_1 = 2$, $c_2 = \frac{2^2 + 5}{2 \cdot 2} = 2.25$

and $c_3 = \frac{(2.25)^2 + 5}{2(2.25)} = 2.236$.

3. Here $f(x) = x^2 - x - 3$ so $f'(x) = 2x - 1$. Substitution into

$$c_{i+1} = c_i - \frac{f(c_i)}{f'(c_i)} \text{ gives } c_{i+1} = c_i - \frac{c_i^2 - c_i - 3}{2c_i - 1} = \frac{c_i^2 + 3}{2c_i - 1}.$$

Thus, since $c_1 = -1$, $c_2 = \frac{(-1)^2 + 3}{2(-1) - 1} = -1.\overline{3}$ and $c_3 = \frac{(-1.\overline{3})^2 + 3}{2(-1.\overline{3}) - 1} = -1.\overline{30}$.

7. Here $f(x) = 3x^3 - x^2 - 4$ and $f'(x) = 9x^2 - 2x$ so

$$c_{i+1} = c_i - \frac{3c_i^3 - c_i^2 - 4}{9c_i^2 - 2c_i} = \frac{6c_i^3 - c_i^2 + 4}{9c_i^2 - 2c_i}. \text{ Then since } c_1 = 1,$$

$$c_2 = \frac{6(1)^3 - (1)^2 + 4}{9(1)^2 - 2(1)} = \frac{9}{7} \approx 1.286$$

and $c_3 \approx \frac{6(1.286)^3 - (1.286)^2 + 4}{9(1.286)^2 - 2(1.286)} \approx 1.227$.

9. Since $f(x) = 100 - \frac{1}{x}$, $f'(x) = x^{-2}$. Thus,

$$c_{i+1} = c_i - \frac{100 - 1/c_i}{1/c_i^2} = 2c_i - 100c_i^2. \text{ Since } c_1 = 1, c_2 = -98, \text{ and}$$

$c_3 = -960,204$. Note that these values do <u>not</u> appear to converge to the zero $1/100$. This may be expected since $f(x)$ does not satisfy the criteria of Theorem 9.2.1 when $c_1 = 1$.

11. We will check that $f(x) = x^4 - 5$ satisfies the hypotheses of Theorem 9.2.1. First note that $f(x)$ is continuous and $f(1) < 0$ and $f(2) > 0$. Thus, $f(x)$ must have a zero α in the interval $1 \leq x \leq 2$. Thus, we may take $a = 1$ and $b = 2$. We proceed now to check hypotheses (a), (b), and (c) of Theorem 9.2.1.

    (a) Since $f'(x) = 4x^3$, $f'(x) \neq 0$ for $1 \leq x \leq 2$.

    (b) Since $f''(x) = 12x^2$, $f''(x)$ is continuous and non-zero for $1 \leq x \leq 2$.

    (c) Note that $f'(1) = 4$ and $f'(2) = 32$. Thus, $|f'(x)|$ is smaller at the endpoint 1. On letting $c = 1$, we have $\left|\frac{f(c)}{f'(c)}\right| = \left|\frac{f(1)}{f'(1)}\right| = \left|\frac{-4}{4}\right| = 1$.

    Then since $b - a = 2 - 1 = 1$, $\left|\frac{f(c)}{f'(c)}\right| \leq b - a$.

Consequently, $f(x) = x^4 - 5$ satisfies the criteria for Theorem 9.2.1 on the interval $1 \leq x \leq 2$. Newton's Method will converge for any choice of c_1 such that $1 \leq c_1 \leq 2$. In particular, it must converge if $c_1 = 1$.

13. Since $f(x) = x^2 - 5$, $f(2) = -1 < 0$, and $f(3) = 4 > 0$. Hence, a zero of $f(x)$ must lie between 2 and 3. The choice $c_1 = 2$ then gives an initial error $E_1 \le 1$. In this case $f'(x) = 2x$ and $f''(x) = 2$. Since $|f''(x)| \le 2$ on $2 \le x \le 3$ and $|f'(x)| = |2x| \ge 4 > 0$ on $2 \le x \le 3$, we may choose $M = 2$ and $m = 4$. Then substitution into $E_{n+1} \le E_n^2 \frac{M}{2m}$ gives $E_{n+1} \le E_n^2 \frac{1}{4}$. Since $E_1 \le 1$,

$$E_2 \le (1)^2 \frac{1}{4} = \frac{1}{4} \text{ and } E_3 \le \left(\frac{1}{4}\right)^2 \frac{1}{4} = \frac{1}{64}.$$

15. Since $f(x) = x^2 - x - 3$, $f(-2) = 3 > 0$ and $f(-1) = -1 < 0$. Hence, a zero of $f(x)$ must lie between -2 and -1. The choice of $c_1 = -1$ then gives an initial error $E_1 \le 1$. In this case $f'(x) = 2x - 1$ and $f''(x) = 2$. Since $|f''(x)| \le 2$ on $-2 \le x \le -1$, we may choose $M = 2$. Since $f'(x) = 2x - 1 < 0$ when $-2 \le x \le -1$, we have $|f'(x)| = |2x - 1| = 1 - 2x$ if $-2 \le x \le 1$. Consequently, $|f'(x)| \ge 3 > 0$ if $-2 \le x \le -1$. We then choose $m = 3$. Substitution into $E_{n+1} \le E_n^2 \frac{M}{2m}$ then gives $E_{n+1} \le E_n^2 \frac{1}{3}$. Since $E_1 \le 1$, $E_2 \le (1)^2 \frac{1}{3} = \frac{1}{3}$ and $E_3 \le \left(\frac{1}{3}\right)^2 \frac{1}{3} = \frac{1}{27}.$

17. Since $f(x) = x^4 - 5$, $f(1) = -4 < 0$ and $f(2) = 9 > 0$, a zero of $f(x)$ must lie in the interval $1 \le x \le 2$. The choice of $c_1 = 1$ then gives an initial error $E_1 \le 1$. Here $f'(x) = 4x^3$ and $f''(x) = 12x^2$ so $|f''(x)| = 12x^2 \le 48$ on $1 \le x \le 2$. Moreover, $|f'(x)| = 4x^3 \ge 4 > 0$ on $1 \le x \le 2$. Thus, we choose $M = 48$ and $m = 4$. Substitution into

$E_{n+1} \le E_n^2 \frac{M}{2m}$ gives $E_{n+1} \le E_n^2 \, 6$. Hence since $E_1 \le 1$, $E_2 \le 6$, and

$E_3 \le 6^2 \, 6 = 216$. Thus, even though Newton's Method must converge in view of Exercise 11, our error bounds increase! This should not be too surprising. It merely indicates that the actual errors will decrease but our <u>bounds</u> on the errors do not indicate this fact.

19. Since $\sqrt{a}$ is zero of $f(x) = x^2 - a$, we will apply Newton's Method to this function. Since $f'(x) = 2x$, substitution into

$$c_{i+1} = c_i - \frac{f(c_i)}{f'(c_i)} \text{ gives } c_{i+1} = c_i - \frac{c_i^2 - a}{2c_i} = \frac{c_i^2 + a}{2c_i}.$$

<u>Exercises 9.3</u>: Bisection Method (page 394)

3. Since $3^3 = 27 < 31$ and $4^3 = 64 > 31$, we have $3 < \sqrt[3]{31} < 4$. Since $(3.5)^3 = 42.875 > 31$ and $3^3 < 31$, $3 < \sqrt[3]{31} < 3.5$. Since $(3.25)^3 = 34.328 > 31$ and $3^3 < 31$, $3 < \sqrt[3]{31} < 3.25$. Since $(3.125)^3 \approx 30.516 < 31$ and $(3.25)^3 > 31$,

$3.125 < \sqrt[3]{31} < 3.25.$ Thus, $\sqrt[3]{31} \approx 3.1875$ is an approximation to within 1/10 of the exact value.

Exercises 9.4: Trapezoid Rule (page 398)

1. Here h = 1.

i	x_i	$f(x_i)$	term
0	1	1	.5
1	2	.5	.5
2	3	$.\overline{3}$	$.\overline{3}$
3	4	.25	.25
4	5	.2	.1

Hence, $\int_1^2 \frac{dx}{x} \approx 1(.5 + .5 + .\overline{3} + .25 + .1) = 1.68\overline{3}.$

3. Here h = .5 so,

i	x_i	$f(x_i)$	term
0	1	1	.5
1	.5	.8	.8
2	1	.5	.5
3	1.5	.3077	.3077
4	2	.2	.1

$\int_0^2 \frac{dx}{1 + x^2} \approx .5(.5 + .8 + .5 + .3077 + .1) \approx 1.1038.$

5. In this case h = .25.

i	x_i	$f(x_i)$	term
0	0	1	.5
1	.25	$.\overline{3}$	$.\overline{3}$
2	.5	$.\overline{1}$	$.\overline{1}$
3	.75	$.\overline{037}$	$.\overline{037}$
4	1	.0123456	.0061728

Hence, $\int_0^1 81^{-x} \, dx \approx .25(.5 + .\overline{3} + .\overline{1} + .\overline{037} + .0061728) \approx .2469.$

7. In this case h = .25.

i	x_i	$f(x_i)$	term
0	0	.5	.25
1	.25	$.8\overline{8}$	$.8\overline{8}$
2	.5	1.6	1.6
3	.75	$2.\overline{90}$	$2.\overline{90}$
4	1	$5.\overline{3}$	$2.\overline{6}$

Hence, $\int_0^1 \frac{16^x}{x+2} dx \approx .25(.25 + .8\overline{8} + 1.6 + 2.\overline{90} + 2.\overline{6}) \approx 2.079.$

9. Here $h = \pi/4.$

i	x_i	$f(x_i)$	term
0	π	0	0
1	$5\pi/4$	$-2\sqrt{2}/5\pi$	$-2\sqrt{2}/5\pi$
2	$3\pi/2$	$-2/3\pi$	$-2/3\pi$
3	$7\pi/4$	$-2\sqrt{2}/7\pi$	$-2\sqrt{2}/7\pi$
4	2π	0	0

Thus, $\int_\pi^{2\pi} \frac{\sin x}{x} dx \approx \frac{\pi}{4}(-\frac{2\sqrt{2}}{5\pi} - \frac{2}{3\pi} - \frac{2\sqrt{2}}{7\pi}) \approx -\frac{1}{2}(\frac{\sqrt{2}}{5} + \frac{1}{3} + \frac{\sqrt{2}}{7}) \approx -\frac{1}{2}(\frac{12\sqrt{2}}{35} + \frac{1}{3}).$

15. In this Exercise we approximated $\int_1^5 \frac{dx}{x}$. Since $f(x) = x^{-1}$, $f'(x) = -x^{-2}$ and $f''(x) = 2x^{-3}$. Consequently, $|f''(x)| = 2|x^{-3}| \leq 2$ for $1 \leq x \leq 5$. Thus, we may make $M = 2$. Substitution into the error formula $E_n \leq \frac{b-a}{12} Mh^2$ gives $E_4 \leq \frac{5-1}{12} \cdot 2 \cdot 1^2$ so $E_4 \leq \frac{2}{3}$.

17. In Exercise 5 we approximated the integral $\int_0^1 81^{-x} dx$. Thus, $f(x) = 81^{-x}$, $f'(x) = -81^{-x} \ln 81$, and $f''(x) = 81^{-x} (\ln 81)^2$. Then since $|f''(x)| = |81^{-x}| (\ln 81)^2 \leq (\ln 81)^2 < 5^2 = 25$ for $0 \leq x \leq 1$ we may take $M = 25$. Substitution into the error formula $E_n \leq \frac{b-a}{12} Mh^2$ gives $E_4 \leq \frac{1}{12} \cdot 25 \cdot (\frac{1}{4})^2 = \frac{25}{192}$ so $E_4 \leq .1302$.

19. In this case $h = (b-a)/10 = 3/10$. Since $f(x) = \ln x^2 = 2 \ln x$, $f'(x) = 2x^{-1}$, and $f''(x) = -2x^{-2}$. Then since $|f''(x)| = |2x^{-2}| \leq 2$ for $1 \leq x \leq 4$ we may take $M = 2$. Substitution into the error formula $E_n \leq \frac{b-a}{12} Mh^2$ gives $E_{10} \leq \frac{3}{12} \cdot 2 \cdot (\frac{3}{10})^2 = \frac{9}{200}$.

21. In Exercise 1 we were to evaluate $\int_1^5 \frac{dx}{x}$. Thus, $f(x) = x^{-1}$, $f'(x) = -x^{-2}$, and $f''(x) = 2x^{-2}$. Since $|f''(x)| \leq |2x^{-2}| \leq 2$ for $1 \leq x \leq 5$ we may take $M = 2$. Substitution into $E_n \leq \frac{b-a}{12} Mh^2$ gives $E_n \leq \frac{2}{3} h^2$. Since $h = \frac{b-a}{n} = \frac{4}{n}$ we may write this inequality in terms of the unknown, n. Then, $E_n \leq \frac{2}{3} \cdot \frac{16}{n^2} = \frac{32}{3n^2}$. In part (a) we seek n large enough to ensure $E_n < \frac{1}{100}$. If we pick n so that $\frac{32}{3n^2} < \frac{1}{100}$, then clearly $E_n < \frac{1}{100}$. Solution of this inequality gives $32(100) < 3n^2$ or $n^2 > \frac{3200}{3} = 1066\ 2/3$. Since $32^2 = 1024$ and

$33^2 = 1089$, $n = 33$ is the smallest value of n for which we are assured that $E_n < \frac{1}{100}$. For part (b) we must pick n so that $\frac{32}{3n^2} < \frac{1}{1000}$ or $n^2 > \frac{32000}{3}$ $= 10666\ 2/3$. Since $103^2 = 10609$ and $104^2 = 10816$, we take $n = 104$.

23. In Exercise 5 we were to approximate $\int_0^1 81^{-x}\,dx$. Thus, we take $f(x) = 81^{-x}$ and so $f''(x) = 81^{-x}\,(\ln 81)^2$. Then since $\ln 81 < 5$, we have $|f''(x)| = 81^{-x}$ $(\ln 81)^2 \leq (\ln 81)^2 < 25$ so we may take $M = 25$. Then, $E_n \leq \frac{b-a}{12} Mh^2$ becomes $E_n \leq \frac{1}{12}\,25\,\frac{1}{n^2}$. Thus, the error will be less than $1/100$ if we select n so large that $\frac{25}{12n^2} < \frac{1}{100}$ or, $n^2 > \frac{2500}{12} = 208\ 1/3$. Since $14^2 = 196$ and $15^2 = 225$ we may select $n = 15$.

25. In general the arc length L of $y = f(x)$ between $x = a$ and $x = b$ is given by $L = \int_a^b \sqrt{1 + (f'(x))^2}\,dx$. In this case then $L = \int_0^\pi \sqrt{1 + \cos^2 x}\,dx$, and since $h = \frac{\pi}{4}$,

i	x_i	$f(x_i)$	term
0	0	$\sqrt{2}$	$\sqrt{2}/2$
1	$\pi/4$	$\sqrt{3}/2$	$\sqrt{3}/2$
2	$\pi/2$	1	1
3	$3\pi/4$	$\sqrt{3}/2$	$\sqrt{3}/2$
4	π	$\sqrt{2}$	$\sqrt{2}/2$

Thus, $L \approx \frac{\pi}{4}\,(\sqrt{2}/2 + \sqrt{3}/2 + 1 + \sqrt{3}/2 + \sqrt{2}/2) = \frac{\pi}{4}\,(1 + \sqrt{2} + \sqrt{6})$.

27. Since the volume is given by $V = \int_0^8 A(x)\,dx$ we may take $n = 4$ and so $h = 2$. Then,

i	x_i	$A(x_i)$	term
0	0	0	0
1	2	4	4
2	4	3.5	3.5
3	6	2	2
4	8	.5	.25

and so $V \approx 2(0 + 4 + 3.5 + 2 + .25) = 19.5\ m^3$.

Exercises 9.5: Simpson's Rule (page 404)

1. Here h = 1 and

i	x_i	$f(x_i)$	term
0	1	1	1
1	2	.5	2
2	3	$.\overline{3}$	$.\overline{6}$
3	4	.25	1
4	5	.2	.2

Thus, $\int_1^5 \frac{dx}{x} \approx \frac{1}{3} (1 + 2 + .\overline{6} + 1 + .2) \approx 1.622$.

3. In this case h = 1/2 and

i	x_i	$f(x_i)$	term
0	0	1	1
1	.5	.8	3.2
2	1	.5	1
3	1.5	.3077	1.2308
4	2	.2	.2

Thus, $\int_0^2 \frac{dx}{1 + x^2} \approx \frac{1}{6} (1 + 3.2 + 1 + 1.2308 + .2) \approx 1.1051$.

5. In this case h = 1/4.

i	x_i	$f(x_i)$	term
0	0	1	1
1	.25	$.\overline{3}$	$1.\overline{3}$
2	.5	$.\overline{1}$	$.2\overline{2}$
3	.75	$.\overline{037}$	$.\overline{148}$
4	1	.01234	.01234

Thus, $\int_0^1 81^{-x} dx \approx \frac{1}{12} (1 + 1.\overline{3} + .2\overline{2} + .\overline{148} + .01234) \approx .2263$.

7. In this case h = 1/4.

i	x_i	$f(x_i)$	term
0	0	1/2	.5
1	.25	8/9	$3.\overline{5}$
2	.5	8/5	3.2
3	.75	32/11	$11.6\overline{36}$
4	1	16/3	5.3

Thus, $\int_0^1 \frac{16^x}{x + 2} dx \approx \frac{1}{12} (.5 + 3.\overline{5} + 3.2 + 11.6\overline{36} + 5.\overline{3}) \approx 2.0188$.

9. Here h = $\pi/4$ and

i	x_i	$f(x_i)$	term
0	π	0	0
1	$5\pi/4$	$-4/5\pi\sqrt{2}$	$-16/5\pi\sqrt{2}$
2	$3\pi/2$	$-2/3\pi$	$-4/3\pi$
3	$7\pi/4$	$-4/7\pi\sqrt{2}$	$-16/7\pi\sqrt{2}$
4	2π	0	0

Thus, $\int_\pi^{2\pi} \dfrac{\sin x}{x}\, dx \approx \dfrac{\pi}{12}\left(-\dfrac{16}{5\pi\sqrt{2}} - \dfrac{4}{3\pi} - \dfrac{16}{7\pi\sqrt{2}}\right) \approx -\dfrac{1}{3}\left(\dfrac{1}{3} + \dfrac{48}{35\sqrt{2}}\right)$

$\approx -\dfrac{1}{3}\left(\dfrac{1}{3} + \dfrac{24\sqrt{2}}{35}\right)$.

15. In Exercise 1 we approximated $\int_1^5 \dfrac{dx}{x} = \ln 5$. Thus, $f(x) = x^{-1}$ and so

$f^{(4)}(x) = 24x^{-5}$. Thus, since $\left|f^{(4)}(x)\right| = \left|24x^{-5}\right| \leq 24$ for $1 \leq x \leq 5$ we may

take M = 24. Substitution into $E_n \leq \dfrac{h^4(b-a)}{180}$ M with n = 4 and h = 1 gives

$E_n \leq \dfrac{4}{180}\, 24 = \dfrac{24}{45}$. In Exercise 15 of Section 9.4 we found that the correspond-

ing error bound using the trapezoid rule was 2/3.

17. In Exercise 5 we approximated $\int_0^1 81^{-x}\, dx$. Thus, since n = 4, h = 1/4. More-

over, since $f(x) = 81^{-x}$, $f^{(4)}(x) = 81^{-x}(\ln 81)^4$. Thus, since $\left|f^{(4)}(x)\right|$

$= \left|81^{-x}(\ln 81)^4\right| \leq (\ln 81)^4 < 5^4$ we may take M = 5^4. Substitution into

$E_n \leq \dfrac{h^4(b-a)}{180}$ M yields $E_4 \leq \dfrac{5^4}{4^4 \cdot 180} \approx .01356$. In Exercise 17 of Section 9.4

we found that the error in the trapezoid rule was bounded by .2552.

19. Since $f(x) = \ln x^2 = 2 \ln x$, $f^{(4)}(x) = -12x^{-4}$. Thus, $\left|f^{(4)}(x)\right| = 12x^{-4} \leq 12$

for $1 \leq x \leq 4$. Consequently, we may select M = 4. Substitution into

$E_n \leq \dfrac{h^4(b-a)}{180}$ M gives $E_{10} \leq \dfrac{(\frac{3}{10})^4 (3)}{180}\, 12 = \dfrac{81}{50,000} = .00162$. In Exercise 19 of

Section 9.4 we found that the corresponding error for the trapezoid rule was

bounded by $\dfrac{9}{200} = .045$.

21. In Exercise 1 we were to approximate $\int_1^5 \dfrac{dx}{x} = \ln 5$. Since $f(x) = x^{-1}$,

$f^{(4)}(x) = 24x^{-5}$. Thus, $\left|f^{(4)}(x)\right| = 24|x|^{-5} \leq 24$ for $1 \leq x \leq 5$, we may take

M = 24. Substitution of this along with h = $\dfrac{4}{n}$ into $E_n \leq \dfrac{h^4(b-a)}{180}$ M yields

$E_n \leq \dfrac{(\frac{4}{n})^4 4}{180} 24$ or, $E_n \leq \dfrac{24 \cdot 4^5}{180n^4}$. If we select n large enough that $\dfrac{24 \cdot 4^5}{180n^4} < \dfrac{1}{100}$.

The error will be less than 1/100. Thus, we seek n so that

$n^4 > \dfrac{24 \cdot 4^5 \cdot 100}{180} = 13{,}653 \ 1/3$. Since $10^4 = 10{,}000$ and $11^4 = 14{,}641$, n = 11

fits the inequality. However, since we must select even n, we must take n = **12**. In part (b) we must select n large enough so that

$\dfrac{24 \cdot 4^5}{180n^4} < \dfrac{1}{1000}$ or, $n^4 > 136{,}533 \ 1/3$. Since $19^4 = 130{,}321$ and $20^4 = 160{,}000$,

we select n = 20. In Exercise 21 of Section 9.4 we found that to obtain an accuracy of 1/100 using the trapezoid rule we must use n = 33. To obtain an accuracy of 1/1000 we needed n = 104.

25. The volume is given by $V = \displaystyle\int_0^{\pi/2} 2\pi x \sin x \, dx$. Thus, if we take n = 2, we have

$h = \pi/4$ and

i	x_i	$f(x_i)$	term
0	0	0	0
1	$\pi/4$	$\sqrt{2}\pi^2/4$	$\sqrt{2}\pi^2$
2	$\pi/2$	π^2	π^2

Hence, $V \approx \dfrac{\pi}{12} (\sqrt{2}\pi^2 + \pi^2) = \dfrac{\pi^3}{12} (1 + \sqrt{2})$. Since $f(x) = 2\pi x \sin x$,

$f^{(4)}(x) = 2\pi (-4 \cos x + x \sin x)$. Then since $f^{(5)}(x) = 2\pi (5 \sin x + x \cos x)$

> 0 for $0 < x \leq \pi/2$, $f^{(4)}(x)$ is increasing for $0 \leq x \leq \pi/2$. Thus, we may take

$M = f^{(4)}(\pi/2) = \pi^2$. Substitution into $E_n \leq \dfrac{h^4(b - a)}{180} M$ gives

$E_n \leq \dfrac{(\pi/4)^4 \pi/2}{180} \pi^2 = \dfrac{\pi^7}{180 \cdot 4^4 \cdot 2} = \dfrac{\pi^7}{92{,}160}$.

27. The work done by a force $f(x)$ in moving an object from a to b is

$W = \displaystyle\int_a^b f(x) \, dx$. In this case, we have $f(0) = 20$, $f(5) = 27$, $f(10) = 26$,

$f(15) = 23$, $f(20) = 20$, $f(25) = 18$, $f(30) = 12$. Thus, if we apply Simpson's Rule with n = 6, h = 30/6 = 5, we get

i	x_i	$f(x_i)$	term
0	0	20	20
1	5	27	108
2	10	26	52
3	15	23	92
4	20	20	40
5	25	18	72
6	30	12	12

Thus, $\int_a^b f(x)\,dx \approx \frac{5}{3}(20 + 108 + 52 + 92 + 40 + 72 + 12) = 660$ kg m.

Exercises 9.6: Linear Approximations (page 409)

1. Here $f(x) = x^2 - 6x - 2$ so $f(a) = f(0) = -2$. Since $f'(x) = 2x - 6$, $f'(a) = f'(0) = -6$. Thus, substitution into $f(x) \approx f(a) + f'(a)(x - a)$ yields $x^2 - 6x - 2 \approx -2 - 6x$.

3. Here $f(x) = x^2 - 6x - 2$ so $f(-1) = f(a) = 5$. Since $f'(x) = 2x - 6$, $f'(a) = f'(-1) = -8$. Thus, substitution into $f(x) \approx f(a) + f'(a)(x - a)$ gives $x^2 - 6x - 2 \approx 5 - 8(x + 1)$.

5. Since $f(x) = -2 + 7$ and $f'(x) = -2$, $f(a) = f(0) = 7$ and $f'(a) = f'(0) = -2$. Thus, on substitution into $f(x) \approx f(a) + f'(a)(x - a)$ we have $-2x + 7 \approx 7 - 2x$.

7. Since $f(x) = -2x + 7$ and $f'(x) = -2$, $f(a) = f(1) = 5$ and $f'(a) = f'(1) = -2$. Thus, on substitution into $f(x) \approx f(a) + f'(a)(x - a)$ we have $-2x + 7 \approx 5 - 2(x - 1)$.

9. Since $f(x) = \ln x$ and $f'(x) = \frac{1}{x}$, $f(a) = f(2) = \ln 2$ and $f'(a) = f'(2) = 1/2$. Thus, $\ln x \approx \ln 2 + 1/2\,(x - 2)$.

11. Since $f(x) = e^x$ and $f'(x) = e^x$, $f(a) = f(0) = 1$ and $f'(a) = f'(0) = 1$. Thus, $e^x \approx 1 + x$.

13. Here $f(x) = \tan x$ and $f'(x) = \sec^2 x$. Thus, $f(a) = f(\pi/4) = \tan \pi/4 = 1$ and $f'(a) = \sec^2 \pi/4 = 2$. Consequently, $\tan x \approx 1 + 2(x - \pi/4)$.

17. Since $f(x) = x \ln x$ and $f'(x) = \ln x + 1$, we have $f(a) = f(1) = 0$ and $f'(a) = f'(1) = 1$. Thus, $x \ln x \approx x - 1$.

19. Since $f(x) = e^{\sin x}$ and $f'(x) = \cos x\, e^{\sin x}$, we have $f(a) = f(0) = 1$ and $f'(a) = f'(0) = 1$. Thus, $e^{\sin x} \approx 1 + x$.

21. Substitution of $x = 1$ into $e^x \approx 1 + x$ yields $e \approx 2$.

25. Theorem 9.6.1 tells us that if M denotes the maximum of $|f''(x)|$ in N, then $|E_1(x)| \le \frac{1}{2} M (x - a)^2$ where $E_1(x)$ is the error at x. In this case

$f(x) = x^2 - 6x - 2$ so $f''(x) = 2$. Thus, we may take $M = 2$. Substitution into the error bound formula above then gives $|E_1(x)| \leq (x + 1)^2$ since $a = -1$. The actual error is the difference between $f(x)$ and its approximation, $5 - 8(x + 1)$. Thus, $|E_1(x)| = |x^2 - 6x - 2 - 5 + 8(x + 1)| = |x^2 + 2x + 1| = (x + 1)^2$. In this case the bound on the error and the error are equal.

27. We again use the inequality $|E_1(x)| \leq \frac{1}{2} M (x - a)^2$ where M is the maximum of

    $|f''(x)|$ in N. In this case, $f(x) = e^x$, $a = 0$. Thus, $f''(x) = e^x$. Since $f''(x) = e^x$ is increasing its maximum e occurs at $x = 1$. Hence, we may take $M = e$. Substitution into the error inequality gives $|E_1(x)| \leq \frac{e}{2} x^2$ for $|x| \leq 1$. Note then that the error is less than or equal to e/2 for all x in the interval $|x| \leq 1$.

29. Since $\frac{\pi}{4} - .01$ is near $\frac{\pi}{4}$ and the trigonometric functions are easily calculated

    at $\pi/4$, we will use a linear approximation to $f(x) = \tan x$ at $a = \pi/4$. Since $f(x) = \tan x$ and $f'(x) = \sec^2 x$, $f(a) = f(\pi/4) = 1$ and $f'(a) = f'(\pi/4) = 2$. Hence, $\tan x \approx 1 + 2(x - \pi/4)$. Substitution of $x = \pi/4 - .01$ gives $\tan (\pi/4 - .01) \approx 1 + 2(\pi/4 - .01 - \pi/4) = .98$.

31. Since the linear approximation for $a = 1$ to $\ln x$ is given by $\ln x \approx x - 1$, we

    have $x \ln x \approx x^2 - x$. Thus, $\int_{.9}^{1} x \ln x\, dx \approx \int_{.9}^{1} (x^2 - x)\, dx$. Note that

    $$|\int_{.9}^{1} x \ln x\, dx - \int_{.9}^{1} (x^2 - x)\, dx| \leq \int_{.9}^{1} |x \ln x - x^2 + x|\, dx$$

    $$\leq \int_{.9}^{1} |x| |\ln x - x + 1|\, dx. \text{ Then since } |\ln x - x + 1| \leq \frac{1}{2} M (x - 1)^2$$

    $$\leq \frac{1}{2} (.9)^{-2}(x - 1)^2 \text{ we have } |\int_{.9}^{1} x \ln x\, dx - \int_{.9}^{1} (x^2 - x)\, dx|$$

    $$\leq \frac{1}{2} (.9)^{-2} \int_{.9}^{1} x(x - 1)^2\, dx \leq .00019.$$

<u>Exercises 9.7:</u> Second and Higher Degree Approximations; Taylor's Theorem (page 418)

1. Since $f(x) = x^4 - x + 7$, we have the following:

   $f(x) = x^4 - x + 7$ $f(a) = f(0) = 7$

   $f'(x) = 4x^3 - 1$ $f'(a) = f'(0) = -1$

   $f''(x) = 12x^2$ $f''(a) = f''(0) = 0$

   $f'''(x) = 24x$ $f'''(a) = f'''(0) = 0$

   $f^{(4)}(x) = 24$ $f^{(4)}(a) = f^{(4)}(0) = 24$

   $f^{(5)}(x) = 0$

Substitution into $f(x) \approx f(a) + f'(a)(x - a) + \frac{1}{2} f''(a)(x - a)^2$
$+ \frac{1}{3!} f'''(a)(x - a)^3 + \frac{1}{4!} f^{(4)}(a)(x - a)^4$ gives $x^4 - x + 7 \approx 7 - x + x^4$. Note
that here the approximation is exact and so $E_4(x) = 0$. The same result can be
obtained by use of the inequality $|E_4(x)| \leq \frac{1}{5!} M_5 |x - a|^5$ where M_5 is a bound
on $|f^{(5)}(x)|$ for x in N. Since $f^{(5)}(x) = 0$ for all x, we may take $M_5 = 0$.
Thus, $|E_4(x)| \leq 0$ and $E_4(x) = 0$.

3. Since $f(x) = 2x^5 - 3x^4 + x^3 - 3$, we have the following:

$f(x) = 2x^5 - 3x^4 + x^3 - 3$ $f(a) = f(0) = -3$

$f'(x) = 10x^4 - 12x^3 + 3x^2$ $f'(a) = f'(0) = 0$

$f''(x) = 40x^3 - 36x^2 + 6x$ $f''(a) = f''(0) = 0$

$f'''(x) = 120x^2 - 72x + 6$ $f'''(a) = f'''(0) = 6$

$f^{(4)}(x) = 240x - 72$ $f^{(4)}(x) = -72$

$f^{(5)}(x) = 240$

Substitution into $f(x) \approx f(a) = f'(a)(x - a) + \frac{1}{2} f''(a)(x - a)^2$
$+ \frac{1}{3!} f'''(a)(x - a)^3 + \frac{1}{4!} f^{(4)}(a)(x - a)^4$ gives

$2x^5 - 3x^4 + x^3 - 3 \approx -3 + x^3 - 3x^4$. Use of the inequality $|E_4(x)| \leq \frac{1}{5!} M_5$

$|x - a|^5$ where $M_5 \geq |f^{(5)}(x)|$ for x in N will yield a bound on the error.

Here $f^{(5)}(x) = 240$ so we may take $M_5 = 240$. Thus, $|E_4(x)| \leq \frac{240}{5!} |x|^5 = 2|x|^5$.

Note that since $|E_4(x)| = |f(x) - (f(a) + f'(a)(x - a) + \frac{f''(a)}{2!} (x - a)^2$

$+ \frac{f'''(a)}{3!} (x - a)^3 + \frac{f^4(a)}{4!} (x - a)^4)| = |2x^5 - 3x^4 + x^3 - 3 - (-3 + x^3 - 3x^4)|$

$= 2|x|^5$. The error is _exactly_ $2|x|^5$ for all x.

5. Since $f(x) = \sin x$, we have:

$f(x) = \sin x$ $f(a) = f(0) = 0$

$f'(x) = \cos x$ $f'(a) = f'(0) = 1$

$f''(x) = -\sin x$ $f''(a) = f''(0) = 0$

$f'''(x) = -\cos x$ $f'''(a) = f'''(0) = -1$

$f^{(4)}(x) = \sin x$ $f^{(4)}(a) = f^{(4)}(0) = 0$

$f^{(5)}(x) = \cos x$

Thus, $\sin x \approx 0 + 1x + \frac{0}{2!} x^2 - \frac{1}{3!} x^3 + \frac{0}{4!} x^4$ or, $\sin x \approx x - \frac{1}{3!} x^3$. Since

$|f^5(x)| = |\cos x| \leq 1$ for all x, we may take $M_5 = 1$. Substitution into $|E_4(x)| \leq \dfrac{M_5}{5!}|x - a|^5$ gives $|E_4(x)| \leq \dfrac{1}{5!}|x|^5$ for all x.

7. Since $f(x) = e^x$, we have:

$f(x) = e^x$ $\qquad\qquad$ $f(a) = f(0) = 1$

$f'(x) = e^x$ $\qquad\qquad$ $f'(a) = f'(0) = 1$

$f''(x) = e^x$ $\qquad\qquad$ $f''(a) = f''(0) = 1$

$f'''(x) = e^x$ $\qquad\qquad$ $f'''(a) = f'''(0) = 1$

$f^{(4)}(x) = e^x$ $\qquad\qquad$ $f^{(4)}(a) = f^{(4)}(0) = 1$

$f^{(5)}(x) = e^x$

Thus, $e^x \approx 1 + x + \dfrac{1}{2!}x^2 + \dfrac{1}{3!}x^3 + \dfrac{1}{4!}x^4$. Since $|f^{(5)}(x)| = |e^x| = e^x$ is increasing its maximum on the interval $|x| \leq 1$ occurs at $x = 1$. Thus, we may take $M_5 = e^1 = e$ for $|x| \leq 1$. Then, substitution into

$|E_4(x)| \leq \dfrac{M_5}{5!}|x - a|^5$ gives $|E_4(x)| \leq \dfrac{e}{5!}|x|^5$ if $|x| \leq 1$.

9. Since $f(x) = (x + 2)^{1/3}$, we have:

$f(x) = (x + 2)^{1/3}$ $\qquad$ $f(a) = f(0) = 2^{1/3}$

$f'(x) = \dfrac{1}{3}(x + 2)^{-2/3}$ $\qquad$ $f'(a) = f'(0) = \dfrac{1}{3 \cdot 2^{2/3}}$

$f''(x) = -\dfrac{2}{3^2}(x + 2)^{-5/3}$ $\qquad$ $f''(a) = f''(0) = -\dfrac{1}{3^2 \cdot 2^{2/3}}$

$f'''(x) = \dfrac{2 \cdot 5}{3^3}(x + 2)^{-8/3}$ $\qquad$ $f'''(a) = f'''(0) = \dfrac{5}{3^3 \cdot 2^{5/3}}$

$f^{(4)}(x) = -\dfrac{2 \cdot 5 \cdot 8}{3^4}(x + 2)^{-11/3}$ $\qquad$ $f^{(4)}(a) = f^{(4)}(0) = -\dfrac{5 \cdot 8}{3^4 \cdot 2^{8/3}}$

$f^{(5)}(x) = \dfrac{2 \cdot 5 \cdot 8 \cdot 11}{3^5}(x + 2)^{-14/3}$

Thus, $\sqrt[3]{x + 2} \approx 2^{1/3} + \dfrac{1}{3 \cdot 2^{2/3}}x - \dfrac{1}{3^2 \cdot 2^{2/3} \cdot 2!}x^2 + \dfrac{5}{3^3 \cdot 2^{5/3} \cdot 3!}x^3$

$- \dfrac{5 \cdot 8}{3^4 \cdot 2^{8/3} \cdot 4!}x^4$. Since $|f^{(5)}(x)| = \dfrac{2 \cdot 5 \cdot 8 \cdot 11}{3^5}|x + 2|^{-14/3}$, we have

$|f^{(5)}(x)| \leq \dfrac{2 \cdot 5 \cdot 8 \cdot 11}{3^5}$ for $|x| \leq 1$ and we can take $M_5 = \dfrac{2 \cdot 5 \cdot 8 \cdot 11}{3^5}$.

Then substitution into $|E_4(x)| \leq \dfrac{M_5}{5!}|x - a|^5$ yields

$$|E_4(x)| \le \frac{2 \cdot 5 \cdot 8 \cdot 11}{3^5 \cdot 5!} |x|^5 \le \frac{22}{3^6} |x|^5 \text{ if } |x| \le 1.$$

11. Since $f(x) = \dfrac{1}{x-3} = (x-3)^{-1}$, we have

$$f(x) = (x-3)^{-1} \qquad\qquad\qquad f(0) = -1/3$$

$$f'(x) = -(x-3)^{-2} \qquad\qquad\quad f'(0) = -1/3^2$$

$$f''(x) = 2(x-3)^{-3} \qquad\qquad\quad f''(0) = -2/3^3$$

$$f'''(x) = -6(x-3)^{-4} \qquad\qquad f'''(0) = -6/3^4$$

$$f^{(4)}(x) = 24(x-3)^{-5} \qquad\qquad f^{(4)}(0) = -24/3^5$$

$$f^{(5)}(x) = -120(x-3)^{-6}$$

Thus, $\dfrac{1}{x-3} \approx -\dfrac{1}{3} - \dfrac{1}{3^2} x - \dfrac{1}{3^3} x^2 - \dfrac{1}{3^4} x^3 - \dfrac{1}{3^5} x^4$. Since $\left| f^{(5)}(x) \right|$

$= \left| \dfrac{120}{(x-3)^6} \right| \le \dfrac{120}{2^6} = \dfrac{15}{8}$ and we can take $M_5 = 15/8$. Then substitution into

$|E_4(x)| \le \dfrac{M_5}{5!} |x-a|^5$ gives $|E_4(x)| \le \dfrac{15}{8 \cdot 5!} |x|^5 = \dfrac{|x|^5}{64}$ if $|x| \le 1$.

13. Since $f(x) = e^{(x^2)}$, we have

$$f(x) = e^{(x^2)} \qquad\qquad\qquad\qquad f(0) = 1$$

$$f'(x) = 2xe^{(x^2)} \qquad\qquad\qquad\quad f'(0) = 0$$

$$f''(x) = 2e^{(x^2)} + 4x^2 e^{(x^2)} \qquad\qquad f''(0) = 2$$

$$f'''(x) = 12xe^{(x^2)} + 8x^3 e^{(x^2)} \qquad\quad f'''(0) = 0$$

$$f^{(4)}(x) = 12e^{(x^2)} + 48x^2 e^{(x^2)} + 16x^4 e^{(x^2)} \qquad f^{(4)}(0) = 12$$

$$f^{(5)}(x) = 120xe^{(x^2)} + 160x^3 e^{(x^2)} + 32x^5 e^{(x^2)}$$

Thus, $e^{(x^2)} \approx 1 + x^2 + \dfrac{1}{2} x^4$. Since $f^{(5)}(x)$ is an increasing function

$(f^{(6)}(x) \ge 0$ for all $x)$ we have $\left| f^{(5)}(x) \right| \le \left| f^{(5)}(1) \right| = 312e$ for $|x| < 1$.

Thus, we may take $M_5 = 312e$. Substitution into $|E_4(x)| \le \dfrac{M_5}{5!} |x-a|^5$ gives

$|E_4(x)| \le \dfrac{13}{5} |x|^5$ for $|x| < 1$.

17. $\cos \frac{\pi x}{2}$. In this case

$f(x) = \cos \frac{\pi x}{2}$ $\qquad$ $f(1) = 0$

$f'(x) = -\frac{\pi}{2} \sin \frac{\pi x}{2}$ $\qquad$ $f'(1) = -\pi/2$

$f''(x) = -(\frac{\pi}{2})^2 \cos \frac{\pi x}{2}$ $\qquad$ $f''(1) = 0$

$f'''(x) = (\frac{\pi}{2})^3 \sin \frac{\pi x}{2}$ $\qquad$ $f'''(1) = (\pi/2)^3$

$f^{(4)}(x) = (\frac{\pi}{4})^4 \cos \frac{\pi x}{2}$ $\qquad$ $f^{(4)}(1) = 0$

$f^{(5)}(x) = -(\frac{\pi}{2})^5 \sin \frac{\pi x}{2}$

Thus, $\cos \frac{\pi x}{2} \approx -\frac{\pi}{2}(x-1) + (\frac{\pi}{2})^3 \frac{(x-1)^3}{3!} \approx -\frac{\pi}{2}(x-1) + \frac{\pi^3}{48}(x-1)^3$. Since

$|f^{(5)}(x)| = (\frac{\pi}{2})^5 |\sin \frac{\pi x}{2}| \leq (\frac{\pi}{2})^5$, we may take $M_5 = (\frac{\pi}{2})^5$. Then substitution into

$|E_4(x)| \leq \frac{M_5}{5!} |x-a|^5$ gives $|E_4(x)| \leq \frac{\pi^5}{2^5 \cdot 5!} |x-1|^5$.

19. In this case

$f(x) = x^{1/2}$ $\qquad$ $f(1) = 1$

$f'(x) = \frac{1}{2} x^{-1/2}$ $\qquad$ $f'(1) = 1/2$

$f''(x) = -\frac{1}{4} x^{-3/2}$ $\qquad$ $f''(1) = -1/4$

$f'''(x) = \frac{3}{8} x^{-5/2}$ $\qquad$ $f'''(1) = 3/8$

$f^{(4)}(x) = -\frac{15}{16} x^{-7/2}$ $\qquad$ $f^{(4)}(1) = -15/16$

$f^{(5)}(x) = \frac{105}{32} x^{-9/2}$

Thus, $x^{1/2} \approx 1 + \frac{1}{2}(x-1) - \frac{1}{8}(x-1)^2 + \frac{1}{16}(x-1)^3 - \frac{5}{128}(x-1)^4$. Since $f^{(5)}(x)$ is a decreasing function we have $|f^{(5)}(x)| \leq |f^{(5)}(1/2)|$ for $1/2 \leq x \leq 3/2$. Thus, $|f^{(5)}(x)| \leq \frac{105}{32}(2)^{9/2}$ for $1/2 \leq x \leq 3/2$ and we may take $M_5 = 105(2)^{9/2}/32$. Then substitution into $|E_4(x)| \leq \frac{M_5}{5!}|x-a|^5$ gives $|E_4(x)| \leq \frac{105(2)^{9/2}}{32 \cdot 5!}|x-1|^5$ or, $|E_4(x)| \leq \frac{7(2)^{1/2}}{16}|x-1|^5$ for $1/2 \leq x \leq 3/2$.

21. Since $f(x) = x \ln x$, we have

$f(x) = x \ln x$ $f(1) = 0$

$f'(x) = \ln x + 1$ $f'(1) = 1$

$f''(x) = x^{-1}$ $f''(1) = 1$

$f'''(x) = -x^{-2}$ $f'''(1) = -1$

$f^{(4)}(x) = 2x^{-3}$ $f^{(4)}(1) = 2$

$f^{(5)}(x) = -6x^{-4}$

Thus, $x \ln x \approx (x - 1) + \frac{1}{2}(x - 1)^2 - \frac{1}{6}(x - 1)^3 + \frac{1}{12}(x - 1)^4$. Since

$\left| f^{(5)}(x) \right| \leq 6 \cdot 2^4 = 96$, we can take $M_5 = 96$. Thus, $\left| E_4(x) \right| \leq \frac{96}{5!} \left| x - 1 \right|^5$

$= \frac{4}{5} \left| x - 1 \right|^5$.

27. In Exercise 13 we found that $e^{(x^2)} \approx 1 + x^2 + \frac{x^4}{2}$ where $\left| E_4(x) \right| \leq \frac{13}{5} \left| x \right|^5$. By

using this approximation one can write $\int_0^{.01} e^{x^2} dx \approx \int_0^{.01} (1 + x^2 + \frac{x^4}{2}) \, dx$.

In order to compute a bound on the error made in using this approximation note

that $e^{x^2} = x + x^2 + \frac{1}{2} x^4 + E_4(x)$ where $\left| E_4(x) \right| \leq \frac{13}{5} \left| x \right|^5$. Thus,

$\int_0^{.01} e^{x^2} dx = \int_0^{.01} (x + x^2 + \frac{1}{2} x^4) \, dx + \int_0^{.01} E_4(x) \, dx$ and the error is

$\int_0^{.01} E_4(x) \, dx \leq \int_0^{.01} \frac{13}{5} \left| x \right|^5 \, dx \leq \frac{13}{5} \frac{x^6}{6} \Big|_0^{.01} \leq 4.3 \times 10^{-13}$.

Technique Review Exercises, Chapter 9 (page 421)

1. Since $\alpha = \sqrt{3}$ is the positive zero of $f(x) = x^2 - 3$, we will apply Newton's
Method to this function. Then since $f'(x) = 2x$, substitution into

$c_{i+1} = c_i - \dfrac{f(c_i)}{f'(c_i)}$ gives $c_{i+1} = c_i - \dfrac{c_i^2 - 3}{2c_i}$ or on simplification

$c_{i+1} = \dfrac{c_i^2 + 3}{2c_i}$.

2. Substitution of $c_1 = 2$ into $c_{i+1} = \dfrac{c_i^2 + 3}{2c_i}$ gives $c_2 = \frac{7}{4} = 1.75$. Then

$c_3 = \dfrac{(1.75)^2 + 3}{2(1.75)} \approx 1.732$. Since $f(x) = x^2 - 3$, $f'(x) = 2x$ and $f''(x) = 2$.

Thus, we may take $M = 2$. Since $f'(x) = 2x > 2$ for $1 < x < 2$ and $1 < \sqrt{3} < 2$

we can take $m = 2$. Substitution into $E_{n+1} \leq E_n^2 \dfrac{M}{2m}$ then gives $E_{n+1} \leq \frac{1}{2} E_n^2$.

Since $E_1 \leq 1$, we have $E_2 \leq \frac{1}{2}$ and $E_3 \leq \frac{1}{2} \left(\frac{1}{2}\right)^2 = \frac{1}{8} = .125$.

3. Since $(1.5)^2 = 2.25 < 3$ and $2^2 = 4 > 3$, $1.5 < \sqrt{3} < 2$. Since $(1.75)^2 = 3.0625$ > 3 and $(1.5)^2 < 3$, $1.5 < \sqrt{3} < 1.75$. Since $(1.625)^2 \approx 2.640 < 3$ and $(1.75)^2 > 3$, $1.625 < \sqrt{3} < 1.75$. Thus, $\sqrt{3} \approx 1.6875$ is an approximation to within $1/10$ of the exact value.

4. Here $h = .25$ so

i	x_i	$f(x_i)$	term
0	0	1	.5
1	.25	2	2
2	.5	4	4
3	.75	8	8
4	1	16	8

Thus, $\int_0^1 16 \, x \, dx \approx .25 \, (.5 + 2 + 4 + 8 + 8) = 5.625$.

5. Here $h = .25$ and

i	x_i	$f(x_i)$	term
0	0	1	1
1	.25	2	8
2	.5	4	8
3	.75	8	32
4	1	16	16

Thus, $\int_0^1 16^x \, dx \approx \frac{.25}{3} (1 + 8 + 8 + 32 + 16) = 5.41\overline{6}$.

6. (a) The error E_n is bounded by $E_n \leq \frac{b-a}{12} Mh^2$ where $\left|f''(x)\right| \leq M$ for $a \leq x \leq b$.

In this case $f(x) = 16^x$, $f'(x) = 16^x \ln 16$, and $f''(x) = 16^x (\ln 16)^2$. Since $f''(x)$ is increasing and positive, the maximum of $\left|f''(x)\right|$ for $0 \leq x \leq 1$ is $f''(1) = 16 (\ln 16)^2$. Thus, since $\ln 16 < 3$, we can take $M = 16 (3)^2 = 144$. Then, on substitution of $h = .25$, $M = 144$, $b = 1$, and $a = 0$, we have $E_4 \leq \frac{1}{12} (144)(.25)^2$. Thus, $E_4 \leq .75$.

(b) The error E_n is bounded by $E_n \leq \frac{h^4 (b-a)}{180} M$ where $\left|f^{(4)}(x)\right| < M$ for $a \leq x \leq b$. In this case $f(x) = 16^x$, so $f^{(4)}(x) = 16^x (\ln 16)^4$. Since $f^{(4)}(x)$ is an increasing and positive function, the maximum of $\left|f^{(4)}(x)\right|$ for $0 \leq x \leq 1$ is $f^{(4)}(1) = 16 (\ln 16)^4$. Thus, since $\ln 16 < 3$, we can take $M = 16 (3)^4 = 1296$. Then, on substitution of $h = .25$, $M = 1296$, $b = 1$, and $a = 0$ above, get $E_4 \leq \dfrac{(.25)^4}{180} \ 1296 \approx .028.$

7. Since $f(x) = e^{-x}$, we have the following:

$f(x) = e^{-x}$ $\qquad\qquad$ $f(a) = f(0) = 1$

$f'(x) = -e^{-x}$ $\qquad\qquad$ $f'(a) = f'(0) = -1$

$f''(x) = e^{-x}$ $\qquad\qquad$ $f''(a) = f''(0) = 1$

$f'''(x) = -e^{-x}$ $\qquad\qquad$ $f'''(a) = f'''(0) = -1$

$f^{(4)}(x) = e^{-x}$ $\qquad\qquad$ $f^{(4)}(a) = f^{(4)}(0) = 1$

$f^{(5)}(x) = -e^{-x}$

Substitution into $f(x) \approx f(a) + f'(a)(x - a) + \dfrac{1}{2} f''(a)(x - 2)^2 + \dfrac{1}{3!} f'''(a)$

$(x - a)^3 + \dfrac{1}{4!} f^{(4)}(a)(x - a)^4$ gives $e^{-x} \approx 1 - x + \dfrac{x^2}{2} - \dfrac{x^3}{3!} + \dfrac{x^4}{4!}.$ Since

$\left| f^{(5)}(x) \right| = e^{-x}$ is a decreasing function, its maximum on the interval $|x| \leq 1$

occurs at $x = -1$. Thus, we may take $M_5 = e^1 = e$ for $|x| \leq 1$. Then substitu-

tion into $\left| E_4(x) \right| \leq \dfrac{M_5}{5!} |x - a|^5$ gives $\left| E_4(x) \right| \leq \dfrac{e}{5!} |x|^5$ if $|x| \leq 1.$

Additional Exercises, Chapter 9 (page 421)

1. In this case $f(x) = x^2 - 6$ and $f'(x) = 2x$. Thus, on substitution into

$c_{i+1} = c_i - \dfrac{f(c_i)}{f'(c_i)}$ we obtain $c_{i+1} = c_i - \dfrac{c_i^2 - 6}{2c_i} = \dfrac{c_i^2 + 6}{2c_i}.$ Then since

$c_1 = 2$, $c_2 = \dfrac{4 + 6}{4} = 2.5$, and $c_3 = \dfrac{(2.5)^2 + 6}{2(2.5)} = 2.45.$

3. In this case $f(x) = x^4 - 2x - 3$ and $f'(x) = 4x^3 - 2$. Thus,

$c_{i+1} = c_i - \dfrac{c_i^4 - 2c_i - 3}{4c_i^3 - 2} = \dfrac{3(c_i^4 + 1)}{4c_i^3 - 2}.$ Since $c_1 = 1$, we have $c_2 = 3$ and

$c_3 = \dfrac{3(82)}{106} \approx 2.321.$

5. Since $f(x) = x^5 - 2$, $f'(x) = 5x^4$. Thus, $c_{i+1} = c_i - \dfrac{c_i^5 - 2}{5c_i^4} = \dfrac{4c_i^5 + 2}{5c_i^4}.$

Since $c_1 = 1$, $c_2 = \dfrac{6}{5} = 1.2$ and $c_3 = \dfrac{4(1.2)^5 + 2}{5(1.2)^4} \approx 1.153.$

7. In this exercise $f(x) = x^2 + 3x - 1$. Since $f(0) = -1 < 0$ and $f(1) = 3 > 0$, $f(x)$ has a zero α such that $0 < \alpha < 1$. Then $f'(x) = 2x + 3$ and $f''(x) = 2$ are both non-zero and continuous on the interval $0 < x < 1$. Since $f'(0) = 3$ and

$f'(1) = 5$, $|f'(x)|$ is smaller at the endpoint 0. Finally, since $|\frac{f(0)}{f'(0)}| = \frac{1}{3} < 1 - 0 = 1$, we can apply Theorem 9.2.1 to see that Newton's Method converges.

9. Again we will apply Theorem 9.2.1. Since $f(x) = x^3 - 2x^2 - 1$, we have $f(2) = -1 < 0$, and $f(3) = 8 > 0$. Thus, $f(x)$ has a zero α with $2 < \alpha < 3$. Both $f'(x) = 3x^2 - 4x$ and $f''(x) = 6x - 4$ are continuous and non-zero for $2 \le x \le 3$. Since $f'(2) = 4$ and $f'(3) = 15$, $|f'(x)|$ is smaller at the endpoint 2. Then since $|\frac{f(2)}{f'(2)}| = \frac{1}{4} < 3 - 2 = 1$, Newton's Method must converge.

11. Since $2^2 = 4$ and $3^2 = 9$, $2 < \sqrt{5} < 3$. Since $(2.5)^2 > 5$, $2 < \sqrt{5} < 2.5$. Since $(2.25)^2 > 5$, $2 < \sqrt{5} < 2.25$. Finally, since $(2.125)^2 < 5$, $2.125 < \sqrt{5} < 2.25$. Thus, $\sqrt{5} \approx 2.1875$ is an approximation that is within $1/10$ of the exact value.

13. Since $1^3 < 4$ and $2^3 > 4$, $1 < \sqrt[3]{4} < 2$. Since $(1.5)^3 < 4$ and $2^3 > 4$, $1.5 < \sqrt[3]{4} < 2$. Since $(1.75)^3 > 4$ and $(1.5)^3 < 4$, $1.5 < \sqrt[3]{4} < 1.75$. Finally, since $(1.625)^3 > 4$ and $(1.5)^3 < 4$, $1.5 < \sqrt[3]{4} < 1.625$. Thus, $\sqrt[3]{4} \approx 1.5625$ is an approximation to $\sqrt[3]{4}$ that is within $1/10$ of the exact value.

15. In this case $h = \frac{1}{4} = .25$.

i	x_i	$f(x_i)$	term
0	0	1	.5
1	.25	.8	.8
2	.5	$.\overline{6}$	$.\overline{6}$
3	.75	.5714	.5714
4	1	.5	.25

Thus, $\int_0^1 \frac{dx}{1+x} \approx .25 \,(.5 + .8 + .\overline{6} + .5714 + .25) \approx .6970$.

17. Again $h = \frac{1}{4} = .25$.

i	x_i	$f(x_i)$	term
0	0	1	.5
1	.25	$e^{.0625}$	$e^{.0625}$
2	.5	$e^{.25}$	$e^{.25}$
3	.75	$e^{.5625}$	$e^{.5625}$
4	1	e	$e/2$

Thus, $\int_0^1 e^{x^2}\,dx \approx .25 \,(.5 + e^{.0625} + e^{.25} + e^{.5625} + .5e)$.

19. In this case $f(x) = (1 + x)^{-1}$ so, $f''(x) = 2(1 + x)^{-3}$. Since $f''(x)$ is a decreasing function on $0 \le x \le 1$, $|f''(x)| < f''(0) = 2$ for $0 \le x \le 1$. Thus, we may take $M = 2$. Since $h = .25$, substitution into $E_n \le \frac{b - a}{12} Mh^2$ gives

$E_4 \le \frac{1}{12} (2)(.25)^2$ so $E_4 \le .01042$.

21. Here $f(x) = e^{x^2}$ so $f''(x) = 2e^{x^2}(1 + 2x^2)$, an increasing function. Thus, $|f''(x)| \le f''(1) = 6e$ for $0 \le x \le 1$ and we can take $M = 6e$. Substitution into $E_n \le \frac{b - a}{12} Mh^2$ gives $E_n \le \frac{1}{12} (6e) h^2$. Then since $h = \frac{1}{n}$, we must select n so large that $\frac{e}{2n^2} < \frac{1}{1000}$ or, $n^2 > \frac{1000e}{2} \approx 1359$. Since $36^2 = 1296$ and $(37)^2$

$= 1369$, $n = 37$ is the smallest possible choice for n.

23. Here $h = .25$ and

i	x_i	$f(x_i)$	term
0	0	1	1
1	.25	.8	3.2
2	.5	$.\overline{6}$	$1.\overline{3}$
3	.75	.5714	2.2857
4	1	.5	.5

Thus, $\int_0^1 \frac{dx}{1 + x} \approx \frac{.25}{3} (1 + 3.2 + 1.\overline{3} + 2.2857 + .5) \approx .6933$.

25. $\int_0^1 e^{x^2} dx$. Again $h = .25$ and

i	x_i	$f(x_i)$	term
0	0	1	1
1	.25	$e^{.0625}$	$4e^{.0625}$
2	.5	$e^{.25}$	$2e^{.25}$
3	.75	$e^{.5625}$	$4e^{.5625}$
4	1	e	e

Thus, $\int_0^1 e^{x^2} dx \approx \frac{.25}{3} (1 + 4e^{.0625} + 2e^{.25} + 4e^{.5625} + e)$.

27. In this case $f(x) = (1 + x)^{-1}$ so $f^{(4)}(x) = 24 (1 + x)^{-5}$, a decreasing function. Thus, $|f^{(4)}(x)| \le |f^{(4)}(0)| = 24$ for $0 \le x \le 1$. Thus, we may take $M = 24$. Substitution into $E_n \le \frac{h^4(b - a)}{180} M$ gives $E_4 \le \frac{(.25)^4}{180} 24$. Hence, $E_4 \le .000521$.

29. Since $f(x) = e^{x^2}$, $f^{(4)}(x) = 4e^{x^2}(4x^4 + 12x^2 + 3)$, an increasing function. Thus, $|f^{(4)}(x)| \le f^{(4)}(1) = 76e$ for $0 \le x \le 1$ and we can take $M = 76e$. Substitution into $E_n \le \frac{h^4(b - a)}{180} M$ gives $E_n \le \frac{76e}{180n^4}$ since $h = \frac{1}{n}$. Thus, we select n

so large that $\dfrac{76e}{180n^4} < \dfrac{1}{1000}$ or, $n^4 > \dfrac{76e}{180} 1000 \approx 1148$. Since $6^4 = 1296$, $n = 6$

is a suitable choice for n. Since $5^4 = 625$, 6 is the smallest possible choice

for n.

31. Since $f(x) = \cos x$ and $f'(x) = -\sin x$, we have $f(\pi/2) = 0$ and $f'(\pi/2) = -1$.

Thus, the desired linear approximation is $\cos x \approx -(x - \pi/2)$.

33. In this case $f'(x) = \dfrac{5}{6} x^{-1/6}$ does not exist at $x = 0$. Thus, there is no

linear approximation in this case.

35. In this case $|f''(x)| = |-\cos x| \leq 1$ for all x. Thus, we may take $M = 1$. Sub-

stitution into $|E_1(x)| \leq \dfrac{1}{2} M (x - a)^2$ gives $|E_1(x)| \leq \dfrac{1}{2} (x - \pi/2)^2$.

37. $f(x) = \cos x$ $\qquad\qquad\qquad f(\pi/2) = 0$

$f'(x) = -\sin x$ $\qquad\qquad\qquad f'(\pi/2) = -1$

$f''(x) = -\cos x$ $\qquad\qquad\qquad f''(\pi/2) = 0$

$f'''(x) = \sin x$ $\qquad\qquad\qquad f'''(\pi/2) = 1$

$f^{(4)}(x) = \cos x$ $\qquad\qquad\qquad f^{(4)}(\pi/2) = 0$

Thus, $\cos x \approx -(x - \pi/2) + \dfrac{1}{3!} (x - \pi/2)^3$ is the desired approximation.

39. $f(x) = \cos^2 x$ $\qquad\qquad\qquad f(0) = 1$

$f'(x) = -2 \cos x \sin x$ $\qquad\qquad\qquad f'(0) = 0$

$f''(x) = 2(\sin^2 x - \cos^2 x)$ $\qquad\qquad f''(0) = -2$

$f'''(x) = 8 \sin x \cos x$ $\qquad\qquad\qquad f'''(0) = 0$

$f^{(4)}(x) = 8(\cos^2 x - \sin^2 x)$ $\qquad\quad f^{(4)}(0) = 8$

Thus, $\cos^2 x \approx 1 - \dfrac{2}{2!} x^2 + \dfrac{8}{4!} x^4 = 1 - x^2 + \dfrac{1}{3} x^4$ is the desired approximation.

41. $f(x) = x^4 + 3x^2 + 1$ $\qquad\qquad\qquad f(2) = 29$

$f'(x) = 4x^3 + 6x$ $\qquad\qquad\qquad f'(2) = 44$

$f''(x) = 12x^2 + 6$ $\qquad\qquad\qquad f''(2) = 54$

$f'''(x) = 24x$ $\qquad\qquad\qquad f'''(2) = 48$

$f^{(4)}(x) = 24$ $\qquad\qquad\qquad f^{(4)}(2) = 24$

Thus, $x^4 + 3x^2 + 1 \approx 29 + 44(x - 2) + \dfrac{54}{2!} (x - 2)^2 + \dfrac{48}{3!} (x - 2)^3 + \dfrac{24}{4!} (x - 2)^4$

or, $x^4 + 3x^2 + 1 \approx 29 + 44(x - 2) + 27(x - 2)^2 + 8(x - 2)^3 + (x - 2)^4$.

43. Since $f(x) = \cos x$, $f^{(5)}(x) = -\sin x$. Since $|f^{(5)}(x)| \leq 1$ for all x, we can

take $M_5 = 1$. Then substitution into $|E_n(x)| \leq \dfrac{1}{(n + 1)!} M_{n+1} |x - a|^{n+1}$ gives

$|E_4(x)| \leq \dfrac{1}{120} |x - \pi/2|^5$ as a bound on the error.

45. Since $f(x) = \cos^2 x$, $f^{(5)}(x) = -32 \sin x \cos x = -16 \sin 2x$. Thus, $|f^{(5)}(x)| \leq 16$ for all x and we can take $M_5 = 16$. Substitution into $|E_n(x)| \leq \frac{1}{(n+1)!} M_{n+1} |x - a|^{n+1}$ gives $|E_4(x)| \leq 16x^5/120 = 2x^5/15$.

CHAPTER 10: IMPROPER INTEGRALS, INFINITE SEQUENCES, AND SERIES

Exercises 10.1: An Extension of the Limit Concept (page 431)

1. Division of both numerator and denominator by x^4 gives

   $\lim\limits_{x \to \infty} \frac{3x^4 - x^2 + 1}{6 + 40x^3} = \lim\limits_{x \to \infty} \frac{3 - 1/x^2 + 1/x^4}{6/x^4 + 40/x}$. Since the numerator approaches 3 while

   the denominator goes to 0, the limit does not exist. Since the fraction increases without bound, we may write

   $\lim\limits_{x \to \infty} \frac{3x^4 - x^2 + 1}{6 + 40x^3} = \infty$.

5. In this case $\cos x$ continually oscillates between -1 and 1 while x increases without bound. Thus, $\lim\limits_{x \to \infty} \frac{\cos x}{x} = 0$.

7. Since 7^x increases without bound, $\lim\limits_{x \to \infty} \frac{5}{7^x} = 0$.

9. Division of numerator and denominator by x^3 gives $\lim\limits_{x \to \infty} \frac{1}{x^3 - x^2} = \lim\limits_{x \to \infty} \frac{1/x^3}{1 - 1/x}$

   $= 0$.

11. Since $\sin x$ continually assumes all values between -1 and 1 as x increases without bound, this limit does not exist.

13. Division of numerator and denominator by x gives $\lim\limits_{x \to \infty} \frac{5x + 3}{x - 70} = \lim\limits_{x \to \infty} \frac{5 + 3/x}{1 - 70/x}$

    $= 5$.

15. $\lim\limits_{x \to \infty} [\frac{x^2 - 9}{x + 3} - x] = \lim\limits_{x \to \infty} [\frac{(x + 3)(x - 3)}{x + 3} - x] = \lim\limits_{x \to \infty} [x - 3 - x] = -3$.

17. Division of numerator and denominator by n yields $\lim\limits_{n \to \infty} \frac{10n}{n - 6} = \lim\limits_{n \to \infty} \frac{10}{1 - 6/n} = 10$.

19. Division of numerator and denominator by n^3 gives

    $\lim\limits_{n \to \infty} \frac{2n^3 - 6n + 1}{4 - n^3} = \lim\limits_{n \to \infty} \frac{2 - 6/n^2 + 1/n^3}{4/n^3 - 1} = -2$.

21. Since $\lim\limits_{n \to \infty} [\frac{1}{n} - \frac{1}{n^2}] = \lim\limits_{n \to \infty} \frac{n - 1}{n^2}$, we divide numerator and denominator by n^2.

    Then $\lim\limits_{n \to \infty} [\frac{1}{n} - \frac{1}{n^2}] = \lim\limits_{n \to \infty} \frac{1/n - 1/n^2}{1} = 0$.

23. Since $1 - n < 0$ if $n > 1$, the product $n(1 - n)$ is negative if $n > 1$. Moreover, the product $|n(1 - n)|$ increases without bound as n increases. Thus, $\lim [n(1 - n)] = -\infty$.

25. $\lim\limits_{x\to\infty} \dfrac{\sqrt{4x^2 - 1}}{5x} = \lim\limits_{x\to\infty} \sqrt{\dfrac{4}{25} - \dfrac{1}{25x^2}} = 2/5$.

27. $\lim\limits_{n\to\infty} \dfrac{\sqrt[3]{n^6 - 3n}}{n^2 - 7} = \lim\limits_{n\to\infty} \dfrac{\sqrt[3]{1 - 3/n^5}}{1 - 7/n^2} = 1$.

29. $\lim\limits_{n\to\infty} f_n(r) = \lim\limits_{n\to\infty} \dfrac{nr}{1 + (n - 1)\,r} = \lim\limits_{n\to\infty} \dfrac{nr}{1 + nr - r}$. Division of numerator and

denominator by n then yields $\lim\limits_{n\to\infty} f_n(r) = \lim\limits_{n\to\infty} \dfrac{r}{1/n + r - r/n} = 1$. This result

indicates that as a test is made longer its reliability approaches 1.

$$\lim\limits_{r\to 1} f_n(r) = \lim\limits_{r\to 1} \dfrac{nr}{(1 + (n - 1)\,r)} = \dfrac{\lim\limits_{r\to 1} nr}{\lim\limits_{r\to 1} (1 + (n - 1)\,r)} = \dfrac{n}{1 + (n - 1)} = 1.$$

This indicates that as the reliability of the original test approaches 1 the reliability of a test n times as long also approaches 1.

Exercises 10.2: 1'Hospital's Rule (page 436)

1. Both $f(x) = 1 - \cos x$ and $g(x) = x^2$ satisfy the hypothesis of 1'Hospital's Rule. That is, (a) $\lim\limits_{x\to 0} (1 - \cos x) = 0$ and $\lim\limits_{x\to 0} x^2 = 0$.

   (b) $f'(x) = \sin x$ and $g'(x) = 2x$ exist in a neighborhood of 0.

   (c) $g'(x) = 2x$ is non-zero in a neighborhood of 0, except at 0 itself. Thus, $\lim\limits_{x\to 0} \dfrac{1 - \cos x}{x^2} = \lim\limits_{x\to 0} \dfrac{\sin x}{2x}$. We may apply 1'Hospital's Rule again since both

   $f(x) = \sin x$ and $g(x) = 2x$ satisfy the hypotheses. Thus,

   $\lim\limits_{x\to 0} \dfrac{1 - \cos x}{x^2} = \lim\limits_{x\to 0} \dfrac{\sin x}{2x} = \lim\limits_{x\to 0} \dfrac{\cos x}{2} = \dfrac{1}{2}$.

3. Both $f(x) = \sin 3x$ and $g(x) = x$ satisfy the hypothesis of 1'Hospital's Rule.

   Thus, $\lim\limits_{x\to 0} \dfrac{\sin 3x}{x} = \lim\limits_{x\to 0} 3 \cos 3x = 3$.

5. Since both $f(x) = x + 2$ and $g(x) = x^2 - 4$ satisfy the hypothesis of 1'Hospital's

   Rule, we have $\lim\limits_{x\to -2} \dfrac{x + 2}{x^2 - 4} = \lim\limits_{x\to -2} \dfrac{1}{2x} = -\dfrac{1}{4}$.

7. Both $f(x) = -5x^3 + 7x$ and $g(x) = 3 - x^3$ satisfy the hypothesis of the second form of 1'Hospital's Rule, Theorem 10.2.3. That is,

   (a) $\lim\limits_{x\to\infty} (-5x^3 + 7x) = -\infty$ and $\lim\limits_{x\to\infty} (3 - x^3) = -\infty$.

(b) Both $f'(x) = -15x^2 + 7$ and $g'(x) = -3x^2$ exist for all values of x.

(c) $g'(x) = -3x^2$ is non-zero for large values of x.

Thus, $\lim\limits_{x \to \infty} \dfrac{-5x^3 + 7x}{3 - x^3} = \lim\limits_{x \to \infty} \dfrac{-15x^2}{-3x^2} = 5$.

9. Both $f(t) = \sqrt{t^2 + 12} - 2t$ and $g(t) = t - 2$ satisfy the hypothesis of l'Hospi-

tal's Rule. Thus, $\lim\limits_{t \to 2} \dfrac{\sqrt{t^2 + 12} - 2t}{t - 2} = \lim\limits_{t \to 2} \dfrac{t(t^2 + 12)^{-1/2} - 2}{1} = \dfrac{2}{4} - 2 = -\dfrac{3}{2}$.

11. Again both $f(x) = \ln(x + 1)$ and $g(x) = x$ satisfy the hypothesis of l'Hospi-

tal's Rule, so $\lim\limits_{x \to \infty} \dfrac{\ln(x + 1)}{x} = \lim\limits_{x \to \infty} \dfrac{\frac{1}{x + 1}}{1} = 0$.

15. Both $f(x) = 10x^3$ and $g(x) = e^x$ satisfy the conditions of l'Hospital's Rule.

Thus, $\lim\limits_{x \to \infty} \dfrac{10x^3}{e^x} = \lim\limits_{x \to \infty} \dfrac{30x^2}{e^x}$. Again $\lim\limits_{x \to \infty} 30x^2 = \infty$ and $\lim\limits_{x \to \infty} e^x = \infty$. Another

application of l'Hospital's Rule yields $\lim\limits_{x \to \infty} \dfrac{30x^2}{e^x} = \lim\limits_{x \to \infty} \dfrac{60x}{e^x}$. Another applica-

tion of l'Hospital's Rule gives $\lim\limits_{x \to \infty} \dfrac{60x}{e^x} = \lim\limits_{x \to \infty} \dfrac{60}{e^x} = 0$. Thus, $\lim\limits_{x \to \infty} \dfrac{10x^3}{e^x} = 0$.

19. The functions $f(x) = \ln|\cos x|$ and $g(x) = \ln\left|\dfrac{x + 1}{x - 1}\right|$ satisfy the hypothesis

of l'Hospital's Rule. That is,

(a) $\lim\limits_{x \to 0} \ln|\cos x| = \ln\left(\lim\limits_{x \to 0} |\cos x|\right) = \ln 1 = 0$ and,

$\lim\limits_{x \to 0} \ln\left|\dfrac{x + 1}{x - 1}\right| = \ln\left(\lim\limits_{x \to 0}\left|\dfrac{x + 1}{x - 1}\right|\right) = \ln 1 = 0$.

(b) $f'(x) = \tan x$ and $g'(x) = \dfrac{1}{x + 1} - \dfrac{1}{x - 1} = \dfrac{-2}{x^2 - 1}$ both exist in a neighbor-

hood of 0.

(c) $g'(x) \neq 0$ in a neighborhood of zero. Thus,

$\lim\limits_{x \to 0} \dfrac{\ln|\cos x|}{\ln\left|\dfrac{x + 1}{x - 1}\right|} = \lim\limits_{x \to 0} \dfrac{\tan x}{\dfrac{-2}{x^2 - 1}} = \dfrac{0}{2} = 0$.

21. Since $\lim\limits_{x \to 0} \ln|x| = -\infty$ and $\lim\limits_{x \to 0} \cos x = 1$, l'Hospital's Rule does <u>not</u> apply.

However, since the numerator decreases without bound as the denominator goes

to 1, $\lim\limits_{x \to 0} \dfrac{\ln|x|}{\cos x} = -\infty$.

23. The functions $f(x) = \arcsin 3x$ and $g(x) = x$ satisfy the hypothesis of

l'Hospital's Rule. Thus, $\lim\limits_{x \to 0} \dfrac{\arcsin 3x}{x} = \lim\limits_{x \to 0} \dfrac{3}{\sqrt{1 - 9x^2}} = 3$.

29. In its given form, it is not easy to see whether l'Hospital's Rule applies or

not. After dividing numerator and denominator by x we see that it does apply

and we get:

$$\lim\limits_{x \to 0} \frac{\ln |x|}{1 + x^{-1} \ln |x|} = \lim\limits_{x \to 0} \frac{1/x}{x^{-1}x^{-1} - x^{-2} \ln |x|} = \lim\limits_{x \to 0} \frac{x}{1 - \ln |x|} = 0.$$

31. Since $\lim\limits_{x \to 0} \cot 2x = \infty$ and $\lim\limits_{x \to 0} \cot 3x = \infty$, we may apply l'Hospital's Rule. Then,

$\lim\limits_{x \to 0} \dfrac{\cot 2x}{\cot 3x} = \lim\limits_{x \to 0} \dfrac{-2 \csc^2 2x}{-3 \csc^2 3x}$, or, $\lim\limits_{x \to 0} \dfrac{\cot 2x}{\cot 3x} = \lim\limits_{x \to 0} \dfrac{2 \sin^2 3x}{3 \sin^2 2x}$. Another applica-

tion of l'Hospital's Rule then gives $\lim\limits_{x \to 0} \dfrac{\cot 2x}{\cot 3x} = \lim\limits_{x \to 0} \dfrac{12 \sin 3x \cos 3x}{12 \sin 2x \cos 2x}$. One

final application of l'Hospital's Rule yields

$$\lim\limits_{x \to 0} \frac{\cot 2x}{\cot 3x} = \lim\limits_{x \to \infty} \frac{3 \cos^2 3x - 3 \sin^2 3x}{2 \cos^2 2x - 2 \sin^2 2x} = \frac{3}{2}.$$

33. The method of Section 10.1 indicates that we should divide both numerator and

denominator by x^2. Thus, $\lim\limits_{x \to 0} \dfrac{8x^2 + 3x + 1}{100x^2 + 2x} = \lim\limits_{x \to 0} \dfrac{8 + 3/x + 1/x^2}{100 + 2/x} = \dfrac{2}{25}$. By

l'Hospital's Rule, $\lim\limits_{x \to 0} \dfrac{8x^2 + 3x + 1}{100x^2 + 2x} = \lim\limits_{x \to 0} \dfrac{16x + 3}{200x + 2} = \lim\limits_{x \to 0} \dfrac{16}{200} = \dfrac{2}{25}$.

Exercises 10.3:   Extensions of l'Hospital's Rule (page 439)

1. Since $\lim\limits_{x \to 0} (\csc x - \cot x) = \lim\limits_{x \to 0} (\dfrac{1}{\sin x} - \dfrac{\cos x}{\sin x}) = \lim\limits_{x \to 0} \dfrac{1 - \cos x}{\sin x}$, we apply

l'Hospital's Rule to this last limit. $\lim\limits_{x \to 0} \dfrac{1 - \cos x}{\sin x} = \lim\limits_{x \to 0} \dfrac{\sin x}{\cos x} = 0$. Con-

sequently, $\lim\limits_{x \to 0} (\csc x - \cot x) = 0$.

3. Again l'Hospital's Rule can not be applied directly so an algebraic manipula-

tion is first used.

$$\lim\limits_{x \to 0} (\ln |x| - e^x \ln |x|) = \lim\limits_{x \to 0} (1 - e^x) \ln |x| = \lim\limits_{x \to 0} \frac{\ln |x|}{\dfrac{1}{1 - e^x}}.$$

Now the functions $f(x) = \ln |x|$ and $g(x) = \dfrac{1}{1 - e^x}$ satisfy the hypothesis of

l'Hospital's Rule. Thus, $\lim\limits_{x \to 0} \dfrac{\ln |x|}{\dfrac{1}{1 - e^x}} = \lim\limits_{x \to 0} \dfrac{\dfrac{1}{x}}{(1 - e^x)^{-2} e^x} = \lim\limits_{x \to 0} [\dfrac{(1 - e^x)^2}{xe^x}]$.

This last limit is not evident, but one may apply l'Hospital's Rule again.

Then, $\lim\limits_{x \to 0} \left[\dfrac{(1 - e^x)^2}{xe^x}\right] = \lim\limits_{x \to 0} \dfrac{2(1 - e^x)(-e^x)}{e^x + xe^x} = \dfrac{0}{1} = 0.$ Thus,

$\lim\limits_{x \to 0} (\ln |x| - e^x \ln |x|) = 0.$

7. $\lim\limits_{x \to 0} \left(\dfrac{1}{x} - \dfrac{1}{xe^{ax}}\right) = \lim\limits_{x \to 0} \dfrac{e^{ax} - 1}{xe^{ax}}$ and we can apply l'Hospital's Rule to the second

limit. Thus, $\lim\limits_{x \to 0} \left(\dfrac{1}{x} - \dfrac{1}{xe^{ax}}\right) = \lim\limits_{x \to 0} \dfrac{ae^{ax}}{e^{ax} + axe^{ax}} = a.$

9. As this limit stands, l'Hospital's Rule does not apply. However, if we note

that $x^2 \ln |x| = \dfrac{\ln |x|}{1/x^2}$, we may write $\lim\limits_{x \to 0} x^2 \ln |x| = \lim\limits_{x \to 0} \dfrac{\ln |x|}{1/x^2}$ and apply

l'Hospital's Rule to this second limit. Thus,

$\lim\limits_{x \to 0} \dfrac{\ln |x|}{x^{-2}} = \lim\limits_{x \to 0} \dfrac{x^{-1}}{-2x^{-3}} = \lim\limits_{x \to 0} -\dfrac{1}{2} x^2 = 0.$

11. Again an algebraic manipulation will put this limit in a form to which l'Hospital's Rule may be applied.

$\lim\limits_{x \to 0} (\sin x) \ln |x| = \lim\limits_{x \to 0} \dfrac{\ln |x|}{\csc x} = \lim\limits_{x \to 0} \dfrac{\dfrac{1}{x}}{-\csc x \cot x} = \lim\limits_{x \to 0} -\dfrac{\sin x \tan x}{x}.$ We

again may apply l'Hospital's Rule. Thus, $\lim\limits_{x \to 0} -\dfrac{\sin x \tan x}{x}$

$= \lim\limits_{x \to 0} -\dfrac{\cos x \tan x + \sin x \sec^2 x}{1} = 0.$ So $\lim\limits_{x \to 0} (\sin x) \ln |x| = 0.$

13. L'Hospital's Rule applies when we express $\tan x = \dfrac{\sin x}{\cos x}.$ We get

$\lim\limits_{x \to \pi/2} \dfrac{(\frac{\pi}{2} - x) \sin x}{\cos x} = \lim\limits_{x \to \pi/2} \dfrac{(\frac{\pi}{2} - x) \cos x - \sin x}{-\sin x} = \dfrac{-1}{-1} = 1.$

15. Let $y = |x|^{2/x}.$ Then $\ln y = \dfrac{2 \ln |x|}{x}.$ We first find $\lim\limits_{x \to \infty} \ln y.$

$\lim\limits_{x \to \infty} \ln y = \lim\limits_{x \to \infty} \dfrac{2 \ln |x|}{x} = \lim\limits_{x \to \infty} \dfrac{2/x}{1} = 0,$ so $\lim\limits_{x \to \infty} y = 1.$

19. $\lim\limits_{x \to 0} x (\ln |x|)^2 = \lim\limits_{x \to 0} \dfrac{(\ln |x|)^2}{x^{-1}} = \lim\limits_{x \to 0} \dfrac{(2 \ln |x|) x^{-1}}{-x^{-2}} = \lim\limits_{x \to 0} \dfrac{-2 \ln |x|}{x^{-1}}$

$= \lim\limits_{x \to 0} \dfrac{-2x^{-1}}{-x^{-1}} = \lim\limits_{x \to 0} 2x = 0.$

21. Let $y = (\frac{x}{x + 1})^x$. Then $\ln y = x \ln (\frac{x}{x + 1}) = x \ln (\frac{1}{1 + x^{-1}}) = -x \ln (1 + x^{-1})$.

But $\lim\limits_{x \to \infty} \ln y = \lim\limits_{x \to \infty} \frac{-\ln (1 + x^{-1})}{x^{-1}}$. Apply l'Hospital's Rule to get

$\lim\limits_{x \to \infty} \frac{\frac{x^{-2}}{1 + x^{-1}}}{-x^{-2}} = \lim\limits_{x \to \infty} \frac{-1}{1 + x^{-1}} = -1$. So $\lim\limits_{x \to \infty} y = \lim\limits_{x \to \infty} (\frac{x}{x + 1})^x = e^{-1}$.

25. $\lim\limits_{x \to \infty} x(3^{1/x} - 2^{1/x}) = \lim\limits_{x \to \infty} \frac{3^{1/x} - 2^{1/x}}{x^{-1}} = \lim\limits_{x \to \infty} \frac{-x^{-2} 3^{1/x} \ln 3 + x^{-2} 2^{1/x} \ln 2}{-x^{-2}}$

$= \lim\limits_{x \to \infty} (3^{1/x} \ln 3 - 2^{1/x} \ln 2) = \ln 3 - \ln 2 = \ln (3/2)$.

Exercises 10.4: Improper Integrals (page 448)

1. Since $f(x) = x^{-1/2}$ is discontinuous at $x = 0$, this is an important integral.

Thus, $\int_0^1 x^{-1/2} dx = \lim\limits_{p \to 0} \int_p^1 x^{-1/2} dx = \lim\limits_{p \to 0} 2x^{1/2} \Big|_p^1 = \lim\limits_{p \to 0} 2(1 - p^{1/2}) = 2$.

3. Since $f(x) = \frac{1}{x^2}$ is discontinuous at $x = 0$, this is an improper integral. Thus,

$\int_{-1}^0 \frac{dx}{x^2} = \lim\limits_{q \to 0} \int_{-1}^{-q} \frac{dx}{x^2} = \lim\limits_{q \to 0} - x^{-1} \Big|_{-1}^{-q} = \lim\limits_{q \to 0} (\frac{1}{q} - 1)$. This limit does not exist

so the original integral diverges.

5. $\int_{-\infty}^0 \frac{dx}{x - 5} = \lim\limits_{a \to -\infty} \int_a^0 \frac{dx}{x - 5} = \lim\limits_{a \to -\infty} \ln |x - 5| \Big|_a^0 = \lim\limits_{a \to -\infty} (\ln 5 - \ln |a - 5|)$.

Since $\lim\limits_{a \to -\infty} \ln |a - 5| - \infty$, this limit does not exist and the integral diverges.

7. Since $f(x) = (x - 3)^{-3/5}$ is discontinuous at $x = 3$, this is an improper integral. Thus, we write

$\int_0^3 (x - 3)^{-3/5} dx = \lim\limits_{q \to 0} \int_0^{3-q} (x - 3)^{-3/5} dx$

$= \lim\limits_{q \to 0} \frac{5}{2} (x - 3)^{2/5} \Big|_0^{3-q} = \lim\limits_{q \to 0} \frac{5}{2} [(-q)^{2/5} - (-3)^{2/5}] = - \frac{5 \cdot 3^{2/5}}{2}$.

9. $\int_{-\infty}^0 \frac{e^x}{1 + e^{2x}} dx = \lim\limits_{a \to -\infty} \int_a^0 \frac{e^x}{1 + e^{2x}} dx = \lim\limits_{a \to -\infty} \arctan e^x \Big|_a^0$

$= \lim\limits_{a \to -\infty} (\pi/4 - \arctan e^a) = \pi/4$.

11. The function $f(x) = \frac{1}{\sqrt{9 - x^2}}$ is discontinuous at $x = 3$. Thus,

$\int_3^0 \dfrac{dx}{\sqrt{9 - x^2}} = \lim\limits_{q \to 0} \int_{3-q}^0 \dfrac{dx}{\sqrt{9 - x^2}}$. To evaluate this last integral we will use the

trigonometric substitution $x = 3 \sin \theta$, $-\pi/2 \le \theta \le \pi/2$. Then $dx = 3 \cos \theta \, d\theta$

and $\sqrt{9 - x^2} = \cos \theta$. Thus, $\int \dfrac{dx}{\sqrt{9 - x^2}} = \int \dfrac{3 \cos \theta \, d\theta}{\cos \theta} = 3 \theta + C$. Since

$\theta = \arcsin \dfrac{x}{3}$, $\int \dfrac{dx}{\sqrt{9 - x^2}} = 3 \arcsin \dfrac{x}{3} + C$. Consequently,

$\int_3^0 \dfrac{dx}{\sqrt{9 - x^2}} = \lim\limits_{q \to 0} 3 \arcsin \dfrac{x}{3} \Big|_{3-q}^0 = \lim\limits_{q \to 0} 3(\arcsin 0 - \arcsin \dfrac{3 - q}{3})$

$= 3(\arcsin 0 - \arcsin 1) = -\dfrac{3\pi}{2}$.

13. The function $f(x) = (x - 3)^{-5/3}$ is discontinuous at $x = 3$. Thus, we write

$\int_0^4 (x - 3)^{-5/3} \, dx = \int_0^3 (x - 3)^{-5/3} \, dx + \int_3^4 (x - 3)^{-5/3} \, dx$

$= \lim\limits_{q \to 0} \int_0^{3-q} (x - 3)^{-5/3} \, dx + \lim\limits_{p \to 0} \int_{3+p}^4 (x - 3)^{-5/3} \, dx$

$= \lim\limits_{q \to 0} -\dfrac{3}{2} (x - 3)^{-2/3} \Big|_0^{3-q} + \lim\limits_{p \to 0} -\dfrac{3}{2} (x - 3)^{-2/3} \Big|_{3+p}^4$

$= \lim\limits_{q \to 0} -\dfrac{3}{2} [(-q)^{-2/3} - (-3)^{-2/3}] + \lim\limits_{p \to 0} -\dfrac{3}{2} (1 - p^{-2/3})$. Since the limit

$\lim\limits_{q \to 0} (-q)^{-2/3}$ does not exist the limit $\lim\limits_{q \to 0} -\dfrac{3}{2} [(-q)^{-2/3} - (-3)^{-2/3}]$ does not

exist and the original integral is divergent.

15. $\int_{-\infty}^{-3} \dfrac{t}{1 - t^2} \, dt = \lim\limits_{a \to -\infty} \int_a^{-3} \dfrac{t}{1 - t^2} \, dt = \lim\limits_{a \to -\infty} (-\dfrac{1}{2} \ln |1 - t^2| \Big|_a^{-3})$

$= \lim\limits_{a \to -\infty} -\dfrac{1}{2} (\ln 8 - \ln |1 - a^2|)$. Since $\lim\limits_{a \to -\infty} \ln |1 - a^2|$ does not exist, the

integral is divergent.

17. Since the function $f(x) = \dfrac{1}{3 - x}$ is discontinuous at $x = 3$, we first write

$\int_0^\infty \dfrac{dx}{3 - x} = \int_0^3 \dfrac{dx}{3 - x} + \int_3^4 \dfrac{dx}{3 - x} + \int_4^\infty \dfrac{dx}{3 - x}$. Note that the limit of integration

4 is merely a convenience. Any $x > 3$ could be used. Thus,

$\int_0^3 \dfrac{dx}{3 - x} = \lim\limits_{q \to 0} \int_0^{3-q} \dfrac{dx}{3 - x} + \lim\limits_{p \to 0} \int_{3+p}^4 \dfrac{dx}{3 - x} + \lim\limits_{b \to \infty} \int_4^b \dfrac{dx}{3 - x}$

$$= \lim_{q \to 0} - \ln |3 - x| \Big|_0^{3-q} + \lim_{p \to 0} - \ln |3 - x| \Big|_{3+p}^4 + \lim_{b \to \infty} - \ln |3 - x| \Big|_4^b$$

$$= \lim_{q \to 0} (-\ln |q| + \ln 3) + \lim_{p \to 0} (-\ln |p|) + \lim_{b \to \infty} (-\ln |3 - b|).$$

Since $\lim_{q \to 0} \ln |q|$ does not exist, the original integral is divergent.

19. $\int_\infty^0 e^{-x} \, dx = \lim_{a \to \infty} \int_a^0 e^{-x} \, dx = \lim_{a \to \infty} - e^{-x} \Big|_a^0 = \lim_{a \to \infty} (-1 + e^{-a}) = -1.$

21. $\int_6^\infty \frac{dt}{t^2 - 9} = \lim_{b \to \infty} \int_6^b \frac{dt}{t^2 - 9} = \lim_{b \to \infty} \frac{1}{6} \ln \left|\frac{t - 3}{t + 3}\right| \Big|_6^b$

$$= \lim_{b \to \infty} \frac{1}{6} \left(\ln \left|\frac{b - 3}{b + 3}\right| - \ln \frac{1}{3}\right) = -\frac{1}{6} \ln \frac{1}{3} = \frac{1}{6} \ln 3.$$

23. $\int_2^\infty \frac{dx}{x\sqrt{x^2 + 4}} = \lim_{b \to \infty} \int_2^b \frac{dx}{x\sqrt{x^2 + 4}}$. To evaluate the integral $\int \frac{dx}{x\sqrt{x^2 + 4}}$, we will

use the trigonometric substitution $x = 2 \tan \theta$, $-\pi/2 < \theta < \pi/2$. Then,

$dx = 2 \sec^2 \theta \, d\theta$ and $\sqrt{x^2 + 4} = 2 \sec \theta$. Consequently,

$$\int \frac{dx}{x\sqrt{x^2 + 4}} = \int \frac{2 \sec^2 \theta \, d\theta}{(2 \tan \theta)(2 \sec \theta)} = \frac{1}{2} \int \frac{\sec \theta}{\tan \theta} \, d\theta = \frac{1}{2} \int \frac{1}{\sin \theta} \, d\theta$$

$$= \frac{1}{2} \int \csc \theta \, d\theta = -\frac{1}{2} \ln |\csc \theta + \cot \theta| + C. \text{ Since } \sec \theta = \frac{1}{2}\sqrt{x^2 + 4} \text{ and}$$

$\tan \theta = \frac{x}{2}$, we have $\csc \theta = \frac{\sec \theta}{\tan \theta} = \frac{1}{x}\sqrt{x^2 + 4}$ and, $\cot \theta = \frac{1}{\tan \theta} = \frac{2}{x}$. Thus,

$$\int \frac{dx}{x\sqrt{x^2 + 4}} = -\frac{1}{2} \ln \left(\frac{\sqrt{x^2 + 4}}{x} + \frac{2}{x}\right) + C, \text{ and,}$$

$$\int_2^\infty \frac{dx}{x\sqrt{x^2 + 4}} = -\frac{1}{2} \lim_{b \to \infty} \ln \left(\frac{\sqrt{x^2 + 4}}{x} + \frac{2}{x}\right) \Big|_2^b$$

$$= -\frac{1}{2} \lim_{b \to \infty} \left(\ln \left(\frac{\sqrt{b^2 + 4}}{b} + \frac{2}{b}\right) - \ln (\sqrt{2} + 1)\right). \text{ We now concentrate on calculating}$$

the limit $\lim_{b \to \infty} \ln \left(\frac{\sqrt{b^2 + 4}}{b} + \frac{2}{b}\right)$. Since $\ln x$ is continuous,

$$\lim_{b \to \infty} \ln \left(\frac{\sqrt{b^2 + 4}}{b} + \frac{2}{b}\right) = \ln \left(\lim_{b \to \infty} \left(\frac{\sqrt{b^2 + 4}}{b} + \frac{2}{b}\right)\right) = \ln \left(\lim_{b \to \infty} \left(\sqrt{1 + 4/b^2} + \frac{2}{b}\right)\right)$$

$$= \ln 1 = 0. \text{ Thus, } \int_2^\infty \frac{dx}{x\sqrt{x^2 + 4}} = \frac{1}{2} \ln (\sqrt{2} + 1).$$

25. Since $f(x) = \dfrac{1}{(x - 3)(x + 2)\, x}$ is discontinuous at $x = 3$ and $x = 0$, we write

$$\int_0^\infty \frac{dx}{(x - 3)(x + 2)\, x} = \int_0^2 \frac{dx}{(x - 3)(x + 2)\, x} + \int_2^3 \frac{dx}{(x - 3)(x + 2)\, x}$$

$$+ \int_3^4 \frac{dx}{(x - 3)(x + 2)\, x} + \int_4^\infty \frac{dx}{(x - 3)(x + 2)\, x}$$

$$= \lim_{p \to 0} \int_p^2 \frac{dx}{(x - 3)(x + 2)\, x} + \lim_{q \to 0} \int_2^{3-q} \frac{dx}{(x - 3)(x + 2)\, x}$$

$$+ \lim_{r \to 0} \int_{3+r}^4 \frac{dx}{(x - 3)(x + 2)\, x} + \lim_{b \to \infty} \int_4^b \frac{dx}{(x - 3)(x + 2)\, x}.$$

To evaluate the integral $\int \dfrac{dx}{(x - 3)(x + 2)\, x}$ we will use partial fractions.

Thus, we seek constants a, b, and c such that

$$\frac{1}{(x - 3)(x + 2)\, x} = \frac{a}{x - 3} + \frac{b}{x + 2} + \frac{c}{x} \text{ or, } 1 = ax(x + 2) + bx(x - 3)$$

$+ c(x - 3)(x + 2)$. If $x = 3$, we have $1 = 15a$ or $a = 1/15$. If $x = -2$, we get $1 = 10b$ or $b = 1/10$. Finally, if $x = 0$, $1 = -6c$, so $c = -1/6$. Consequently,

$$\int \frac{dx}{(x - 3)(x + 2)\, x} = \int \frac{1/15}{x - 3}\, dx + \int \frac{1/10}{x + 2}\, dx - \int \frac{1/6}{x}\, dx$$

$= \dfrac{1}{15} \ln |x - 3| + \dfrac{1}{10} \ln |x + 2| - \dfrac{1}{6} \ln |x| + C$. Returning to the improper

integral we obtain $\int_0^\infty \dfrac{dx}{(x - 3)(x + 2)\, x} = \lim\limits_{p \to 0} \left(\dfrac{1}{15} \ln |x - 3| + \dfrac{1}{10} \ln |x + 2|\right.$

$- \dfrac{1}{6} \ln |x| \Big) \Big|_p + \lim\limits_{q \to 0} \left(\dfrac{1}{15} \ln |x - 3| + \dfrac{1}{10} \ln |x + 2| - \dfrac{1}{6} \ln |x|\right) \Big|_2^{3-q}$

$+ \lim\limits_{r \to 0} \left(\dfrac{1}{15} \ln |x - 3| + \dfrac{1}{10} \ln |x + 2| - \dfrac{1}{6} \ln |x|\right) \Big|_{3+r}^4$

$+ \lim\limits_{b \to \infty} \left(\dfrac{1}{15} \ln |x - 3| + \dfrac{1}{10} \ln |x + 2| - \dfrac{1}{6} \ln |x|\right) \Big|_4^b$. Let's center our atten-

tion on the first of these limits.

$\lim\limits_{p \to 0} \left(\dfrac{1}{15} \ln |x - 3| + \dfrac{1}{10} \ln |x + 2| - \dfrac{1}{6} \ln |x|\right) \Big|_p^2$

$= \lim\limits_{p \to 0} \left(\dfrac{1}{10} \ln 4 - \dfrac{1}{6} \ln 2 - \dfrac{1}{15} \ln |p - 3| - \dfrac{1}{10} \ln |p + 2| - \dfrac{1}{6} \ln |p|\right).$

Then since $\lim\limits_{p \to 0} \ln |p|$ does not exist, the integral $\int_0^2 \dfrac{dx}{(x - 3)(x + 2)\, x}$ does

not exist. Consequently, the original integral diverges.

27. By definition $\int_{-\infty}^{\infty} \dfrac{r}{r^2 + 9}\, dr = \int_{-\infty}^{0} \dfrac{r}{r^2 + 9}\, dr + \int_{0}^{\infty} \dfrac{r}{r^2 + 9}\, dr$. Thus,

$$\int_{-\infty}^{\infty} \frac{r}{r^2 + 9}\, dr = \lim_{a \to -\infty} \int_{a}^{0} \frac{r}{r^2 + 9}\, dr + \lim_{b \to \infty} \int_{0}^{b} \frac{r}{r^2 + 9}\, dr$$

$$= \lim_{a \to -\infty} \frac{1}{2} \ln |r^2 + 9| \Big|_{a}^{0} + \lim_{b \to \infty} \frac{1}{2} \ln |r^2 + 9| \Big|_{0}^{b}$$

$$= \lim_{a \to -\infty} \frac{1}{2} (\ln 9 - \ln |a^2 + 9|) + \lim_{b \to \infty} \frac{1}{2} (\ln 9 - \ln |b^2 + 9|).$$

Since $\lim_{a \to -\infty} \ln |a^2 + 9|$ does not exist, the integral diverges.

29. Since $\tan \theta$ is discontinuous at $\theta = \pi/2$, the integral is improper. Thus,

$$\int_{0}^{\pi/2} \tan \theta\, d\theta = \lim_{p \to 0} \int_{0}^{\pi/2 - p} \tan \theta\, d\theta = \lim_{p \to 0} - \ln |\cos \theta| \Big|_{0}^{\pi/2 - p}$$

$$= \lim_{p \to 0} - \ln |\cos (\pi/2 - p)|.$$ Since this limit does not exist, the integral

diverges.

31. $\int_{-\infty}^{\infty} \dfrac{x}{\sqrt{3x^2 + 2}}\, dx = \int_{-\infty}^{0} \dfrac{x}{\sqrt{3x^2 + 2}}\, dx + \int_{0}^{\infty} \dfrac{x}{\sqrt{3x^2 + 2}}\, dx$

$$= \lim_{a \to -\infty} \int_{a}^{0} \frac{x}{\sqrt{3x^2 + 2}}\, dx + \lim_{b \to \infty} \int_{0}^{b} \frac{x}{\sqrt{3x^2 + 2}}\, dx = \lim_{a \to -\infty} \frac{1}{3} (3x^2 + 2)^{1/2} \Big|_{a}^{0}$$

$$+ \lim_{b \to \infty} \frac{1}{3} (3x^2 + 2)^{1/2} \Big|_{0}^{b} = \lim_{a \to -\infty} (\frac{2^{1/2}}{3} - \frac{1}{3} (3a^2 + 2)^{1/2})$$

$$+ \lim_{b \to \infty} (\frac{1}{3} (3b^2 + 2)^{1/2} - \frac{2^{1/2}}{3}).$$ Since $\lim_{a \to -\infty} \frac{1}{3} (3a^2 + 2)^{1/2}$ does not exist, the

integral diverges.

33. Since $1/\sqrt{a^2 - x^2}$ is discontinuous at a, the integral is improper. Thus,

$$\int_{0}^{a} \frac{1}{\sqrt{a^2 - x^2}}\, dx = \lim_{p \to 0} \int_{0}^{a - p} \frac{1}{\sqrt{a^2 - x^2}}\, dx = \lim_{p \to 0} \arcsin \frac{x}{a} \Big|_{0}^{a - p}$$

$$= \lim_{p \to 0} \arcsin \frac{a - p}{a} = \arcsin 1 = \pi/2.$$

35. Since $1/\sqrt{x}\,(x + 1)$ is discontinuous at $x = 0$, the integral is improper. Thus,

$$\int_{0}^{1} \frac{dx}{\sqrt{x}\,(x + 1)} = \lim_{p \to 0} \int_{p}^{1} \frac{dx}{\sqrt{x}\,(x + 1)} = \lim_{p \to 0} \int_{p}^{1} \frac{2\, d\sqrt{x}}{(\sqrt{x})^2 + 1}$$

$$= \lim_{p \to 0} 2 \arctan \sqrt{x} \Big|_{p}^{1} = \lim_{p \to 0} 2(\arctan 1 - \arctan p) = \pi/2.$$

37. Since the function $f(x)$ is discontinuous at $x = 2$, we write

$$\int_1^3 f(x)\ dx = \int_1^2 f(x)\ dx + \int_2^3 f(x)\ dx = \lim_{q \to 0} \int_1^{2-q} f(x)\ dx + \lim_{p \to 0} \int_{2+p}^3 f(x)\ dx.$$

Then since $f(x) = 2x$ if $x < 2$ and $f(x) = x^3$ if $x > 2$, $\int_1^3 f(x)\ dx$

$$= \lim_{q \to 0} \int_1^{2-q} 2x\ dx + \lim_{p \to 0} \int_{2+p}^3 x^3\ dx = \lim_{q \to 0} x^2 \Big|_1^{2-q} + \lim_{p \to 0} \frac{1}{4} x^4 \Big|_{2+p}^3$$

$$= \lim_{q \to 0} [(2 - q)^2 - 1] + \lim_{p \to 0} \frac{1}{4} [3^4 - (2 + p)^4] = 3 + \frac{1}{4} (3^4 - 2^4) = \frac{77}{4}. \quad \text{Note}$$

that the value of $f(x)$ at the point of discontinuity did **not** enter into the calculation.

41. $\int_0^\infty 1000 e^{-.06t}\ dt = \lim_{b \to \infty} \int_0^b e^{-.06t}\ dt = \lim_{b \to \infty} -\frac{1000}{.06} e^{-.06t} \Big|_0^b$

$$= \lim_{b \to \infty} -\frac{1000}{.06} (e^{-.06b} - 1) \approx \frac{1000}{.06} \approx \$16,666.67.$$

43. $L(1) = \int_0^\infty e^{-tx}\ dx = \lim_{b \to \infty} 1000 \int_0^b e^{-tx}\ dx = \lim_{b \to \infty} \frac{-1}{t} e^{-tx} \Big|_0^b = \lim_{b \to \infty} (\frac{-e^{-tb}}{t} + \frac{1}{t}) = \frac{1}{t}$

Technique Review Exercises, Chapter 10 (page 450)

1. (a) $\lim_{x \to 2} \dfrac{x^2 + 3x - 10}{2x^2 - 10x + 12} = \lim_{x \to 2} \dfrac{(x - 2)(x + 5)}{2(x - 3)(x - 2)} = \lim_{x \to 2} \dfrac{x + 5}{2(x - 3)} = -\dfrac{7}{2}$

   (b) $\lim_{x \to 2} \dfrac{x^2 + 3x - 10}{2x^2 - 10x + 11} = \dfrac{0}{-1} = 0$

   (c) $\lim_{x \to \infty} \dfrac{x^2 + 3x - 10}{2x^2 - 10x + 11} = \lim_{x \to \infty} \dfrac{1 + 3/x - 10/x^2}{2 - 10/x + 11/x^2} = \dfrac{1}{2}.$

2. (a) Since both $\lim_{x \to 0} (1 - \cos 3x) = 0$ and $\lim_{x \to 0} x^2 = 0$, we may apply l'Hospital's

   Rule. Then, $\lim_{x \to 0} \dfrac{1 - \cos 3x}{x^2} = \lim_{x \to 0} \dfrac{3 \sin 3x}{2x}$. l'Hospital's Rule may be applied

   again. Thus, $\lim_{x \to 0} \dfrac{1 - \cos 3x}{x^2} = \lim_{x \to 0} \dfrac{3 \sin 3x}{2x} = \lim_{x \to 0} \dfrac{9 \cos 3x}{2} = \dfrac{9}{2}.$

   (b) l'Hospital's Rule is not applicable directly. However, if we write

   $\lim_{x \to 0} 2x \ln\ x = \lim_{x \to 0} \dfrac{\ln |x|}{\frac{1}{2x}}$, then l'Hospital's Rule is applicable to the last

   limit. Hence, $\lim_{x \to 0} 2x \ln |x| = \lim_{x \to 0} \dfrac{\ln |x|}{\frac{1}{2x}} = \lim_{x \to 0} \dfrac{\frac{1}{x}}{-\frac{1}{2x^2}} = \lim_{x \to 0} (-2x) = 0.$

3. (a) $\lim\limits_{x\to\infty} \dfrac{e^x}{\ln x} = \lim\limits_{x\to\infty} \dfrac{e^x}{x^{-1}} = \lim\limits_{x\to\infty} x e^x = \infty$

   (b) Let $y = (\ln x)^{1/x}$, then $\ln y = \dfrac{\ln (\ln x)}{x}$ and $\lim\limits_{x\to\infty} \ln y = \lim\limits_{x\to\infty} \dfrac{\ln (\ln x)}{x}$

   $= \lim\limits_{x\to\infty} \dfrac{\frac{x^{-1}}{\ln x}}{1} = \lim\limits_{x\to\infty} \dfrac{1}{x \ln x} = 0.$ So, $\lim\limits_{x\to\infty} y = \lim\limits_{x\to\infty} (\ln x)^{\frac{1}{x}} = 1.$

4. $\displaystyle\int_1^\infty x e^{-x^2}\, dx = \lim\limits_{b\to\infty} \int_1^b x e^{-x^2}\, dx = \lim\limits_{b\to\infty} -\frac{1}{2} e^{-x^2} \Big|_1^b = \lim\limits_{b\to\infty} (-\frac{1}{2} e^{-b^2} + \frac{1}{2e}) = \frac{1}{2e}.$

5. Since the integrand is not defined at $x = -1$, this integral is improper. Thus,

   we write $\displaystyle\int_{-2}^0 \dfrac{dx}{(x + 1)^2} = \int_{-2}^{-1} \dfrac{dx}{(x + 1)^2} + \int_{-1}^0 \dfrac{dx}{(x + 1)^2}$

   $= \lim\limits_{p\to 0} \displaystyle\int_{-2}^{-1-p} \dfrac{dx}{(x + 1)^2} + \lim\limits_{p\to 0} \int_{-1+p}^0 \dfrac{dx}{(x + 1)^2}$

   $= \lim\limits_{p\to 0} -(x + 1)^{-1} \Big|_{-2}^{-1-p} + \lim\limits_{p\to 0} -(x + 1)^{-1} \Big|_{-1+p}^0$

   $= \lim\limits_{p\to 0} (\frac{1}{p} - 1) + \lim\limits_{p\to 0} (-1 + \frac{1}{p}).$ Since these limits do not exist, the integral

   is divergent.

Additional Exercises, Chapter 10 (page 450)

1. $\lim\limits_{x\to\infty} \dfrac{7x^3 + 3x}{4x^2 - 10x^3} = \lim\limits_{x\to\infty} \dfrac{7 + 3/x^2}{4/x - 10} = -\dfrac{7}{10}$

3. $\lim\limits_{x\to\infty} \dfrac{x^4 - x^3}{x^2 + 5} = \lim\limits_{x\to\infty} \dfrac{1 - 1/x}{1/x^2 + 5/x^4}.$ Since $\lim\limits_{x\to\infty} (1 - 1/x) = 1$ and $\lim\limits_{x\to\infty} (1/x^2 + 5/x^4)$

   $= 0$, this limit does not exist.

5. $\lim\limits_{x\to\infty} \dfrac{3x^2 + 7x}{2x^2 - 10} = \lim\limits_{x\to\infty} \dfrac{3 + 7/x}{2 - 10/x^2} = \dfrac{3}{2}$

9. Since $\lim\limits_{x\to 0} \ln (\cos x) = 0$ and $\lim\limits_{x\to 0} x = 0$, we can apply l'Hospital's Rule. Then,

   $\lim\limits_{x\to 0} \dfrac{\ln (\cos x)}{x} = \lim\limits_{x\to 0} \dfrac{-\sin x}{\cos x} = 0.$

11. Since $\lim\limits_{x\to -2} (x^2 + 3x + 1) = -1$ and $\lim\limits_{x\to -2} (x^2 - x - 6) = 0$, this limit does not

   exist. Note that l'Hospital's Rule is not applicable.

13. $\lim\limits_{x\to\infty} \dfrac{\ln (\ln x)}{x} = \lim\limits_{x\to\infty} \dfrac{\frac{x^{-1}}{\ln x}}{1} = \lim\limits_{x\to\infty} \dfrac{1}{x \ln x} = 0$

17. $\lim\limits_{x \to \pi/2} \dfrac{(2x/\pi) - \sin x}{x - \pi/2} = \lim\limits_{x \to \pi/2} \dfrac{2/\pi - \cos x}{1} = \dfrac{2}{\pi}$

21. Let $y = x^{5/x}$, then $\lim\limits_{x \to \infty} \ln y = \lim\limits_{x \to \infty} \dfrac{5 \ln x}{x} = \lim\limits_{x \to \infty} \dfrac{5/x}{1} = 0$

27. As x gets large $\sin x$ takes all values from -1 to 1 and so $x^{\sin x}$ takes arbitrarily large values and also positive values near zero. Thus, the limit does not exist.

33. $\lim\limits_{x \to 0} [\dfrac{1}{x} - \dfrac{1}{\sin x}] = \lim\limits_{x \to 0} \dfrac{\sin x - x}{x \sin x}$. Then since $\lim\limits_{x \to 0} (x \sin x) = 0$ and

$\lim\limits_{x \to 0} (\sin x - x) = 0$, we can apply l'Hospital's Rule. Then,

$\lim\limits_{x \to 0} \dfrac{\sin x - x}{x \sin x} = \lim\limits_{x \to 0} \dfrac{\cos x - 1}{x \cos x + \sin x}$. Since both $\lim\limits_{x \to 0} (\cos x - 1) = 0$ and

$\lim\limits_{x \to 0} (x \cos x + \sin x) = 0$, we can again apply l'Hospital's Rule. Then

$\lim\limits_{x \to 0} \dfrac{\cos x - 1}{x \cos x + \sin x} = \lim\limits_{x \to 0} \dfrac{-\sin x}{-x \sin x + 2 \cos x} = \dfrac{0}{2} = 0.$

35. $\int_0^1 \dfrac{dx}{x\sqrt{x}} = \lim\limits_{p \to 0} \int_p^1 \dfrac{dx}{x\sqrt{x}} = \lim\limits_{p \to 0} -2x^{-1/2} \Big|_p^1 = \lim\limits_{p \to 0} (-2 + \dfrac{2}{\sqrt{p}})$. Since this limit does not exist the integral diverges.

37. $\int_1^\infty \dfrac{dx}{1 + x^2} = \lim\limits_{b \to \infty} \int_1^b \dfrac{dx}{1 + x^2} = \lim\limits_{b \to \infty} \arctan x \Big|_1^b$

$= \lim\limits_{b \to \infty} (\arctan b - \pi/4) = \pi/2 - \pi/4 = \pi/4.$

39. Since the integrand is discontinuous at $x = 1$, this is an improper integral.

Thus, we write $\int_0^2 \dfrac{dx}{x^2 - 1} = \int_0^1 \dfrac{dx}{x^2 - 1} + \int_1^2 \dfrac{dx}{x^2 - 1}$

$= \lim\limits_{p \to 0} \int_0^{1-p} \dfrac{dx}{x^2 - 1} + \lim\limits_{p \to 0} \int_{1+p}^2 \dfrac{dx}{x^2 - 1} = \lim\limits_{p \to 0} \dfrac{1}{2} \ln \left|\dfrac{x - 1}{x + 1}\right| \Big|_0^{1-p}$

$+ \lim\limits_{p \to 0} \dfrac{1}{2} \ln \left|\dfrac{x - 1}{x + 1}\right| \Big|_{1+p}^2 = \lim\limits_{p \to 0} \dfrac{1}{2} \ln \left|\dfrac{-p}{2 - p}\right| + \lim\limits_{p \to 0} (\dfrac{1}{2} \ln \dfrac{1}{3} - \dfrac{1}{2} \ln \left|\dfrac{p}{2 + p}\right|).$

Since these limits do not exist, the integral diverges.

CHAPTER 11: INFINITE SEQUENCES AND SERIES

Exercises 11.1: Infinite Sequences (page 459)

1. On division of numerator and denominator by n^2, we have

$\lim\limits_{n \to \infty} \dfrac{2n^2 + 3}{3n^2 + 7} = \lim\limits_{n \to \infty} \dfrac{2 + 3/n^2}{3 + 7/n^2} = \dfrac{2}{3}$. Thus, the sequence converges with limit 2/3.

3. Again we divide by n to the highest power it occurs, n^3. Thus,

$$\lim_{n \to \infty} \frac{2n^2 + 3n}{n^3 + 5} = \lim_{n \to \infty} \frac{2/n + 3/n^2}{1 + 5/n^3} = 0.$$ The sequence converges with limit 0.

5. $\lim_{n \to \infty} \dfrac{n^4 - 100n}{n^3 + 50n^2} = \lim_{n \to \infty} \dfrac{1 - 100/n^3}{1/n + 50/n^2} = \infty.$ Thus, the sequence diverges.

7. $\lim_{n \to \infty} (1 + \dfrac{(-1)^n}{n}) = 1 + \lim_{n \to \infty} \dfrac{(-1)^n}{n}.$ Since $(-1)^n$ oscillates between 1 and −1 for

all n, $\lim_{n \to \infty} \dfrac{(-1)^n}{n} = 0.$ Consequently, $\lim_{n \to \infty} (1 + \dfrac{(-1)^n}{n}) = 1$, and the sequence

converges with limit 1.

9. Since $\sin (n\pi) = 0$ for all n, $\lim_{n \to \infty} \sin (n\pi) = 0$. Thus, the sequence converges

with limit 0.

11. Since $(-1)^n = \begin{cases} 1 \text{ if n is even} \\ -1 \text{ if n is odd} \end{cases}$ we have $\dfrac{1 + (-1)^n}{2} = \begin{cases} 1 \text{ if n is even} \\ 0 \text{ if n is odd} \end{cases}.$ Thus,

the sequence continually oscillates between 0 and 1. Hence, $\lim_{n \to \infty} \dfrac{1 + (-1)^n}{2}$

does not exist and the sequence diverges.

13. The value of the limit $\lim_{n \to \infty} \dfrac{\ln n}{n}$ is not apparent. However, we may apply

l'Hospital's Rule to evalute $\lim_{x \to \infty} \dfrac{\ln x}{x}.$ Thus, $\lim_{x \to \infty} \dfrac{\ln x}{x} = \lim_{x \to \infty} \dfrac{1}{x} = 0.$ Conse-

quently, since $\dfrac{\ln x}{x}$ is arbitrarily small for sufficiently large x,

$\lim_{n \to \infty} \dfrac{\ln n}{n} = 0.$ The sequence converges with limit zero.

15. Since $\ln (1/n) = -\ln (n)$, we have $\lim_{n \to \infty} \dfrac{\ln (1/n)}{n^2} = - \lim_{n \to \infty} \dfrac{\ln n}{n}.$ Then on using

l'Hospital's Rule, we get $\lim_{n \to \infty} \dfrac{\ln (1/n)}{n^2} = - \lim_{n \to \infty} 1/n = 0.$

19. Since $(-1)^n$ oscillates between the values of 1 and −1, the terms of the given
sequence are 1, 2, 1/3, 4, 1/5, 6, . . . This sequence clearly diverges.

21. Since the terms of the sequence have the same value, 5, for all $n \geq 1000$, the
sequence converges with limit 5.

25. (a) $\dfrac{n}{n + 1}$ is the n-th term.

(b) Since $\dfrac{n}{n + 1} < 1$ any number B such that $B \geq 1$ is an upper bound. Thus, 1
and 2 are upper bounds.

(c) The least upper bound is 1.

(d) Since $\frac{n}{n+1} \geq \frac{1}{2}$ for all $n \geq 1$, any number $B \leq \frac{1}{2}$ is a lower bound. Thus, $\frac{1}{2}$ and 0 are lower bounds.

(e) The greatest lower bound is $\frac{1}{2}$.

(f) The sequence converges with limit 1 since $\lim_{n \to \infty} \frac{n}{n+1} = 1$.

27. (a) The n-th term is 5.

(b) Any number $B \geq 5$ is an upper bound so 5 and 6 are upper bounds.

(c) 5 is the least upper bound.

(d) Any number $B \leq 5$ is a lower bound so 5 and 4 are lower bounds.

(e) 5 is the greatest lower bound.

(f) Since $\lim_{n \to \infty} 5 = 5$ the sequence converges with limit 5.

29. (a) The n-th term is given by either of the following rules

$$a_n = \begin{cases} 1 & n \text{ odd} \\ 2 & n \text{ even} \end{cases} \quad n \geq 1 \text{ or,}$$

$$a_n = \frac{3}{2} + (-1)^n \frac{1}{2} \quad n \geq 1.$$

(b) Any number $B \geq 2$ is an upper bound so 2 and 5 are upper bounds.

(c) The least upper bound is 2.

(d) Any number $B \leq 1$ is a lower bound so 1 and 0 are lower bounds.

(e) The greatest lower bound is 1.

(f) The sequence diverges since $\lim_{n \to \infty} a_n$ does not exist.

31. (a) The n-th term is given by

$$a_n = \begin{cases} \dfrac{1}{2^{(n-1)/2}} & \text{for } n \text{ odd} \\[2ex] -\dfrac{1}{2^{(n-2)/2}} & \text{for } n \text{ even} \end{cases} \quad n \geq 0$$

(b) Since $a_n \leq 1$ for all n, any number $B \geq$ is an upper bound so 1 and 2 are.

(c) The least upper bound is 1.

(d) Since $a_n \geq -1$ for all n, any number $B \leq -1$ is a lower bound so -1 and -2 are.

(e) The greatest lower bound is -1.

(f) Since $\lim_{n \to \infty} a_n = 0$, the sequence converges with limit 0.

33. (a) n is the n-th term.

(b) There is no upper bound.

(c) There is no least upper bound.

(d) Since $a_n = n \geq 1$ for all $n \geq 1$, any number $B \leq 1$ is a lower bound, so 1 and 0 are lower bounds.

(e) The greatest lower bound is 1.

(f) The sequence diverges since $\lim_{n \to \infty} a_n = \lim_{n \to \infty} n$ does not exist.

35. (a) The n-th term is given by

$$a_n = \begin{cases} 1 & n \text{ odd} \\ \dfrac{n}{n+2} & n \text{ even} \end{cases} \qquad n \geq 1.$$

(b) Since $a_n \leq 1$ for all $n \geq 1$, any number $B \geq 1$ is an upper bound so 1 and 2 are.

(c) the least upper bound is 1.

(d) Since $a_n \geq \frac{1}{2}$ for all $n \geq 1$, any number $B \leq \frac{1}{2}$ is a lower bound so $\frac{1}{2}$ and 0 are.

(e) The greatest lower bound is $\frac{1}{2}$.

(f) Since $\lim_{n \to \infty} a_n = 1$, the sequence converges with limit 1.

37. Since the initial population is N, the population at the end of the first ten year period is 1.2 N. The population at the end of the second ten year period is then $(1.2)(1.2) N = (1.2)^2 N$. In general, the population at the end of the n-th ten year period is $(1.2)^n N$. The sequence we seek is thus

$$\{N, (1.2) N, (1.2)^2 N, (1.2)^3 N, \ldots, (1.2)^n N, \ldots\}. \text{ Since}$$

$\lim_{n \to \infty} (1.2)^n N = N \lim_{n \to \infty} (1.2)^n$ does not exist, the sequence diverges.

Exercises 11.2: Infinite Series (page 466)

1. The n-th partial sum of this series is $s_n = \underbrace{1 + 1 + \ldots + 1}_{n \text{ terms}} = n.$

Since $\lim_{n \to \infty} s_n = \lim_{n \to \infty} n$ does not exist, the series diverges. Note that one may also employ the Divergence Test. Since $\lim_{n \to \infty} a_n = \lim_{n \to \infty} 1 = 1 \neq 0$, the series must diverge.

3. This is a geometric series with $a = 5$ and $r = \frac{1}{3}$. Since $|r| = \frac{1}{3} < 1$, the series converges with sum $S = \dfrac{a}{1 - r} = \dfrac{5}{2/3} = 15/2.$

5. This is a geometric series with $a = 1$ and $r = \frac{6}{5}$. Since $|r| = 6/5 \geq 1$, the series diverges.

7. If we write $\displaystyle\sum_{i=0}^{\infty} (-1)^i (\frac{1}{3})^i = \sum_{i=0}^{\infty} (-\frac{1}{3})^i = 1 + (-\frac{1}{3}) + (-\frac{1}{3})^2 + (-\frac{1}{3})^3 + \ldots$

we see that this series is a geometric series with $a = 1$ and $r = -\frac{1}{3}$. Since

$|r| = \frac{1}{3} < 1$, the series converges with sum $S = \frac{a}{1-r} = \frac{1}{1+\frac{1}{3}} = \frac{3}{4}$.

9. This series is a geometric series with $a = \frac{1}{11}$ and $r = -\frac{9}{7}$. Since $|r| = \frac{9}{7} \geq 1$, the series diverges.

11. The n-th term of this series is $a_n = \frac{n}{4n-1}$. Since $\lim_{n\to\infty} a_n = \lim_{n\to\infty} \frac{n}{4n-1}$

   $= \frac{1}{4} \neq 0$, the Divergence Test indicates that the series diverges.

13. Since the series can be written as $1 - 1 + 1 - 1 + 1 - 1 + \ldots$, it is clear that the even partial sums are 0 and the odd partial sums are 1. Consequently, the sequence of partial sums diverges and the series diverges.

15. We can write this series as $\sum\limits_{n=1}^{\infty} (r^2)^n$. So it is a geometric series with

   $a = r^2$ and sum $\frac{r^2}{1-r^2}$ if $r^2 < 1$. Thus, $\sum\limits_{n=1}^{\infty} (r^2)^n = \frac{r^2}{1-r^2}$ if $r^2 < 1$ and the

   series diverges if $r^2 \geq 1$.

17. This series is a geometric series with $a = 1$ and $r = 2x$. Note that since

   $|x| < \frac{1}{2}$, $|r| = 2|x| < 1$. Thus, the series converges and its sum is

   $S = \frac{a}{1-r} = \frac{1}{1-2x}$. As should be expected, this sum is dependent on the value

   of x.

19. Since $s_n = \frac{n+2}{n-1}$, $\lim_{n\to\infty} s_n = \lim_{n\to\infty} \frac{n+2}{n-1} = 1$. Thus, the series converges with sum

   1.

21. Since $s_n = \frac{(-1)^n}{2n}$, $\lim_{n\to\infty} s_n = \lim_{n\to\infty} \frac{(-1)^n}{2n} = 0$. Thus, the series converges with sum

   0.

23. The n-th partial sum of this series is $s_n = (1 - \frac{1}{2}) + (\frac{1}{2} - \frac{1}{3}) + (\frac{1}{3} - \frac{1}{4})$

   $+ \ldots + (\frac{1}{n-1} - \frac{1}{n}) + (\frac{1}{n} - \frac{1}{n+1})$. On cancelling the recurring terms, we have

   $s_n = 1 - \frac{1}{n+1}$. Since $\lim_{n\to\infty} (1 - \frac{1}{n+1}) = 1$, the series converges and its sum is

   1. The series is a telescopic series due to the collapsing or telescoping of
   the terms of its n-th partial sum.

25. The ball initially drops 6 feet. On the first rebound it rises $(.7)6$ feet. It

   next falls $(.7)6$ feet and rebounds $(.7)^2 6$ feet. Continuing in the same fashion,
   we see that the total distance traveled is
   $D = 6 + 2(.7)6 + 2(.7)^2 6 + 2(.7)^3 6 + \ldots + 2(.7)^n 6 + \ldots$ Thus,

$D = 6 + 12(.7) + 12(.7)^2 + 12(.7)^3 + \ldots + 12(.7)^n + \ldots$ In order to find the sum of this series we note that this series is <u>almost</u> a geometric series. In fact, if the first term were 12 it would be a geometric series with $a = 12$ and $r = .7$. Thus, we write $D = 12 + 12(.7) + 12(.7)^2 + \ldots + 12(.7)^n + \ldots - 6$.

Using the sum formula for geometric series gives $D = \dfrac{12}{1 - .7} - 6 = 40 - 6$

$= 34$ feet.

27. Initially the bank may loan out $M - m$ dollars, keeping the required m dollars in reserve. The loan of $M - m$ dollars will cause $\dfrac{R}{100}(M - m)$ dollars to be redeposited. This $\dfrac{R}{100}(M - m)$ dollars can then be loaned while m dollars is still kept in reserve. The loan of $\dfrac{R}{100}(M - m)$ dollars ultimately causes a deposit of $\dfrac{R}{100}[\dfrac{R}{100}(M - m)] = (\dfrac{R}{100})^2(M - m)$ dollars which can in turn be loaned. Continuing in this fashion we find that

$A = (M - m) + \dfrac{R}{100}(M - m) + (\dfrac{R}{100})^2(M - m) + \ldots + (\dfrac{R}{100})^n(M - m) + \ldots$ is an

infinite series representing the total amount ultimately on loan. Since this is a geometric series with $a = M - m$ and $r = \dfrac{R}{100}$, $A = \dfrac{a}{1 - r} = \dfrac{M - m}{1 - \dfrac{R}{100}}$

$= \dfrac{100(M - m)}{100 - R}$.

29. If we factor $\dfrac{1}{2}$ out of each term of this series, we have

$\dfrac{1}{2} + \dfrac{1}{8} + \dfrac{1}{32} + \ldots + \dfrac{1}{2^{2n+1}} + \ldots = \dfrac{1}{2}(1 + \dfrac{1}{4} + \dfrac{1}{16} + \ldots + \dfrac{1}{2^{2n}} + \ldots)$

$= \dfrac{1}{2}(1 + \dfrac{1}{4} + \dfrac{1}{16} + \ldots + \dfrac{1}{(2^2)^n} + \ldots) = \dfrac{1}{2}(1 + \dfrac{1}{4} + \dfrac{1}{16} + \ldots + (\dfrac{1}{4})^n + \ldots)$.

Now the series in parenthesis is recognized as a geometric series with $a = 1$ and $r = \dfrac{1}{4}$. Thus, $\dfrac{1}{2} + \dfrac{1}{8} + \dfrac{1}{32} + \ldots + \dfrac{1}{2^{2n+1}} + \ldots = \dfrac{1}{2}\dfrac{1}{1 - \dfrac{1}{4}} = \dfrac{2}{3}$.

Exercises 11.3: Series of Positive Terms (page 472)

1. The n-th term of this series is $\dfrac{1}{\sqrt[3]{n}} = \dfrac{1}{n^{1/3}}$. Thus, this is a k-series with

$k = \dfrac{1}{3} \le 1$. The series then diverges.

3. This series is a k-series with $k = 2 > 1$. Thus, this k-series converges. The series to be tested is simply this k-series with one term removed. Thus, the original series also converges.

5. This is simply the series $\sum\limits_{n=1}^{\infty} \frac{1}{n^2 + 1}$ with its first term removed. To test

$\sum\limits_{n=1}^{\infty} \frac{1}{n^2 + 1}$, we will compare it with the known convergent k-series $\sum\limits_{n=1}^{\infty} \frac{1}{n^2}$.

Since $\frac{1}{n^2 + 1} \leq \frac{1}{n^2}$ for all n, the comparison test indicates that $\sum\limits_{n=1}^{\infty} \frac{1}{n^2 + 1}$

converges. Thus, the original series also converges.

7. This is the series $\sum\limits_{n=1}^{\infty} \frac{1}{100\,n + 1}$. We will apply the integral test with

$f(x) = 1/(100\,x + 1)$. That function satisfies all the criteria of the integral

test. Thus, the series converges or diverges as the integral

$\int_{1}^{\infty} \frac{dx}{100\,x + 1}$ converges or diverges.

$\int_{1}^{\infty} \frac{dx}{100\,x + 1} = \lim\limits_{b \to \infty} \int_{1}^{b} \frac{dx}{100\,x + 1} = \lim\limits_{b \to \infty} \frac{1}{100} \ln (100\,x + 1)\ \Big|_{1}^{b} = \infty.$

Consequently, the series diverges.

9. This is the series $\sum\limits_{n=0}^{\infty} \frac{1}{2 + 3n}$. To determine whether this series converges or

diverges, we will apply the integral test to $\sum\limits_{n=1}^{\infty} \frac{1}{2 + 3n}$. The function

$f(x) = \frac{1}{2 + 3x}$ satisfies all the criteria of this test. Thus, the series con-

verges or diverges as the integral $\int_{1}^{\infty} \frac{dx}{2 + 3x}$ converges or diverges.

$\int_{1}^{\infty} \frac{dx}{2 + 3x} = \lim\limits_{b \to \infty} \int_{1}^{b} \frac{dx}{2 + 3x} = \lim\limits_{b \to \infty} \frac{1}{3} \ln |2 + 3x|\ \Big|_{1}^{b} = \lim\limits_{b \to \infty} \frac{1}{3} (\ln |2 + 3b| - \ln 5).$

Since this limit does not exist, the integral diverges and consequently the

series diverges.

11. This series may be written as $1 + \sum\limits_{n=1}^{\infty} \frac{1}{3n}$. Since the first term does not affect

the convergence or divergence of the series, we will consider the series

$\sum\limits_{n=1}^{\infty} \frac{1}{3n}$. This series is simply the harmonic series, $\sum\limits_{n=1}^{\infty} \frac{1}{n}$ multiplied by the

non-zero constant $\frac{1}{3}$. Thus, the series $\sum\limits_{n=1}^{\infty} \frac{1}{3n}$ diverges and so does the original

series.

13. This series may be written as $\sum\limits_{n=0}^{\infty} \frac{n + 2}{3 + 10n}$. Since $\lim\limits_{n \to \infty} a_n = \lim\limits_{n \to \infty} \frac{n + 2}{3 + 10\,n}$

$= \frac{1}{10} \neq 0$, the Divergence Test indicates that the series diverges.

15. This is a geometric series with a = 1 and $r = \frac{13}{15}$. Since $|r| = \frac{13}{15} < 1$, the series converges.

17. This is a geometric series with a = 1 and r = 1.0001. Since $|r| = 1.001 \geq 1$, the series diverges.

19. Since $\sum\limits_{i=0}^{\infty} \frac{1}{\sqrt{i+3}} = \frac{1}{\sqrt{3}} + \frac{1}{\sqrt{4}} + \frac{1}{\sqrt{5}} + \ldots = \sum\limits_{i=3}^{\infty} \frac{1}{i^{1/2}}$, we see that this is simply the

   divergent k-series $\sum\limits_{i=1}^{\infty} \frac{1}{i^{1/2}}$ with the first two terms removed. Since the removal

   of a finite number of terms does not affect the convergence or divergence, the original series must also diverge.

21. We will compare this series to the convergent k-series $\sum\limits_{n=1}^{\infty} \frac{1}{n^{4/3}}$. Note that

   $\frac{1}{(n^4+5)^{1/3}} \leq \frac{1}{(n^4)^{1/3}}$ for all n. (This last inequality is equivalent to the

   inequality $n^4 \leq n^4 + 5$.) Consequently, since $\sum\limits_{n=1}^{\infty} \frac{1}{n^{4/3}}$ converges, so does

   $\sum\limits_{n=1}^{\infty} \frac{1}{\sqrt[3]{n^4+5}}$.

23. Since $\lim\limits_{n \to \infty} a_n = \lim\limits_{n \to \infty} \frac{n^2 + 3n}{2 + 5n^2} = \frac{1}{5} \neq 0$, the Divergence Test indicates that this

   series diverges.

25. By combining the fractions over a common denominator, we see that

   $\sum\limits_{n=4}^{\infty} (\frac{1}{n-3} - \frac{1}{n}) = \sum\limits_{n=4}^{\infty} \frac{3}{n(n-3)} = \frac{3}{1 \cdot 4} + \frac{3}{2 \cdot 5} + \frac{3}{3 \cdot 6} + \ldots$ It is then clear

   that this series may also be written as $\sum\limits_{n=1}^{\infty} \frac{3}{n(n+3)}$. We will compare this

   rewritten series with the convergent series $\sum\limits_{n=1}^{\infty} \frac{3}{n^2}$. Since $\frac{3}{n(n+3)} \leq \frac{3}{n^2}$ for

   all $n \geq 1$, the series $\sum\limits_{n=1}^{\infty} \frac{3}{n(n+3)} = \sum\limits_{n=4}^{\infty} (\frac{1}{n-3} - \frac{1}{n})$ must converge.

27. We can apply the integral test with $f(x) = xe^{-x^2}$. Then the series converges or

   diverges as the integral $\int_1^{\infty} xe^{-x^2} dx$ converges or diverges. $\int_1^{\infty} xe^{-x^2} dx$

   $= \lim\limits_{b \to \infty} \int_1^b xe^{-x^2} dx = \lim\limits_{b \to \infty} -\frac{1}{2} \int_1^b e^{-x^2}(-2x\,dx) = \lim\limits_{b \to \infty} -\frac{1}{2} e^{-x^2} \Big|_1^b$

$= \lim_{b\to\infty} -\frac{1}{2}(e^{-b^2} - \frac{1}{2}e^{-1}) = \frac{1}{4e}$. Since the integral converges, the series must also converge.

29. We will apply the integral test with $f(x) = \frac{1}{x \ln x}$. Then the series converges or diverges as the integral $\int_2^\infty \frac{dx}{x \ln x}$ converges or diverges.

$\int_2^\infty \frac{dx}{x \ln x} = \lim_{b\to\infty} \int_2^b \frac{dx}{x \ln x} = \lim_{b\to\infty} \int_2^b \frac{\frac{dx}{x}}{\ln x} = \lim_{b\to\infty} \ln|\ln x| \Big|_2^b$

$= \lim_{b\to\infty} (\ln|\ln b| - \ln|\ln 2|)$. Since $\lim_{b\to\infty} \ln|\ln b|$ does not exist, the integral diverges and so does the series.

31. We will apply the integral test with $f(x) = \frac{x}{2x^2 - 1}$. Then the series converges or diverges as the integral $\int_1^\infty \frac{x}{2x^2 - 1} dx$ converges or diverges.

$\int_1^\infty \frac{x}{2x^2 - 1} dx = \lim_{b\to\infty} \int_1^b \frac{x}{2x^2 - 1} dx = \lim_{b\to\infty} \frac{1}{4} \ln|2x^2 - 1| \Big|_1^b = \lim_{b\to\infty} \frac{1}{4} \ln|2b^2 - 1|$.

Since this limit does not exist, the integral diverges. Consequently, the series also diverges.

33. We will apply the integral test with $f(x) = \frac{1}{x^3 + x^2}$. Then the series converges or diverges as the integral $\int_1^\infty \frac{dx}{x^3 + x^2}$ converges or diverges.

$\int_1^\infty \frac{dx}{x^3 + x^2} = \lim_{b\to\infty} \int_1^b \frac{dx}{x^3 + x^2} = \lim_{b\to\infty} (\int_1^b -\frac{1}{x} dx + \int_1^b \frac{1}{x^2} dx + \int_1^b \frac{dx}{x + 1})$

$= \lim_{b\to\infty} (-\ln|x| \Big|_1^b - 2/x \Big|_1^b + \ln|x + 1| \Big|_1^b)$

$= \lim_{b\to\infty} (-\ln b + \ln 1 - \frac{2}{b} + 2 + \ln|b + 1| - \ln 2)$

$= \lim_{b\to\infty} (\ln \frac{b + 1}{b} + 0 - \frac{2}{b} + 2 - \ln 2) = 2 - \ln 2$ exists so the series converges.

35. Since $\frac{1}{n + n^2} < \frac{1}{n^2}$ and the series $\sum_{n=1}^\infty \frac{1}{n^2}$ is a convergent k-series, $\sum \frac{1}{n + n^2}$ is a convergent series.

38. Let E_N denote the error in approximating the sum of the series by $\sum_{n=1}^N \frac{1}{n^2}$. By

Exercise 34, $E_N < \int_N^\infty \frac{dx}{x^2}$. If we find N such that $\int_N^\infty \frac{dx}{x^2} < 10^{-3}$ then $E_N < 10^{-3}$.

Since $\int_N^\infty \frac{dx}{x^2} = \lim_{b\to\infty} -\frac{1}{x}\Big|_N^b = \frac{1}{N}$, we will need N such that $\frac{1}{N} < 10^{-3}$ or, $N > 10^3$.

Consequently, N = 1001 is the smallest number of terms that can be added.

Exercises 11.4: The Limit Form of the Comparison Test and the Ratio
Test (page 476)

1. Since $p_n = \frac{n!}{(n-2)!}$, $\lim_{n\to\infty} \frac{p_{n+1}}{p_n} = \lim_{n\to\infty} \frac{(n+1)!}{(n-1)!}/\frac{n!}{(n-2)!} = \lim_{n\to\infty} \frac{(n+1)!}{(n-1)!} \frac{(n-2)!}{n!}$.

Then using the facts that $(n+1)! = (n+1)n!$ and $(n-1)! = (n-1)(n-2)!$,

we have $\lim_{n\to\infty} \frac{p_{n+1}}{p_n} = \lim_{n\to\infty} \frac{n+1}{n-1} = 1$. The Ratio Test thus fails to determine

whether the series converges or not. We must try another test. Since

$\frac{n!}{(n-2)!} = \frac{n(n-1)(n-2)!}{(n-2)!} = n(n-2)$, $\sum_{n=1}^\infty \frac{n!}{(n-2)!} = \sum_{n=1}^\infty n(n-2)$. Then

since $\lim_{n\to\infty} n(n-2)$ does not exist, the Divergence Test indicates that the

series diverges.

3. Since $p_n = \frac{n!}{100^n}$, $\lim_{n\to\infty} \frac{p_{n+1}}{p_n} = \lim_{n\to\infty} \frac{(n+1)!}{100^{n+1}}/\frac{n!}{100^n} = \lim_{n\to\infty} \frac{(n+1)!}{100^{n+1}} \frac{100^n}{n!}$

$= \lim_{n\to\infty} \frac{n+1}{100} = \infty$. Thus, by the Ratio Test the series diverges.

5. Since $p_n = \frac{6^{n-1}}{n5^{n+3}}$, $\lim_{n\to\infty} \frac{p_{n+1}}{p_n} = \lim_{n\to\infty} \frac{6^n}{(n+1)\,5^{n+4}}/\frac{6^{n-1}}{n5^{n+3}}$

$= \lim_{n\to\infty} \frac{6^n}{(n+1)\,5^{n+4}} \frac{n5^{n+3}}{6^{n-1}} = \lim_{n\to\infty} \frac{6n}{5(n+1)} = \frac{6}{5} > 1$. Consequently, by the Ratio

Test the series diverges.

7. We will use the Limit Form of the Comparison Test to compare this series with

the divergent series $\sum_{n=1}^\infty \frac{1}{n} = \sum_{n=1}^\infty a_n$. $\lim_{n\to\infty} \frac{a_n}{p_n} = \lim_{n\to\infty} \frac{1}{n}/\frac{n^3-5}{n^4+7n} = \lim_{n\to\infty} \frac{n^4+7n}{n(n^3-5)}$

$= \lim_{n\to\infty} \frac{n^4+7n}{n^4-5n} = 1$. Since $\lim_{n\to\infty} \frac{a_n}{p_n}$ exists, the series must diverge.

9. We will use the Limit Form of the Comparison Test to compare this series with

the convergent series $\sum_{n=1}^\infty \frac{1}{n^2} = \sum_{n=1}^\infty b_n$. Then, $\lim_{n\to\infty} \frac{p_n}{b_n} = \lim_{n\to\infty} \frac{n^3+3n^2+1}{2n^5+4n^3+6}/\frac{1}{n^2}$

$= \lim\limits_{n\to\infty} \dfrac{n^5 + 3n^4 + n^2}{2n^5 + 4n^3 + 6} = \dfrac{1}{2}$. Since the limit exists, the original series must con-

verge.

11. We will use the Limit Form of the Comparison Test to compare this series with

the convergent series $\sum\limits_{n=1}^{\infty} 1/n^{3/2}$. Then,

$\lim\limits_{n\to\infty} \dfrac{1/n^{3/2}}{1/(n\,(n+2)^{1/2})} = \lim\limits_{n\to\infty} \dfrac{n\,(n+2)^{1/2}}{n^{3/2}} = \lim\limits_{n\to\infty} (\dfrac{n+2}{n})^{1/2} = 1$. Consequently,

the series converges.

13. Since $p_n = \dfrac{n^7 6^{n+2}}{2^{3n}}$, $\lim\limits_{n\to\infty} \dfrac{p_{n+1}}{p_n} = \lim\limits_{n\to\infty} \dfrac{(n+1)^7 6^{n+3}}{2^{3n+3}} / \dfrac{n^7 6^{n+2}}{2^{3n}}$

$= \lim\limits_{n\to\infty} \dfrac{(n+1)^7 6^{n+3}}{2^{3n+3}} \dfrac{2^{3n}}{n^7 6^{n+2}} = \lim\limits_{n\to\infty} \dfrac{6}{2^3} (\dfrac{n+1}{n})^7 = \dfrac{3}{4} < 1$. Consequently, by the

Ratio Test the series converges.

15. Since $\lim\limits_{n\to\infty} 2^{1/n} = 2^0 = 1 \neq 0$, this series diverges by the Divergence Test.

17. This series may be written as $\sum\limits_{n=2}^{\infty} \dfrac{n}{(n+1)(n+2)} = \sum\limits_{n=2}^{\infty} \dfrac{n}{n^2 + 3n + 2}$. We will

thus use the Limit Form of the Comparison Test to compare this series with the

divergent series $\sum\limits_{n=1}^{\infty} \dfrac{1}{n} = \sum\limits_{n=1}^{\infty} a_n$. Then, $\lim\limits_{n\to\infty} \dfrac{a_n}{p_n} = \lim\limits_{n\to\infty} \dfrac{1}{n} / \dfrac{n}{n^2 + 3n + 2}$

$= \lim\limits_{n\to\infty} \dfrac{n^2}{n^2 + 3n + 2} = 1$. Since the limit exists, the original series must

diverge.

19. This series may be written as $\sum\limits_{n=1}^{\infty} \dfrac{1}{\sqrt{n(n+1)(n+2)}}$. Thus, we will use the

Limit Form of the Comparison Test to compare it with the convergent k-series

$\sum\limits_{n=1}^{\infty} \dfrac{1}{n^{3/2}} = \sum\limits_{n=1}^{\infty} b_n$. Then, $\lim\limits_{n\to\infty} \dfrac{p_n}{b_n} = \lim\limits_{n\to\infty} \dfrac{n^{3/2}}{\sqrt{n(n+1)(n+2)}}$

$= \lim\limits_{n\to\infty} \sqrt{\dfrac{n^3}{n^3 + 3n^2 + 2n}} = 1$. Since the limit exists, the original series must

also converge.

21. This series may be written as $\sum\limits_{n=2}^{\infty} \dfrac{n}{2(n-1)\,2n} = \sum\limits_{n=2}^{\infty} \dfrac{n}{4n^2 - 4n}$. Thus, we will

use the Limit Form of the Comparison Test to compare it with the divergent series $\sum\limits_{n=2} a_n = \sum\limits_{n=2}^{\infty} \frac{1}{n}$. Then, $\lim\limits_{n\to\infty} = \frac{a_n}{p_n} = \lim\limits_{n\to\infty} \frac{1}{n} \cdot \frac{4n^2 - 4n}{n} = \lim\limits_{n\to\infty} \frac{4n^2 - 4n}{n^2} = 4$.

Since this limit exists, the original series must diverge.

23. This series may be written as $\sum\limits_{n=1}^{\infty} \frac{2 \cdot 5 \cdot 8 \cdot \ldots \cdot (3n-1)}{2 \cdot 4 \cdot 6 \cdot \ldots \cdot (2n)}$. Then since

$p_n = \frac{2 \cdot 5 \cdot 8 \cdot \ldots \cdot (3n-1)}{2 \cdot 4 \cdot 6 \cdot \ldots \cdot (2n)}$, $\lim\limits_{n\to\infty} p_{n+1}/p_n$

$= \lim\limits_{n\to\infty} \frac{2 \cdot 5 \cdot 8 \cdot \ldots \cdot (3n+2)}{2 \cdot 4 \cdot 6. \cdot \ldots \cdot (2n+2)} / \frac{2 \cdot 5 \cdot 8 \cdot \ldots \cdot (3n-1)}{2 \cdot 4 \cdot 6 \cdot \ldots \cdot (2n)}$

$= \lim\limits_{n\to\infty} \frac{2 \cdot 5 \cdot 8 \cdot \ldots \cdot (3n+2)}{2 \cdot 4 \cdot 6 \cdot \ldots \cdot (2n+2)} \frac{2 \cdot 4 \cdot 6 \cdot \ldots \cdot (2n)}{2 \cdot 5 \cdot 8 \cdot \ldots \cdot (3n-1)}$

$= \lim\limits_{n\to\infty} \frac{3n+2}{2n+2} = \frac{3}{2} > 1$. Thus, the series diverges by the Ratio Test.

25. We will apply the Ratio Test. Since $p_n = \frac{3^{n-1}}{n\,2^n}$,

$\lim\limits_{n\to\infty} \frac{p_{n+1}}{p_n} = \lim\limits_{n\to\infty} \frac{3^n}{(n+1)\,2^{n+1}} \frac{n\,2^n}{3^{n-1}} = \lim\limits_{n\to\infty} \frac{3n}{2(n+1)} = \frac{3}{2} > 1$. Since the limit is

greater than 1, the series diverges.

27. Since $p_n = \frac{n^5 6^n}{(n+1)!}$, $\lim\limits_{n\to\infty} \frac{p_{n+1}}{p_n} = \lim\limits_{n\to\infty} \frac{(n+1)^5 6^{n+1}}{(n+2)!} \frac{(n+1)!}{n^5 6^n}$

$= \lim\limits_{n\to\infty} \frac{6(n+1)^5}{(n+2)\,n^5} = \lim\limits_{n\to\infty} \frac{6}{n+2} (\frac{n+1}{n})^5 = 0 < 1$. Thus, by the Ratio Test, the

series converges.

29. Since $p_n = \frac{n^3 n!}{(3n)!}$, $\lim\limits_{n\to\infty} \frac{p_{n+1}}{p_n} = \lim\limits_{n\to\infty} \frac{(n+1)^3 (n+1)!}{(3n+3)!} \frac{(3n)!}{n^3 n!}$

$= \lim\limits_{n\to\infty} \frac{(n+1)^3 (n+1)}{(3n+3)(3n+2)(3n+1)\,n^3}$

$= \lim\limits_{n\to\infty} (\frac{n+1}{n})^3 \frac{n+1}{(3n+3)(3n+2)(3n+1)} = 0 < 1$. Thus, by the Ratio Test, this

series converges.

31. Since $p_n = \frac{2^n}{(2n)!\,n^2}$, $\lim\limits_{n\to\infty} \frac{p_{n+1}}{p_n} = \lim\limits_{n\to\infty} \frac{2^{n+1}}{(2n+2)!(n+1)^2} \frac{(2n)!\,n^2}{2^n}$

$= \lim\limits_{n\to\infty} \frac{2n^2}{(2n+2)(2n+1)(n+1)^2} = \lim\limits_{n\to\infty} (\frac{n}{n+1})^2 \frac{2}{(2n+2)(2n+1)} = 0 < 1$.

Thus, by the Ratio Test, this series converges.

33. We will apply the Ratio Test. In this case $p_n = \frac{x^n}{n}$, so

$\lim_{n\to\infty} \frac{p_{n+1}}{p_n} = \lim_{n\to\infty} \frac{x^{n+1}}{n+1} \frac{n}{x^n} = \lim_{n\to\infty} (x \frac{n}{n+1}) = x \lim_{n\to\infty} \frac{n}{n+1} = x.$ Now since we are only

concerned with those values of x such that $0 < x < 1$, $\lim_{n\to\infty} \frac{p_{n+1}}{p_n} = x < 1$. Con-

sequently, by the Ratio Test, this series converges for all x in the interval
$0 < x < 1$. Of course the sum of the series is dependent on the particular
value of x used.

35. We already know from Exercise 31 that this series converges for all x in the
interval $0 < x < 1$. It is only necessary to check what happens if $x = 1$.

Then, $\sum_{n=1}^{\infty} \frac{x^n}{n} = \sum_{n=1}^{\infty} \frac{1}{n}$, the divergent harmonic series. Consequently, the series

$\sum_{n=1}^{\infty} \frac{x^n}{n}$ does not converge for all x in the interval $0 < x \le 1$.

37. We will apply the Ratio Test. Since $p_n = \frac{x^n}{n!}$,

$\lim_{n\to\infty} \frac{p_{n+1}}{p_n} = \lim_{n\to\infty} \frac{x^{n+1}}{(n+1)!} \frac{n!}{x^n} = \lim_{n\to\infty} \frac{x}{n+1} = x \lim_{n\to\infty} \frac{1}{n+1} = x \cdot 0 = 0$ for all x.

Thus, since $\lim_{n\to\infty} \frac{p_{n+1}}{p_n} = 0 < 1$ for all x, the series $\sum_{n=0}^{\infty} \frac{x^n}{n!}$ must converge for

$x > 0$.

Exercises 11.5: Series of Positive and Negative Terms (page 482)

1. This series can be written as $\sum_{n=0}^{\infty} \frac{(-1)^{n+1}}{1 + 7n}$. To apply the Alternating Series

Test, note that since $a_n = \frac{1}{1 + 7n}$ and $a_{n+1} = \frac{1}{8 + 7n}$, $0 < a_{n+1} < a_n$ for all n.

Moreover, $\lim_{n\to\infty} a_n = \lim_{n\to\infty} \frac{1}{1 + 7n} = 0$ so the series converges.

3. Since $a_n = \frac{2n}{3 + 10n}$, $\lim_{n\to\infty} a_n = \lim_{n\to\infty} \frac{2n}{3 + 10n} = \frac{1}{5} \ne 0$. Thus, by the Alternating

Series Test, the series must diverge.

5. In this case $a_n = \frac{5n}{2^n}$. L'Hospital's Rule can be used to show

$\lim_{x\to\infty} \frac{5x}{2^x} = \lim_{x\to\infty} \frac{5}{2^x \ln 2} = 0$. Consequently, $\lim_{n\to\infty} a_n = \frac{5n}{2^n} = 0$. Thus, $a_{n+1} < a_n$ for

all n, the series will converge by the Alternating Series Test. Thus, we investigate the statement $a_{n+1} = \dfrac{5n + 5}{2^{n+1}} < \dfrac{5n}{2^n} = a_n$. This inequality is equiva-

lent on cross multiplication to the inequality $2^n(5n + 5) < 2^{n+1} 5n$ or,

$5n + 5 < 10n$ or, $5 < 5n$. Since this last inequality is clearly true for all

$n > 1$, we have $a_{n+1} < a_n$ if $n > 1$. Thus, the series

$-\dfrac{10}{4} + \dfrac{15}{8} - \dfrac{20}{16} + \ldots = -a_2 + a_3 - a_4 + \ldots$ converges and hence the original

series converges.

7. In this case $a_n = \dfrac{n}{2n + 3}$. Thus, $\lim\limits_{n \to \infty} a_n = \lim\limits_{n \to \infty} \dfrac{n}{2n + 3} = \dfrac{1}{2} \neq 0$ and the series

diverges.

9. Here $a_n = \dfrac{1}{\sqrt[3]{n}}$ and so $\lim\limits_{n \to \infty} a_n = 0$. Moreover, $a_{n+1} = \dfrac{1}{\sqrt[3]{n + 1}} < \dfrac{1}{\sqrt[3]{n}} = a_n$ so the

series converges by the Alternating Series Test.

11. Here $a_n = \dfrac{1}{\ln n}$ and so $\lim\limits_{n \to \infty} a_n = 0$. Moreover, $a_{n+1} = \dfrac{1}{\ln(n + 1)} < \dfrac{1}{\ln n} = a_n$ so

the series converges by the Alternating Series Test.

13. Since $a_n = \dfrac{7n + 2}{n^2 + 1}$, we have $\lim\limits_{n \to \infty} a_n = 0$. In addition,

$a_{n+1} = \dfrac{7n + 9}{n^2 + 2n + 2} < \dfrac{7n + 2}{n^2 + 1} = a_n$ since $(7n + 9)(n^2 + 1) < (n^2 + 2n + 2)(7n + 2)$

or, $7n^3 + 9n^2 + 7n + 9 < 7n^3 + 16n^2 + 18n + 4$. Consequently, the series con-

verges by the Alternating Series Test.

15. $\lim\limits_{n \to \infty} a_n = \lim\limits_{n \to \infty} \dfrac{3\sqrt{n}}{2n + 5} = 0$ and $a_{n+1} = \dfrac{3\sqrt{n + 1}}{2n + 7} < \dfrac{3\sqrt{n}}{2n + 5} = a_n$ for $n \geq 3$. Thus, the

series converges by the Alternating Series Test.

17. If S represents the sum of the series and s_n its n-th partial sum, then by

Theorem 11.5.4, $\left|S - s_n\right| < a_{n+1}$. In this case, we want $\left|S - s_n\right| < 5.10^{-2}$, so

we select n so that $a_{n+1} = \dfrac{n + 1}{4^n} < 5.10^{-2}$. Since $5/4^4 < 5/100$, $n = 4$ works.

Hence, $1 - \dfrac{2}{4} + \dfrac{3}{4^2} - \dfrac{4}{4^3} = .625$ is the desired approximation.

19. Since $\dfrac{4}{10^4} < \dfrac{5}{10^4} = 5 \cdot 10^{-4}$, Theorem 10.8.4 indicates that $\left|s_3 - S\right| < 5 \cdot 10^{-4}$

where $s_3 = 1 - \dfrac{1}{10} + \dfrac{2}{10^2} - \dfrac{3}{10^3}$. Thus, $s_3 = 1 - \dfrac{1}{10} + \dfrac{2}{10^2} - \dfrac{3}{10^3} = \dfrac{917}{1000}$ is the

desired approximation.

21. In this case the series may be written as $-1 + \sum\limits_{n=1}^{\infty} \dfrac{(-1)^{n+1} n}{2n + 4}$. Since

$\lim\limits_{n\to\infty} \dfrac{n}{2n + 4} = \dfrac{1}{2} \neq 0$, the series is divergent.

23. In this case the series may be written as $1 + \sum\limits_{n=1}^{\infty} \dfrac{(-1)^n}{\sqrt{2n}}$. To test for absolute

convergence consider the series $1 + \sum\limits_{n=1}^{\infty} \dfrac{1}{\sqrt{2n}}$. We can use the Limit Form of the

Comparison Test to compare this series to the divergent k-series $\sum\limits_{n=1}^{\infty} \dfrac{1}{n^{1/2}}$.

Then, $\lim\limits_{n\to\infty} \dfrac{n^{1/2}}{\sqrt{2n}} = \dfrac{1}{\sqrt{2}} \lim\limits_{n\to\infty} \dfrac{n^{1/2}}{n^{1/2}} = \dfrac{1}{\sqrt{2}}$. Since this limit exists, the series

$1 + \sum\limits_{n=1}^{\infty} \dfrac{1}{\sqrt{2n}}$ diverges and the original series is <u>not</u> absolutely convergent.

However, since $a_{n+1} \dfrac{1}{\sqrt{2n + 2}} < \dfrac{1}{\sqrt{2n}} = a_n$ for all n and $\lim\limits_{n\to\infty} a_n = \lim\limits_{n\to\infty} \dfrac{1}{\sqrt{2n}} = 0$, the

Alternating Series Test indicates that the series $1 + \sum\limits_{n=1}^{\infty} \dfrac{(-1)^n}{\sqrt{2n}}$ does converge.

Since this series converges, but is not absolutely convergent, is is <u>conditionally convergent</u>.

25. We first test for absolute convergence. That is, we test the series

$\sum\limits_{n=1}^{\infty} \dfrac{n^3 + 1}{3n^{9/2} + 7n^2}$. We will use the Limit Form of the Comparison Test to compare

this series with the convergent k-series $\sum\limits_{n=1}^{\infty} \dfrac{1}{n^{3/2}}$. Then,

$\lim\limits_{n\to\infty} \dfrac{n^3 + 1}{3n^{9/2} + 7n^2} / \dfrac{1}{n^{3/2}} = \lim\limits_{n\to\infty} \dfrac{n^{9/2} + n^{3/2}}{3n^{9/2} + 7n^2} = \dfrac{1}{3}$. Since this limit exists, the

series $\sum\limits_{n=1}^{\infty} \dfrac{n^3 + 1}{3n^{9/2} + 7n^2}$ converges, and the original series converges absolutely.

27. Since $\lim\limits_{n\to\infty} a_n = \lim\limits_{n\to\infty} 7^{1/n} = 7^0 = 1$, the series diverges.

29. Since $\lim\limits_{n\to\infty} a_n = \lim\limits_{n\to\infty} \dfrac{1}{\ln(n + 1)} = 0$ and $a_{n+1} = \dfrac{1}{\ln(n + 2)} < \dfrac{1}{\ln(n + 1)} = a_n$, the

series converges by the Alternating Series Test. However, since $\dfrac{1}{\ln(n + 1)}$

$> \dfrac{1}{n + 1}$ and the series $\sum \dfrac{1}{n + 1}$ diverges, the series is not absolutely convergent. Thus, it is conditionally convergent.

31. First note that this series is an alternating series for $-1 \leq x < 0$. We will apply the absolute value form of the Ratio Test, Theorem 10.8.2. Then,

$$\lim_{n \to \infty} \left| \frac{a_{n+1}}{a_n} \right| = \lim_{n \to \infty} \left| \frac{\frac{x^{n+1}}{n+1}}{\frac{x^n}{n}} \right| = \lim_{n \to \infty} \left| x \frac{n}{n+1} \right|. \quad \text{Since } \frac{n}{n+1} > 0,$$

$$\lim_{n \to \infty} \left| \frac{a_{n+1}}{a_n} \right| = \lim_{n \to \infty} \left(|x| \frac{n}{n+1} \right) = |x| \lim_{n \to \infty} \frac{n}{n+1} = |x|. \quad \text{Thus, for all } x \text{ such that}$$

$|x| < 1$, that is $-1 < x < 1$, the series converges and, in fact, converges absolutely. It is only necessary to test the series for one remaining value of x, $x = -1$. If $x = -1$, $\sum\limits_{n=1}^{\infty} \frac{x^n}{n} = \sum\limits_{n=1}^{\infty} \frac{(-1)^n}{n}$, a conditionally convergent series.

Consequently, $\sum\limits_{n=1}^{\infty} \frac{x^n}{n}$ converges for all x in the interval $-1 \leq x < 1$.

33. Again this will be an alternating series if $x < 0$. On applying the absolute value form of the Ratio Test (Theorem 10.8.2), we have

$$\lim_{n \to \infty} \left| \frac{a_{n+1}}{a_n} \right| = \lim_{n \to \infty} \left| \frac{x^{n+1}}{(n+1)!} \frac{n!}{x^n} \right| = \lim_{n \to \infty} |x| \left| \frac{1}{n+1} \right| = |x| \lim_{n \to \infty} \frac{1}{n+1} = |x| \cdot 0 = 0$$

Thus, since $\lim\limits_{n \to \infty} \left| \frac{a_{n+1}}{a_n} \right| = 0 < 1$ for all x, the series does converge for all values of x.

35. Since the series $1 + \frac{1}{2} + \frac{1}{4} + \frac{1}{16} + \frac{1}{32} + \frac{1}{64} + \ldots = \sum\limits_{n=0}^{\infty} \frac{1}{2^n}$ is a convergent

geometric series, the series $1 + \frac{1}{2} - \frac{1}{4} + \frac{1}{16} + \frac{1}{32} - \frac{1}{64} + \ldots$ converges absolutely. Thus, it must also converge.

<u>Exercises 11.6:</u> Testing Series (page 487)

1. We will apply the Limit Form of the Comparison Test to compare this series with

the convergent k-series $\sum\limits_{n=1}^{\infty} \frac{1}{n^{3/2}}$. Then, $\lim\limits_{n \to \infty} \frac{P_n}{a_n} = \lim\limits_{n \to \infty} \frac{10n^3 + 7n^2}{n^{9/2} + 3} / \frac{1}{n^{3/2}}$

$= \lim\limits_{n \to \infty} \frac{10n^{9/2} + 7n^{7/2}}{n^{9/2} + 3} = 10$. Since this limit exists, the series converges.

3. We apply the integral test with the function $f(x) = x^2 e^{-x^3}$. Note that this function satisfies all the conditions for applying the integral test. Then,

$$\int_1^{\infty} x^2 e^{-x^3} \, dx = \lim_{b \to \infty} -\frac{1}{3} \int_1^b e^{-x^3} (-3x^2 \, dx) = \lim_{b \to \infty} -\frac{1}{3} e^{-x^3} \Big|_1^b$$

$= \lim_{b \to \infty} - \frac{1}{3} (e^{-b^3} - e^{-1}) = \frac{1}{3e}$. Since the integral converges, the series also

converges.

5. This series can be written as $\sum_{n=1}^{\infty} \frac{(n + 1)(-1)^{n+1}}{7n}$. Since $\lim_{n \to \infty} a_n = \lim_{n \to \infty} \frac{n + 1}{7n}$

$= \frac{1}{7} \neq 0$, the series must diverge by the Alternating Series Test.

7. On combining the fractions, we get $\sum_{n=1}^{\infty} (\frac{1}{n + 1} - \frac{1}{n}) = \sum_{n=1}^{\infty} \frac{-1}{n(n + 1)}$

$= - \sum_{n=1}^{\infty} \frac{1}{n^2 + n}$. Since $\frac{1}{n^2 + n} < \frac{1}{n^2}$ for all n and the series $\sum_{n=1}^{\infty} \frac{1}{n^2}$ converges,

the Comparison Test indicates that $\sum_{n=1}^{\infty} \frac{1}{n^2 + n}$ also converges. Thus,

$- \sum_{n=1}^{\infty} \frac{1}{n^2 + n} = \sum_{n=1}^{\infty} (\frac{1}{n + 1} - \frac{1}{n})$ converges. Note that this series may also be

treated as a telescopic series.

9. We will apply the Limit Form of the Comparison Test to compare this series with

the convergent k-series $\sum \frac{1}{n^2}$. Thus, $\lim_{n \to \infty} \frac{P_n}{a_n} = \lim_{n \to \infty} \frac{n + 100}{n^3} \cdot n^2$

$= \lim_{n \to \infty} \frac{n^3 + 100n^2}{n^3} = 1$. Since the limit exists, the series must converge.

11. This series may be written as $\sum_{n=1}^{\infty} \frac{1}{1000n + n} = \sum_{n=1}^{\infty} \frac{1}{n(1001)}$. Since this is

simply the divergent harmonic series, $\sum_{n=1}^{\infty} \frac{1}{n}$, multiplied by the non-zero

constant $\frac{1}{1001}$, it must also diverge.

13. Since $\lim_{n \to \infty} \frac{1 - n^2}{50n^2 + 7n + 2} = - \frac{1}{50} \neq 0$, the Divergence Test indicates that the

series diverges.

15. This is the series $\sum_{n=1}^{\infty} \frac{1}{\sqrt{n(n + 1)}}$. We will use the Limit Form of the Comparison

Test to compare it with the divergent series $\sum_{n=1}^{\infty} \frac{1}{n}$. Then,

$\lim_{n \to \infty} \frac{a_n}{P_n} = \lim_{n \to \infty} \frac{1}{\sqrt{n(n + 1)}} \cdot n = \lim_{n \to \infty} \sqrt{\frac{n^2}{n^2 + n}} = 1$. Since this limit exists, the

series must diverge.

17. If we omit the first two terms, this is simply the geometric series $\sum_{n=1}^{\infty} 2(\frac{2}{3})^n$.

Here $a = 2$ and $r = \frac{2}{3}$. Since $|r| = \frac{2}{3} < 1$, the series must converge.

19. By the Ratio Test, $\lim_{n \to \infty} \frac{P_{n+1}}{P_n} = \lim_{n \to \infty} \frac{(n+1)!}{(2n+2)!} \cdot \frac{(2n)!}{n!} = \lim_{n \to \infty} \frac{(n+1)}{(2n+2)(2n+1)}$

$= \lim_{n \to \infty} \frac{n+1}{4n^2 + 6n + 2} = 0 < 1$. Consequently, the series converges.

21. The function $f(x) = \frac{1}{x \ln x}$ is a continuous decreasing function with

$f(n) = \frac{1}{n \ln n}$. Thus, we may apply the integral test.

$\int_{2}^{\infty} \frac{dx}{x \ln x} = \lim_{b \to \infty} \int_{2}^{b} \frac{\frac{dx}{x}}{\ln x} = \lim_{b \to \infty} \ln |\ln x| \Big|_{2}^{b} = \lim_{b \to \infty} (\ln |\ln b| - \ln |\ln 2|)$.

Since this limit does not exist, the integral diverges and so does the series.

23. We use the Limit Form of the Comparison Test to compare this series with the

convergent k-series $\sum_{n=1}^{\infty} \frac{1}{n^{3/2}}$. Then, $\lim_{n \to \infty} \frac{P_n}{b_n} = \lim_{n \to \infty} \frac{1}{\sqrt{n^3 + 1}} \cdot n^{3/2}$

$= \lim_{n \to \infty} \sqrt{\frac{n^3}{n^3 + 1}} = 1$. Since the limit exists, the series converges.

25. This is the series $\sum_{n=1}^{\infty} \frac{(-1)^{n+1} n}{n+2}$. Since $\lim_{n \to \infty} a_n = \lim_{n \to \infty} \frac{n}{n+2} = 1 \neq 0$, the series

must diverge by the Divergence Test.

27. We use the Absolute Value Form of the Ratio Test. Then,

$\lim_{n \to \infty} \frac{P_{n+1}}{P_n} = \lim_{n \to \infty} \frac{(n+1)!}{1 \cdot 3 \cdot 5 \cdot 7 \ldots (2n-1)(2n+1)} \cdot \frac{1 \cdot 3 \cdot 5 \cdot 7 \ldots (2n-1)}{n!}$

$= \lim_{n \to \infty} \frac{n+1}{2n+1} = \frac{1}{2} < 1$. Thus, the series is absolutely convergent.

29. We will use the Ratio Test. Then,

$\lim_{n \to \infty} \frac{P_{n+1}}{P_n} = \lim_{n \to \infty} \frac{2^{n+2}}{3^{n+1}} \cdot \frac{3^n}{2^{n+1}} = \lim_{n \to \infty} \frac{2}{3} = \frac{2}{3} < 1$. Thus, the series converges.

31. We will use the integral test with the function $f(x) = \frac{1}{x(\ln x)^5}$. Note that

this function satisfies the conditions of the integral test. That is, it is a
continuous decreasing function for $x \geq 2$ and $f(n) = p_n$. Then the series con-

verges if and only if the integral $\int_{2}^{\infty} \frac{dx}{x(\ln x)^5}$ converges.

$$\int_2^\infty \frac{dx}{x(\ln x)^5} = \lim_{b\to\infty} \int_2^b \frac{\frac{dx}{x}}{(\ln x)^5} = \lim_{b\to\infty} -\frac{1}{4}(\ln x)^{-4}\Big|_2^b$$

$$= \lim_{b\to\infty} -\frac{1}{4}\left(\frac{1}{(\ln b)^4} - \frac{1}{(\ln 2)^4}\right) = \frac{1}{4(\ln 2)^4}.$$ Since the integral converges, so

does this series.

33. We will apply the Ratio Test.

$$\lim_{n\to\infty} \frac{p_{n+1}}{p_n} = \lim_{n\to\infty} \frac{(n+1)^2}{3^{n+1}}\frac{3^n}{n^2} = \lim_{n\to\infty} \frac{1}{3}\left(\frac{n+1}{n}\right)^2 = \frac{1}{3} < 1.$$ Thus, the series con-

verges.

35. We again apply the Ratio Test. Here,

$$\lim_{n\to\infty} \frac{p_{n+1}}{p_n} = \lim_{n\to\infty} \frac{3^{n+2}}{4^{n+4}}\frac{4^{n+3}}{3^{n+1}} = \lim_{n\to\infty} \frac{3}{4} = \frac{3}{4} < 1.$$ Thus, the series converges.

37. Since $\lim_{n\to\infty} \frac{p_{n+1}}{p_n} = \lim_{n\to\infty} \frac{2^{n+1}}{(2n+2)!}\frac{(2n)!}{2^n} = \lim_{n\to\infty} \frac{2}{(2n+1)(2n+2)} = 0 < 1$, the series

converges by the Ratio Test.

<u>Exercises 11.7:</u> Power Series (page 493)

1. Since $u_n = \frac{x^n}{n!}$, the Ratio Test gives

$$\lim_{n\to\infty} \left|\frac{u_{n+1}}{u_n}\right| = \lim_{n\to\infty} \left|\frac{x^{n+1}}{(n+1)!}\frac{n!}{x^n}\right| = |x|\lim_{n\to\infty} \frac{1}{n+1} = 0.$$ Consequently,

$$\lim_{n\to\infty} \left|\frac{u_{n+1}}{u_n}\right| = 0 < 1$$ for all x, so the series converges for all x. The radius of

convergence is ∞. The derivative series is $1 + x + \frac{x^2}{2!} + \frac{x^3}{3!} + \frac{x^4}{4!} + \ldots$ (the

same series) and also has radius of convergence ∞.

3. Since $u_n = \frac{x^{3(n-1)}}{n!}$, the Ratio Test gives

$$\lim_{n\to\infty} \left|\frac{u_{n+1}}{u_n}\right| = \lim_{n\to\infty} \left|\frac{x^{3n}}{(n+1)!}\frac{n!}{x^{3(n-1)}}\right| = |x^3|\lim_{n\to\infty} \frac{1}{n+1} = 0.$$ Thus, since

$$\lim_{\to\infty} \left|\frac{u_{n+1}}{u_n}\right| = 0 < 1$$ for all x, this series converges for all x and has radius of

convergence ∞. The derivative series $\frac{3x^2}{2!} + \frac{6x^5}{3!} + \frac{9x^8}{4!} + \frac{12x^{11}}{5!} + \ldots$

$+ \frac{3(n-1)x^{3n-4}}{n!} + \ldots$ also has radius of convergence ∞.

5. Since $u_n = \frac{2n+1}{7^n}x^n$, the Ratio Test gives

$\lim\limits_{n \to \infty} \left| \dfrac{u_{n+1}}{u_n} \right| = \lim\limits_{n \to \infty} \left| \dfrac{2n + 3}{7^{n+1}} x^{n+1} \dfrac{7^n}{(2n + 1)x^n} \right| = |x| \lim\limits_{n \to \infty} \dfrac{1}{7} \dfrac{2n + 3}{2n + 1} = \dfrac{1}{7} |x|$. Conse-

quently, the series converges if $\lim\limits_{n \to \infty} \left| \dfrac{u_{n+1}}{u_n} \right| = \dfrac{|x|}{7} < 1$ or $|x| < 7$ and diverges

if $\lim\limits_{n \to \infty} \left| \dfrac{u_{n+1}}{u_n} \right| = \dfrac{|x|}{7} > 1$ or $|x| > 7$. If $|x| = 7$, the Ratio Test does not yield

any information on convergence or divergence. We must then check the points

where $|x| = 7$ separately. If $x = 7$, $\sum\limits_{n=0}^{\infty} \dfrac{2n + 1}{7^n} x^n = \sum\limits_{n=0}^{\infty} \dfrac{2n + 1}{7^n} 7^n$

$= \sum\limits_{n=0}^{\infty} (2n + 1)$, a divergent series. If $x = -7$,

$\sum\limits_{n=0}^{\infty} \dfrac{2n + 1}{7^n} (-7)^n = \sum\limits_{n=0}^{\infty} (-1)^n (2n + 1)$, again a divergent series. Thus, the

series $\sum\limits_{n=0}^{\infty} \dfrac{2n + 1}{7^n} x^n$ converges if and only if $-7 < x < 7$. The radius of con-

vergence is 7. The derivative series, $\dfrac{3}{7} + \dfrac{5 \cdot 2}{7^2} x + \dfrac{7 \cdot 3}{7^3} x^2 + \dfrac{9 \cdot 4}{7^4} x^3 + \dots$

$= \sum\limits_{n=1}^{\infty} \dfrac{(2n + 1)n}{7^n} x^{n-1}$ also has radius of convergence 7.

7. Since $u_n = (-1)^n \dfrac{x^n}{3^n n^{3/2}}$, $\lim\limits_{n \to \infty} \left| \dfrac{u_{n+1}}{u_n} \right| = \lim\limits_{n \to \infty} \left| (-1)^{n+1} \dfrac{x^{n+1}}{3^{n+1}(n + 1)^{3/2}} \cdot \dfrac{3^n n^{3/2}}{(-1)^n x^n} \right|$

$= \lim\limits_{n \to \infty} \left| \dfrac{x}{3} \left(\dfrac{n}{n + 1} \right)^{3/2} \right| = \dfrac{|x|}{3} \lim\limits_{n \to \infty} \left(\dfrac{n}{n + 1} \right)^{3/2} = \dfrac{|x|}{3}$. Thus, by the Ratio Test, the

series converges if $\dfrac{|x|}{3} < 1$, that is, if $|x| < 3$, and diverges if $\dfrac{|x|}{3} > 3$, that

is, if $|x| > 3$. Since no information concerning convergence or divergence is

gained when $|x| = 3$, we must check the points $x = \pm 3$ separately. If $x = 3$,

$\sum\limits_{n=1}^{\infty} (-1)^n \dfrac{x^n}{3^n n^{3/2}} = \sum\limits_{n=1}^{\infty} (-1)^n \dfrac{3^n}{3^n n^{3/2}} = \sum\limits_{n=1}^{\infty} \dfrac{(-1)^n}{n^{3/2}}$. This series converges by the

Alternating Series Test. If $x = -3$, $\sum\limits_{n=1}^{\infty} (-1)^n \dfrac{x^n}{3^n n^{3/2}} = \sum\limits_{n=1}^{\infty} (-1)^n \dfrac{(-3)^n}{3^n n^{3/2}}$

$= \sum\limits_{n=1}^{\infty} \dfrac{1}{n^{3/2}}$, a convergent k-series. Consequently, the original series converges

if and only if $-3 \le x \le 3$, and has radius of convergence 3. The derivative

series, $\sum\limits_{n=1}^{\infty} (-1)^n \dfrac{nx^{n-1}}{3^n n^{3/2}}$ also has radius of convergence 3.

163

11. By use of the Ratio Test we have

$$\lim_{n\to\infty}\left|\frac{u_{n+1}}{u_n}\right| = \lim_{n\to\infty}\left|(-1)^{n+1}\frac{n+1}{3^{n+1}}(x-1)^{n+1}\frac{3^n}{(-1)^n n(x-1)^n}\right| = \lim_{n\to\infty}\left|\frac{n+1}{3n}(x-1)\right|$$

$= \frac{1}{3}|x-1|$. Thus, this series converges if $\frac{1}{3}|x-1| < 1$ or, $|x-1| < 3$ and

diverges if $|x-1| > 3$. Consequently, the series converges if $-2 < x < 4$ and

diverges if $x < -2$ or $x > 4$. We must check the points $x = -2$ and $x = 4$ separ-

ately. If $x = -2$, $\sum_{n=0}^{\infty}(-1)^n\frac{n}{3^n}(x-1)^n = \sum_{n=0}^{\infty}(-1)^n\frac{n}{3^n}(-3)^n = \sum_{n=0}^{\infty}n$. Thus,

the series diverges at $x = -2$. If $x = 4$, $\sum_{n=0}^{\infty}(-1)^n\frac{n}{3^n}(x-1)^n = \sum_{n=0}^{\infty}(-1)^n$

$\frac{n}{3^n}3^n = \sum_{n=0}^{\infty}(-1)^n n$. Then the series also diverges at $x = 4$. In summary then,

this series has radius of convergence 3 and converges if and only if $-2 < x < 4$.

The derivative series $\sum_{n=1}^{\infty}(-1)^n\frac{n^2}{3^n}(x-1)^{n-1}$ also has radius of convergence 3.

13. Use of the Ratio Test yields

$$\lim_{n\to\infty}\left|\frac{u_{n+1}}{u_n}\right| = \lim_{n\to\infty}\left|(-1)^{n+1}\frac{(n+1)!}{(20)^{n+1}}x^{n+1}\frac{(20)^n}{(-1)^n n!\ x^n}\right| = \lim_{n\to\infty}\left|\frac{(n+1)}{20}x\right|$$

$= \frac{x}{20}\lim_{n\to\infty}(n+1) = \infty$ for all x except $x = 0$. Note if $x = 0$, $\lim_{n\to\infty}\left|\frac{u_{n+1}}{u_n}\right| = 0$.

Consequently, this series converges if and only if $x = 0$. Its radius of con-

vergence is 0. In this case the derivative series need not converge to the

derivative function anywhere.

15. Since $u_n = \frac{(x+4)^n}{5^{n+1}n^2}$, we have

$$\lim_{n\to\infty}\left|\frac{u_{n+1}}{u_n}\right| = \lim_{n\to\infty}\left|\frac{(x+4)^{n+1}}{5^{n+2}(n+1)^2}\frac{5^{n+1}n^2}{(x+4)^n}\right| = \lim_{n\to\infty}\left|\frac{(x+4)}{5}\left(\frac{n}{n+1}\right)^2\right| = \frac{|x+4|}{5}.$$

Consequently, this series converges if $\frac{|x+4|}{5} < 1$ and diverges if $\frac{|x+4|}{5} > 1$.

Since the inequality $\frac{|x+4|}{5} < 1$ is equivalent to $|x+4| < 5$ or, $-9 < x < 1$,

we have convergence if $-9 < x < 1$ and divergence if $x < -9$ or $x > 1$. The

points $x = -9$ and $x = 1$ must be checked separately. If $x = -9$,

$\sum_{n=1}^{\infty}\frac{(x+4)^n}{5^{n+1}n^2} = \sum_{n=1}^{\infty}\frac{(-5)^n}{5^{n+1}n^2} = \sum_{n=1}^{\infty}\frac{(-1)^n}{5n^2}$, a series which converges by the Alter-

nating Series Test. If $x = 1$, $\sum_{n=1}^{\infty}\frac{(x+4)^n}{5^{n+1}n^2} = \sum_{n=1}^{\infty}\frac{5^n}{5^{n+1}n^2} = \sum_{n=1}^{\infty}\frac{1}{5n^2} = \frac{1}{5}\sum_{n=1}^{\infty}\frac{1}{n^2}$,

a convergent k-series. Thus, the original series converges if and only if

$-9 \le x \le 1$ and has radius of convergence 5. The derivative series,

$\sum\limits_{n=1}^{\infty} \dfrac{(x + 4)^{n-1}}{5^{n+1}n}$ also has radius of convergence 5.

17. Since $u_k = k^k(x - 5)^k$, we have

$$\lim_{k \to \infty} \left| \frac{u_{k+1}}{u_k} \right| = \lim_{k \to \infty} \left| \frac{(k + 1)^{k+1}(x - 5)^{k+1}}{k^k(x - 5)^k} \right| = \lim_{k \to \infty} (1 + 1/k)^k (k + 1) |x - 5|. \quad \text{Then}$$

since $\lim\limits_{k \to \infty} (1 + 1/k)^k = e \ne 0$, we have $\lim\limits_{k \to \infty} \left| \dfrac{u_{k+1}}{u_k} \right| = \begin{cases} 0 & \text{if } x = 5 \\ \infty & \text{if } x \ne 5. \end{cases}$ Thus, this

series converges only if $x = 5$. It has zero radius of convergence. In this case, the derivative series need not converge to the derivative function anywhere.

19. Since $u_k = \dfrac{(-1)^k(x + 3)^k}{\sqrt{k}}$, we have $\lim\limits_{k \to \infty} \left| \dfrac{u_{k+1}}{u_k} \right| = \lim\limits_{k \to \infty} \left| \dfrac{(-1)^{k+1}(x + 3)^{k+1}\sqrt{k}}{\sqrt{k + 1}\,(-1)^k(x + 3)^k} \right|$

$= \lim\limits_{k \to \infty} \sqrt{\dfrac{k}{k + 1}} \, |x + 3| = |x + 3|$. Then, by the Ratio Test, the series converges

if $|x + 3| < 1$ and diverges if $|x + 3| > 1$. If $|x + 3| = 1$, then $x = -2$, or $x = -4$ and no information is gained so we test these points separately. If

$x = -2$, then $\sum\limits_{k=1}^{\infty} \dfrac{(-1)^k(x + 3)^k}{\sqrt{k}} = \sum\limits_{k=1}^{\infty} \dfrac{(-1)^k}{\sqrt{k}}$. This series converges by the Al-

ternating Series Test. If $x = -4$, then $\sum\limits_{k=1}^{\infty} \dfrac{(-1)^k(x + 3)^k}{\sqrt{k}} = \sum\limits_{k=1}^{\infty} \dfrac{1}{\sqrt{k}}$. This is a

divergent k-series. Thus, the original series converges if and only if

$-4 < x \le -2$ and has radius of convergence 1. The derivative series $\sum\limits_{k=1}^{\infty} (-1)^k\sqrt{k}$

$(x + 3)^{k-1}$ also has radius of convergence 1.

21. Since $u_n = \dfrac{n^3}{x^{2n}}$, $\lim\limits_{n \to \infty} \left| \dfrac{u_{n+1}}{u_n} \right| = \lim\limits_{n \to \infty} \left| \dfrac{(n + 1)^3}{x^{2n+2}} \dfrac{x^{2n}}{n^3} \right| = \lim\limits_{n \to \infty} \left(\dfrac{n + 1}{n} \right)^3 \dfrac{1}{x^2} = \dfrac{1}{x^2}$. Thus,

the given series converges if $\dfrac{1}{x^2} < 1$, that is, if $x > 1$ or $x < -1$. Note that

this series is __not__ a power series and so has no radius of convergence. We test

the points $x = \pm 1$ separately. If $x = 1$, $\sum\limits_{n=0}^{\infty} \dfrac{n^3}{x^{2n}} = \sum\limits_{n=0}^{\infty} n^3$, a divergent series.

If $x = -1$, $\sum\limits_{n=0}^{\infty} \dfrac{n^3}{x^{2n}} = \sum\limits_{n=0}^{\infty} n^3$, a divergent series. Thus, the given series con-

verges only if $x > 1$ or $x < -1$ and diverges otherwise.

25. Since $u_n = \dfrac{n^3}{n!} x^n$, $\lim\limits_{n\to\infty} \left|\dfrac{u_{n+1}}{u_n}\right| = \lim\limits_{n\to\infty} \left|\dfrac{(n+1)^3}{(n+1)!} x^{n+1} \dfrac{n!}{n^3 x^n}\right| = \lim\limits_{n\to\infty} \dfrac{(n+1)^2}{n^3} |x| = 0$

for all x. Consequently, this series converges for all x and has radius of

convergence ∞. The derivative series, $\sum\limits_{n=1}^{\infty} \dfrac{n^4}{n!} x^{n-1}$ also has radius of conver-

gence ∞.

27. This power series can also be viewed as a geometric series with a = 1 and

r = x. Thus, if $|x| < 1$ the series converges with sum $f(x) = \dfrac{a}{1-r} = \dfrac{1}{1-x}$.

29. This series is the derivative series of $\sum\limits_{n=0}^{\infty} x^n$, the series of Exercise 25.

Since $\dfrac{1}{1-x} = \sum\limits_{n=0}^{\infty} x^n$ if $|x| < 1$, $(1-x)^{-2} = \sum\limits_{n=0}^{\infty} nx^{n-1}$ if $|x| < 1$.

31. This series is an indefinite integral of the series of Exercise 25. Then since

$\dfrac{1}{1-x} = \sum\limits_{n=0}^{\infty} x^n$ if $|x| < 1$, $-\ln|1-x| + C = \sum\limits_{n=0}^{\infty} \dfrac{x^{n+1}}{n+1}$ if $|x| < 1$. It is only

necessary to evaluate C to complete the problem. To accomplish this, substi-

tute x = 0. Then, $-\ln 1 + C = 0$ so $C = \ln 1 = 0$. Consequently,

$\sum\limits_{n=0}^{\infty} \dfrac{x^{n+1}}{n+1} = -\ln|1-x|$ if $|x| < 1$.

33. Note that by letting n = k - 2, we have $\sum\limits_{n=0}^{\infty} (n+2)(n+1) x^n = \sum\limits_{k=2}^{\infty} k(k-1)$

x^{k-2}. This last series is clearly the derivative series of $\sum\limits_{k=0}^{\infty} kx^{k-1}$

$= (1-x)^{-2}$ if $|x| < 1$ which was seen in Exercise 27. Thus,

$\sum\limits_{n=0}^{\infty} (n+2)(n+1) x^n = \sum\limits_{k=2}^{\infty} k(k-1) x^{k-2} = 2(1-x)^{-3}$ if $|x| < 1$.

Exercises 11.8: Taylor Series (page 501)

1. Since $f(x) = x^3 + 1$, we have $f'(x) = 3x^2$

$$f''(x) = 6x$$
$$f'''(x) = 6$$
$$f^{(k)}(x) = 0, \ k \geq 4.$$

(a) In this case, a = 0 so f(a) = f(0) = 1

$$f'(a) = f''(a) = 0$$
$$f'''(a) = 6$$
$$f^{(k)}(a) = 0, \ k \geq 4.$$

Thus, the Taylor Series we seek is

$$f(a) + f'(a)(x - a) + \frac{f''(a)}{2!} (x - a)^2 + \ldots = 1 + \frac{6}{3!} x^3 = 1 + x^3.$$

Note that this series is of finite length and converges to the function $f(x)$ for all x.

(b) In this case, a = 1 so, $f(a) = f(1) = 2$

$$f'(a) = f'(1) = 3$$
$$f''(a) = f''(1) = 6$$
$$f'''(a) = f'''(1) = 6$$
$$f^{(k)}(a) = f^{(k)}(1) = 0, \; k \geq 4.$$

Thus, the Taylor Series is

$$f(a) + f'(a)(x - a) + \frac{f''(a)}{2!} (x - a)^2 + \ldots = 2 + 3(x - 1)$$

$$+ \frac{6}{2!} (x - 1)^2 + \frac{6}{3!} (x - 1)^3 = 2 + 3(x - 1) + 3(x - 1)^2 + (x - 1)^3.$$

A simple algebraic manipulation of this series indicates that it converges to $f(x)$ for all x.

(c) Here a = -1, so $f(a) = f(-1) = 0$

$$f'(a) = f'(-1) = 3$$
$$f''(a) = f''(-1) = -6$$
$$f'''(a) = f'''(-1) = 6$$
$$f^{(k)}(a) = f^{(k)}(-1) = 0, \; k \geq 4.$$

Thus, the Taylor Series is

$$f(a) + f'(a)(x - a) + \frac{f''(a)}{2!} (x - a)^2 + \frac{f'''(a)}{3!} (x - a)^3 + \ldots$$

$$= 0 + 3(x + 1) - \frac{6}{2!} (x + 1)^2 + \frac{6}{3!} (x + 1)^3 = 3(x + 1) - 3(x + 1)^2 + (x + 1)^3.$$

Again, a simple algebraic manipulation of this series will yield the function $f(x)$. Consequently, the series converges to $f(x)$ for all x. Uniqueness of power series is guaranteed only for <u>fixed</u> values of a. Each distinct value of a may well yield a distinct power series for $f(x)$ as in this example.

3. Since $f(x) = \tan x$, $f'(x) = \sec^2 x$

$$f''(x) = 2 \sec^2 x \tan x.$$

Thus, $f(a) = f(\pi/4) = 1$

$$f'(a) = f'(\pi/4) = 2$$
$$f''(a) = f''(\pi/4) = 4.$$

Consequently, the first three non-zero terms of the Taylor Series for $f(x)$

$= \tan x$ about $x = \pi/4$ are $1 + 2(x - \pi/4) + 2(x - \pi/4)^2 + \ldots$

7. We already know that $\cos x = 1 - x^2/2! + x^4/4! - x^6/6! + \ldots$ for all x. Thus,

$\cos 2x = 1 - (2x^2)/2! + (2x)^4/4! - (2x)^6/6! + \ldots$ for all x. Consequently,

$\cos 2x = 1 - \dfrac{2^2}{2!} x^2 + \dfrac{2^4}{4!} x^4 - \dfrac{2^6}{6!} x^6 + \ldots$ for all x is a power series for $\cos 2x$ about $x = 0$. By the uniqueness of the power series representation for $f(x)$ about $x = 0$, it must be the desired Taylor Series.

9. We know that $\sin x = x - \dfrac{1}{3!} x^3 + \dfrac{1}{5!} x^5 - \dfrac{1}{7!} x^7 + \ldots$ for all x. Thus,

$x \sin x = x^2 - \dfrac{1}{3!} x^4 + \dfrac{1}{5!} x^6 - \dfrac{1}{7!} x^8 + \ldots$ for all x. By the uniqueness of the power series representation for $f(x)$ about $x = 0$, this series must be the desired Taylor Series.

11. The geometric series $\displaystyle\sum_{n=0}^{\infty} x^n$ converges with sum $1/(x + 1)$, when $|x| < 1$. Thus,

$\dfrac{1}{-x + 1} = \displaystyle\sum_{n=0}^{\infty} (-x)^n$ when $|x| < 1$. By the uniqueness of the power series representation for $f(x)$ about $x = 0$, this series must be the desired Taylor Series.

13. Since $e^x = 1 + x + x^2/2! + x^3/3! + \ldots + x^n/n! + \ldots$ for all x and $\sin x$

$= x - x^3/3! + x^5/5! - x^7/7! + \ldots + x^{2n-1}/(2n - 1)! + \ldots$ for all x, $e^x \sin x$

$= (1 + x + x^2/2! + x^3/3! + \ldots)(x - x^3/3! + x^5/5! - x^7/7! + \ldots)$

$= x + x^2 + (\dfrac{1}{2!} - \dfrac{1}{3!}) x^3 + (\dfrac{1}{3!} - \dfrac{1}{3!}) x^4 + (\dfrac{1}{5!} - \dfrac{1}{2!} \cdot \dfrac{1}{3!} + \dfrac{1}{4!}) x^5$

$+ (\dfrac{1}{5!} - \dfrac{1}{3!} \cdot \dfrac{1}{3!} + \dfrac{1}{5!}) x^6$. Consequently, the terms we seek are

$x + x^2 + (\dfrac{1}{2!} - \dfrac{1}{3!}) x^3 + (\dfrac{1}{5!} - \dfrac{1}{2!} \cdot \dfrac{1}{3!} + \dfrac{1}{4!}) x^5 + (\dfrac{2}{5!} - \dfrac{1}{(3!)^2}) x^6 + \ldots$ Since the series for e^x and the series for $\sin x$ both converge for all x, the product series will also converge for all x.

15. Since a geometric series with $a = 1$ has sum $\dfrac{1}{1 - r}$, if $|r| < 1$, a geometric series with $a = 1$ and $r = -x^2$ has sum $\dfrac{1}{1 + x^2}$. Thus,

$\dfrac{1}{1 + x^2} = 1 - x^2 + x^4 - x^6 + \ldots + (-1)^n x^{2n} + \ldots$ if $|x| < 1$. This then is a power series for $f(x) = 1/(1 + x^2)$. By the uniqueness of power series it must be the Taylor Series for $f(x) = 1/(1 + x^2)$ about $x = 0$.

17. Since $(1 - x)^{-1} = \displaystyle\sum_{n=0}^{\infty} x^n$, for $|x| < 1$, $(1 - x)^{-2} = (\displaystyle\sum_{n=0}^{\infty} x^n)^2$, for $|x| < 1$.

Thus, $(1 - x)^{-2} = (1 + x + x^2 + x^3 + \ldots)(1 + x + x^2 + x^3 + \ldots)$

$= 1 + (1 + 1) x + (1 + 1 + 1) x^2 + (1 + 1 + 1 + 1) x^3 + \ldots$

$= 1 + 2x + 3x^2 + 4x^3 + \ldots = \sum_{n=0}^{\infty} (n + 1) x^n$ if $|x| < 1$, as we were to show.

19. (a) Since the geometric series with $a = 1$ and $r = -x$ has sum $1/(1 + x)$, we have

$\dfrac{1}{1 + x} = 1 - x + x^2 - x^3 + x^4 - \ldots$ for $|x| < 1$. Then, on integrating both

sides, we have $\ln |1 + x| + C = x - \dfrac{1}{2} x^2 + \dfrac{1}{3} x^3 - \dfrac{1}{4} x^4 + \dfrac{1}{5} x^5 + \ldots$ for

$|x| < 1$. In order to evaluate C, substitute $x = 0$. Then, $\ln 1 + C = 0$ so

$C = 0$. Thus, we have $\ln |1 + x| = x - \dfrac{1}{2} x^2 + \dfrac{1}{3} x^3 - \dfrac{1}{4} x^4 + \ldots$ for $|x| < 1$,

as we were to show.

(b) From part (a), we now have $\ln |1 + x| = x - \dfrac{1}{2} x^2 + \dfrac{1}{3} x^3 - \dfrac{1}{4} x^4 + \ldots$ for

$|x| < 1$. Thus, on substitution of $x - 1$ for x, we have

$\ln |x| = (x - 1) - \dfrac{1}{2} (x - 1)^2 + \dfrac{1}{3} (x - 1)^3 - \dfrac{1}{4} (x - 1)^4 + \ldots$ for $|x - 1| < 1$.

Since the inequality $|x - 1| < 1$ is equivalent to $0 < x < 2$, and since $|x| = x$

here, we have $\ln x = (x - 1) - \dfrac{1}{2} (x - 1)^2 + \dfrac{1}{3} (x - 1)^3 - \dfrac{1}{4} (x - 1)^4 + \ldots$ for

$0 < x < 2$.

21. Since $e^x = 1 + x + \dfrac{1}{2!} x^2 + \dfrac{1}{3!} x^3 + \ldots$ for all x,

$e^{-1} = 1 - 1 + \dfrac{1}{2!} - \dfrac{1}{3!} + \dfrac{1}{4!} - \ldots$ Since this is an alternating series, we have

$\left| e^{-1} - (1 - 1 + \dfrac{1}{2!} - \dfrac{1}{3!} + \dfrac{1}{4!} - \ldots + \dfrac{(-1)^n}{n!}) \right| < \dfrac{1}{(n + 1)!}.$ Consequently,

$\left| e^{-1} - (1 - 1 + \dfrac{1}{2!} - \dfrac{1}{3!}) \right| < \dfrac{1}{4!} = \dfrac{1}{24} < \dfrac{1}{20} = 5 \cdot 10^{-2}.$ Thus,

$e^{-1} \approx (1 - 1 + \dfrac{1}{2!} - \dfrac{1}{3!}) = \dfrac{1}{3}$ is an approximation with the desired accuracy.

23. Since $\cos x = 1 - x^2/2! + x^4/4! - x^6/6! + \ldots$ for all x,

$\cos x^3 = 1 - x^6/2! + x^{12}/4! - x^{18}/6! + \ldots$ for all x. Consequently,

$\displaystyle\int_0^1 \cos x^3 \, dx = \int_0^1 dx - \int_0^1 \dfrac{x^6}{2!} \, dx + \int_0^1 \dfrac{x^{12}}{4!} \, dx - \int_0^1 \dfrac{x^{18}}{6!} \, dx + \ldots$

$= x \Big|_0^1 - \dfrac{1}{2! \cdot 7} x^7 \Big|_0^1 + \dfrac{1}{4! \cdot 13} x^{13} \Big|_0^1 - \dfrac{1}{6! \cdot 19} x^{19} \Big|_0^1 + \ldots$

$= 1 - \dfrac{1}{2! \cdot 7} + \dfrac{1}{4! \cdot 13} - \dfrac{1}{6! \cdot 19} + \ldots$ This is an alternating series so

$\left| \displaystyle\int_0^1 \cos x^3 \, dx - (1 - \dfrac{1}{2! \cdot 7} + \dfrac{1}{4! \cdot 13}) \right| < \dfrac{1}{6! \cdot 19} < 5 \cdot 10^{-4}.$ Consequently,

$\displaystyle\int_0^1 \cos x^3 \, dx \approx 1 - \dfrac{1}{2! \cdot 7} + \dfrac{1}{4! \cdot 13} \approx .9318$ is the desired approximation.

1. (a) In this case the n-th term of the sequence is $(-1)^n \frac{n}{n+3}$. Then since

    $\lim_{n \to 0} \frac{n}{n+3} = 1$, the terms of the sequence are alternating close to 1 and -1.

    Thus, the sequence diverges.

    (b) The n-th term of this sequence is $(-1)^{n+1} \frac{3}{2^n}$. Since $\lim_{n \to \infty} \frac{3}{2^n} = 0$, the se-

    quence converges with limit 0.

2. The n-th partial sum is given by

    $s_n = \ln(2) + \ln(\frac{3}{2}) + \ln(\frac{4}{3}) + \ln(\frac{5}{4}) + \ldots + \ln(\frac{n+1}{n})$. Then,

    $s_n = \ln(2) + (\ln(3) - \ln(2)) + (\ln(4) - \ln(3)) + (\ln(5) - \ln(4)) + \ldots$
    $+ (\ln(n) - \ln(n-1)) + (\ln(n+1) - \ln(n)) = \ln(n+1)$. Consequently,

    $\lim_{n \to \infty} s_n = \lim_{n \to \infty} \ln(n+1)$ does not exist and the series diverges.

3. Since $\lim_{n \to \infty} a_n = \lim_{n \to \infty} \frac{n+3}{1000+n} = 1 \neq 0$, the series diverges by the Divergence

    Test.

4. Note that the given series is a geometric series with $a = \frac{2}{9}$ and $r = \frac{1}{3}$. Thus,

    it converges with sum $\dfrac{\frac{2}{9}}{1 - \frac{1}{3}} = \frac{1}{3}$.

5. We will use the Limit Form of the Comparison Test to compare this series with

    the convergent k-series $\sum_{n=1}^{\infty} \frac{1}{n^{3/2}}$. Then,

    $\lim_{n \to \infty} \frac{1/n^{3/2}}{(\sqrt{n}+3)/(1-n+n^2)} = \lim_{n \to \infty} \frac{1-n+n^2}{(\sqrt{n}+3)n^{3/2}} = \lim_{n \to \infty} \frac{1-n+n^2}{n^2 + 3n^{3/2}} = 1$. Consequent-

    ly, the series converges.

6. The function $f(x) = \frac{1}{x(\ln x)^3}$ is a decreasing function with the property that

    $f(n) = \frac{1}{n(\ln n)^3}$. Then on applying the integral test, we get

    $\int_2^{\infty} \frac{dx}{x(\ln x)^3} = \lim_{b \to \infty} \int_2^b \frac{d(\ln x)}{(\ln x)^3} = \lim_{b \to \infty} -\frac{1}{2}(\ln x)^{-2} \Big|_2^b$

    $= \lim_{b \to \infty} (-\frac{1}{2(\ln b)^2} + \frac{1}{2(\ln 2)^2}) = \frac{1}{2(\ln 2)^2}$. Then, since the integral exists,

    the series converges.

7. Application of the absolute value form of the Ratio Test gives

$$\lim_{n\to\infty} \left|\frac{a_{n+1}}{a_n}\right| = \lim_{n\to\infty} \frac{|x+3|^{n+1}}{(n+1)2^{n+1}} \cdot \frac{n \cdot 2^n}{|x+3|^n} = |x+3| \lim_{n\to\infty} \frac{n}{2(n+1)} = \frac{1}{2}|x+3|.$$

Then the series is absolutely convergent if $\frac{1}{2}|x+3| < 1$ and divergent if

$\frac{1}{2}|x+3| > 1$. Consequently, the series is absolutely convergent for

$-5 < x < -1$ and divergent for $x > -1$ or $x < -5$. The test fails when

$\frac{1}{2}|x+3| = 1$, that is for $x = -1$ or $x = -5$. When $x = -1$, we have

$$\sum_{n=1}^{\infty} \frac{(x+3)^n}{n \cdot 2^n} = \sum_{n=1}^{\infty} \frac{1}{n},$$ the divergent harmonic series. When $x = -5$, we have

$$\sum_{n=1}^{\infty} \frac{(x+3)^n}{n \cdot 2^n} = \sum_{n=1}^{\infty} \frac{(-1)^n}{n},$$ the convergent alternating harmonic series. Conse-

quently, the series converges for $-5 \le x < -1$ and diverges for all other values

of x.

8. We first test for absolute convergence. That is, we test the series

$$\sum_{n=30}^{\infty} \frac{1}{\sqrt{n}-5}$$ for convergence. We use the Limit Form of the Comparison Test and

compare the series with the divergent k-series $\sum\limits_{n=30}^{\infty} \frac{1}{n^{1/2}}$. Then since

$$\lim_{n\to\infty} \frac{1/n^{1/2}}{1/(n^{1/2}-5)} = \lim_{n\to\infty} \frac{n^{1/2}-5}{n^{1/2}} = 1,$$ the series $\sum\limits_{n=30}^{\infty} \frac{1}{\sqrt{n}-5}$ diverges and the

original series is not absolutely convergent. We now apply the alternating

series test. Since $a_n = \dfrac{1}{\sqrt{n}-5} > \dfrac{1}{\sqrt{n+1}-5} = a_{n+1}$ and $\lim\limits_{n\to\infty} a_n = \lim\limits_{n\to\infty} \dfrac{1}{\sqrt{n}-5}$

$= 0$, the series $\sum\limits_{n=30}^{\infty} \dfrac{(-1)^n}{\sqrt{n}-5}$ converges. Since the series is not absolutely con-

vergent, it is conditionally convergent.

9. We know that the series $e^x = 1 + x + \dfrac{x^2}{2!} + \dfrac{x^3}{3!} + \ldots + \dfrac{x^n}{n!} + \ldots$ converges for all

x. Consequently, on substitution of $-x^3$, we have

$$e^{-x^3} = 1 + (-x^3) + \frac{(-x^3)^2}{2!} + \frac{(-x^3)^3}{3!} + \ldots + \frac{(-x^3)^n}{n!} + \ldots \text{ for all x,}$$

$$e^{-x^3} = 1 - x^3 + \frac{x^6}{2!} - \frac{x^9}{3!} + \ldots + (-1)^n \frac{x^{3n}}{n} + \ldots \text{ for all x. Moreover, since}$$

this is a power series, we know from the uniqueness of power series that it is

the desired Taylor Series.

10. We know that $\sin x = x - \frac{x^3}{3!} + \frac{x^5}{5!} - \frac{x^7}{7!} + \ldots$ for all x. Thus,

$2 \sin x = 2x - \frac{2}{3!} x^3 + \frac{2}{5!} x^5 - \frac{2}{7!} x^7 + \ldots$ for all x. Since

$\cos x = 1 - \frac{x^2}{2!} + \frac{x^4}{4!} - \frac{x^6}{6!} + \ldots$ for all x, we can add the series to obtain

$2 \sin x + \cos x = 1 + 2x - \frac{1}{2!} x^2 - \frac{2}{3!} x^3 + \frac{1}{4!} x^4 + \frac{2}{5!} x^5 - \frac{1}{6!} x^6 + \frac{2}{7!} x^7 + \ldots$

Moreover, since both the series for 2 sin x and cos x converge for all x, the sum converges to 2 sin x + cos x for all x.

Additional Exercises, Chapter 11 (page 506)

1. Since $\lim_{n \to \infty} (-\frac{1}{2})^n = 0$, $\lim_{n \to \infty} (3 - \frac{1}{(-2)^n}) = 3$.

3. Since $\cos n\pi = \begin{cases} 1 \text{ for n even} \\ -1 \text{ for n odd,} \end{cases}$ this sequence oscillates between 1 and -1. Thus, the sequence diverges.

5. $\lim_{n \to \infty} \frac{n^2 + 5n^3}{13n^3 - 7} = \lim_{n \to \infty} \frac{1/n + 5}{13 - 7/n^3} = 5/13$.

9. This is a geometric series with $a = \frac{1}{5}$ and $r = \frac{1}{8}$. Since $|r| < 1$, the series converges with sum $\frac{1/5}{1 - 1/8} - \frac{1}{5} = \frac{1}{35}$.

11. Since $\lim_{n \to \infty} a_n = \lim_{n \to \infty} (\frac{1}{1000} - \frac{1}{n}) = \frac{1}{1000} \neq 0$, this series diverges by the Divergence Test.

13. This series is just the divergent series $\sum_{n=1}^{\infty} \frac{1}{\sqrt{n}}$ multiplied by 10 with the first 400 terms removed. Since neither of these operations changes a divergent series to a convergent one, the given series diverges.

15. Since $\sum_{n=1}^{\infty} \frac{1}{e^{2n}} = \sum_{n=1}^{\infty} (\frac{1}{e^2})^n$, this is a geometric series with $a = \frac{1}{e^2}$ and $r = \frac{1}{e^2}$.

Since $|r| < 1$, the series converges with sum $\frac{1/e^2}{1 - 1/e^2} = \frac{1}{e^2 - 1}$.

17. Since $\frac{2n + 1}{n^2(n + 1)^2} = \frac{1}{n^2} - \frac{1}{(n + 1)^2}$, this series may be written as

$(1 - \frac{1}{2^2}) + (\frac{1}{2^2} - \frac{1}{3^2}) + (\frac{1}{3^2} - \frac{1}{4^2}) + (\frac{1}{4^2} - \frac{1}{5^2}) + \ldots$ Thus, the n-th partial sum

of the series is $s_n = (1 - \frac{1}{2^2}) + (\frac{1}{2^2} - \frac{1}{3^2}) + (\frac{1}{3^2} - \frac{1}{4^2}) + \ldots + (\frac{1}{n^2} - \frac{1}{(n + 1)^2})$

$= 1 - \dfrac{1}{(n + 1)^2}$. Then since $\lim\limits_{n\to\infty} s_n = \lim\limits_{n\to\infty} (1 - \dfrac{1}{(n + 1)^2}) = 1$, the series con-

verges with sum 1.

19. By the integral test this series converges if and only if the integral

$\displaystyle\int_2^\infty \dfrac{dx}{x^2 - x}$ converges. $\displaystyle\int_2^\infty \dfrac{dx}{x^2 - x} = \lim\limits_{b\to\infty} \int_2^b \dfrac{dx}{x(x - 1)} = \lim\limits_{b\to\infty} \int_2^b (\dfrac{1}{x - 1} - \dfrac{1}{x})\, dx$

$= \lim\limits_{b\to\infty} \ln \left|\dfrac{x - 1}{x}\right| \Big|_2^b = \lim\limits_{b\to\infty} (\ln \left|\dfrac{b - 1}{b}\right| - \ln \dfrac{1}{2}) = \ln 2$. Since the integral con-

verges, so does the series.

21. Apply the integral test with $f(x) = \dfrac{1}{10x + x^{1/2}}$. The series converges or

diverges as the integral $\displaystyle\int_1^\infty \dfrac{dx}{10x + x^{1/2}}$ converges or diverges. Note

$\displaystyle\int_1^\infty \dfrac{dx}{10x + x^{1/2}} = \int_1^\infty \dfrac{x^{-1/2}\, dx}{10x^{1/2} + 1} = \dfrac{1}{5} \int_1^\infty \dfrac{d\,(10x^{1/2})}{10x^{1/2} + 1} = \lim\limits_{b\to\infty} \ln (10x^{1/2} + 1) \Big|_1^b = \infty.$

Consequently, the series diverges.

23. Since $p_n = \dfrac{7n^2 + 1}{n^2 (n + 1)!}$, $\lim\limits_{n\to\infty} \dfrac{p_{n+1}}{p_n} = \lim\limits_{n\to\infty} \dfrac{7(n + 1)^2 + 1}{(n + 1)^2 (n + 2)!} \dfrac{n^2 (n + 1)!}{7n^2 + 1}$

$= \lim\limits_{n\to\infty} \dfrac{(7(n + 1)^2 + 1)\, n^2}{(n + 1)^2 (n + 2)(7n^2 + 1)} = 0 < 1$. Thus, this series converges by the

Ratio Test.

25. We will use the Limit Form of the Comparison Test to compare this series with

the convergent k-series $\displaystyle\sum_{n=1}^\infty \dfrac{1}{n^2}$. Then since

$\lim\limits_{n\to\infty} \dfrac{p_n}{b_n} = \lim\limits_{n\to\infty} \dfrac{n^3 - 3n^2 + 1}{n^5 + 10n^3 - 1} \cdot n^2 = \lim\limits_{n\to\infty} \dfrac{n^5 - 3n^4 + n^2}{n^5 + 10n^3 - 1} = 1$, the given series con-

verges.

27. Since $p_n = \dfrac{2^n}{(n - 1)!\ 3^{n-1}}$, $\lim\limits_{n\to\infty} \dfrac{p_{n+1}}{p_n} = \lim\limits_{n\to\infty} \dfrac{2^{n+1}}{n!\ 3^n} \dfrac{(n - 1)!\ 3^{n-1}}{2^n} = \lim\limits_{n\to\infty} \dfrac{2}{3^n} = 0.$

Thus, by the Ratio Test the given series converges.

29. To test for absolute convergence, we consider the series $\displaystyle\sum_{n=1}^\infty \left|(-1)^n \dfrac{1}{\sqrt{n}}\right|$

$= \displaystyle\sum_{n=1}^\infty \dfrac{1}{\sqrt{n}}$. Since this is a divergent k-series, the given series is not absolute-

ly convergent. However, since $\lim\limits_{n\to\infty} a_n = \lim\limits_{n\to\infty} \dfrac{1}{\sqrt{n}} = 0$, the series does converge.

Consequently, the given series is conditionally convergent.

31. To test the absolute convergence consider the series $\sum\limits_{n=1}^{\infty} (n!)^2/(2n)!$. We will

apply the Ratio Test. $\lim\limits_{n\to\infty} \dfrac{P_{n+1}}{P_n} = \lim\limits_{n\to\infty} \dfrac{((n+1)!)^2}{(2n+2)!} \dfrac{(2n)!}{(n!)^2} = \lim\limits_{n\to\infty} \dfrac{(n+1)^2}{(2n+2)(2n+1)}$

$= \lim\limits_{n\to\infty} \dfrac{n+1}{2(2n+1)} = \dfrac{1}{4} < 1$. Thus, the series is absolutely convergent.

33. To test for absolute convergence consider the series

$\dfrac{1}{7} + \dfrac{1}{3}\left(\dfrac{1}{7}\right)^3 + \dfrac{1}{5}\left(\dfrac{1}{7}\right)^5 + \dfrac{1}{7}\left(\dfrac{1}{7}\right)^7 + \ldots = \sum\limits_{n=1}^{\infty} \dfrac{1}{2n-1}\left(\dfrac{1}{7}\right)^{2n-1}$. Use of the Ratio Test

gives $\lim\limits_{n\to\infty} \dfrac{P_{n+1}}{P_n} = \lim\limits_{n\to\infty} \dfrac{1}{2n+1}\left(\dfrac{1}{7}\right)^{2n+1} 7^{2n-1} (2n-1) = \lim\limits_{n\to\infty} \dfrac{1}{7^2} \dfrac{2n-1}{2n+1} = \dfrac{1}{49} < 1$.

Thus, the series converges absolutely.

35. Since $\lim\limits_{n\to\infty} \dfrac{P_{n+1}}{P_n} = \lim\limits_{n\to\infty} \dfrac{3^{n+1}}{(n+1)!} \dfrac{n!}{3^n} = \lim\limits_{n\to\infty} \dfrac{3}{n+1} = 0 < 1$, the series converges by

the Ratio Test.

37. This series may be written as $\sum\limits_{n=1}^{\infty} (-1)^n \dfrac{5n}{2^n}$. Since

$\lim\limits_{n\to\infty} \left|\dfrac{a_{n+1}}{a_n}\right| = \lim\limits_{n\to\infty} \dfrac{5(n+1)}{2^{n+1}} \dfrac{2^n}{5n} = \lim\limits_{n\to\infty} \dfrac{5(n+1)}{10n} = \dfrac{1}{2} < 1$, this series converges

absolutely by the Ratio Test.

41. Since $\lim\limits_{n\to\infty} \left|\dfrac{a_{n+1}}{a_n}\right| = \lim\limits_{n\to\infty} \left|\dfrac{n+2}{\sqrt{n+1}} (x+3)^{n+1} \dfrac{\sqrt{n}}{(n+1)(x+3)^n}\right|$

$= \lim\limits_{n\to\infty} \dfrac{(n+2)\sqrt{n}}{\sqrt{n+1}(n+1)} |x+3| = |x+3| \lim\limits_{n\to\infty} \dfrac{(1+2/n)}{\sqrt{1+1/n}(1+1/n)} = |x+3|$, the

series will converge for all values of x for which $|x+3| < 1$, and diverge

where $|x+3| > 1$. We must test separately $x = -4$ and $x = -2$ where $|x+3| = 1$.

If $x = -4$, $\sum\limits_{n=1}^{\infty} \dfrac{n+1}{\sqrt{n}} (x+3)^n = \sum\limits_{n=1}^{\infty} (-1)^n \dfrac{n+1}{\sqrt{n}}$. Since $\lim\limits_{n\to\infty} \dfrac{n+1}{\sqrt{n}} = \infty$, this

series diverges. If $x = -2$, $\sum\limits_{n=1}^{\infty} \dfrac{n+1}{\sqrt{n}} (x+3)^n = \sum\limits_{n=1}^{\infty} \dfrac{n+1}{\sqrt{n}}$, again a divergent

series. Consequently, the given series converges if and only if $|x+3| < 1$,

that is, $-4 < x < -2$.

43. Since $a_n = (-1)^n \dfrac{x^{2n+1}}{(2n+1)!}$, $\lim\limits_{n\to\infty} \left|\dfrac{a_{n+1}}{a_n}\right| = \lim\limits_{n\to\infty} \left|\dfrac{x^{2n+3}}{(2n+3)!} \dfrac{(2n+1)!}{x^{2n+1}}\right|$

$= |x|^2 \lim\limits_{n\to\infty} \dfrac{1}{(2n+3)(2n+2)} = 0 < 1$. Consequently, the given series converges

for all x.

45. $\lim\limits_{n\to\infty} \left| \dfrac{x^{n+1}}{(n+1)(n+2)} \dfrac{n(n+1)}{x^n} \right| = \lim\limits_{n\to\infty} |x| \dfrac{n}{n+2} = |x|$. Thus, the series con-

verges absolutely for $|x| < 1$. If $x = 1$, the series becomes

$\sum\limits_{n=1}^{\infty} \dfrac{1}{n(n+1)}$. Since $\dfrac{1}{n(n+1)} < \dfrac{1}{n^2}$ and $\sum\limits_{n=1}^{\infty} 1/n^2$ converges, $\sum\limits_{n=1}^{\infty} \dfrac{1}{n(n+1)}$ con-

verges. When $x = -1$, the series becomes $\sum\limits_{n=1}^{\infty} \dfrac{(-1)^n}{n(n+1)}$, an absolutely convergent

series. Consequently, the series converges for $|x| \le 1$.

47. The series $\sum\limits_{n=0}^{\infty} x^n$ is a geometric series with $a = 1$ and $x = r$. Thus, if

$|r| = |x| < 1$, the series converges with sum $1/(1 - x)$. By the uniqueness of

power series expansions $\sum\limits_{n=0}^{\infty} x^n$ must be the Taylor Series of $1/(1 - x)$ for

$|x| < 1$.

CHAPTER 12: POLAR COORDINATES

Exercises 12.1: Polar Coordinates (page 516)

21. The situation is illustrated below

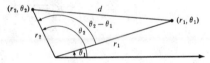

Then a direct application of the "law of cosines" gives

$d^2 = r_1^2 + r_2^2 - 2r_1 r_2 \cos(\theta_2 - \theta_1)$ or since $d > 0$,

$d = \sqrt{r_1^2 + r_2^2 - 2r_1 r_2 \cos(\theta_2 - \theta_1)}$.

Exercises 12.2: Graphs of Polar Equations (page 522)

25. Since Q is on the line $r \cos \theta = a$, the distance $\overline{OQ}$ is $\dfrac{a}{\cos \theta}$. Consequently,

the distance $\overline{OP}$ is $\dfrac{a}{\cos \theta} - b$. Thus, the polar equation we seek is

$r = \dfrac{a}{\cos \theta} - b = a \sec \theta - b$.

Exercises 12.3: Relations between Polar and Rectangular Coordinates
(page 525)

1. Since $x^2 + y^2 = r^2$, this equation becomes $r^2 = 8$.

3. Since $x^2 + y^2 = r^2$ and $x = r \cos \theta$, this equation may be transformed to $r^2 = 4r \cos \theta$. Note that for $r \neq 0$, one may divide both sides of this equation by r. However, such a cancellation does eliminate the solutions $(0, \theta)$ for all θ except when θ is an odd multiple of $\frac{\pi}{2}$.

5. Use of the relations $x^2 + y^2 = r^2$, $x = r \cos \theta$, and $y = r \sin \theta$ yields $r \sin \theta = 2r \cos \theta + r^2$.

7. Since $x = r \cos \theta$, this becomes $r \cos \theta = -1$.

9. On writing this equation as $x^2 + y^2 = 2x - 3$ and using the relations $x^2 + y^2 = r^2$ and $x = r \cos \theta$, we have $r^2 = 2r \cos \theta - 3$.

11. Since $x^2 + y^2 = r^2$, $x = r \cos \theta$, and $y = r \sin \theta$, we obtain $r^4 = 2r^2 \sin \theta \cos \theta = r^2 \sin 2\theta$.

13. Since $x^2 + y^2 = r^2$ and $x = r \cos \theta$, we obtain $r^6 = r^2 \cos^2 \theta$.

15. Since $x^2 + y^2 = r^2$, $x = r \cos \theta$, and $y = r \sin \theta$, we have $r^2 + 7r \cos \theta = 3r \sin \theta$.

17. Since $r = \pm\sqrt{x^2 + y^2}$, we must have either $\sqrt{x^2 + y^2} = -8$ or $-\sqrt{x^2 + y^2} = -8$. Then since $\sqrt{x^2 + y^2}$ is always positive, we must select $-\sqrt{x^2 + y^2} = -8$ or $\sqrt{x^2 + y^2} = 8$ or $x^2 + y^2 = 64$.

19. By writing this equation in the equivalent form $r \cos \theta + r \sin \theta = 1$ and using $x = r \cos \theta$, $y = r \sin \theta$, we obtain $x + y = 1$.

21. We use $y = r \sin \theta$ in order to write $y = -5$.

23. For $r \neq 0$, this equation is equivalent to the equation $r^2 = 2r \cos \theta$. Consequently, since $r^2 = x^2 + y^2$ and $r \cos \theta = x$, we have $x^2 + y^2 = 2x$, $x^2 + y^2 \neq 0$.

25. Since $\sin \theta = \dfrac{y}{\sqrt{x^2 + y^2}}$, and $r^2 = x^2 + y^2$, we have $x^2 + y^2 = \dfrac{y}{\sqrt{x^2 + y^2}}$ or, $(x^2 + y^2)^{3/2} = y$.

27. First write the equation as $r^2 = \cos^2 \theta - \sin^2 \theta$. Then we obtain $x^2 + y^2 = \dfrac{x^2}{x^2 + y^2} - \dfrac{y^2}{x^2 + y^2}$, so $(x^2 + y^2)^2 = x^2 - y^2$.

29. First write the equation as $r + r \cos \theta = 3$. Then we obtain $\pm\sqrt{x^2 + y^2} + x = 3$, or $\pm\sqrt{x^2 + y^2} = 3 - x$. Squaring both sides and simplifying, we get $y^2 = 9 - 6x$.

31. A transformation to polar coordinates will aid in sketching the graph. Use of $x^2 + y^2 = r^2$, $x = r \cos \theta$, and $y = r \sin \theta$ gives $r^4 = 2r^2 \cos \theta \sin \theta$ or, $r^2 = \sin 2\theta$, or $r = 0$.

<u>Exercises 12.4</u>: Area in Polar Coordinates (page 530)

1. A sketch of the graph $r = \cos \theta$ along with the desired region is indicated below.

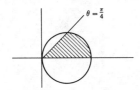

Substitution into the equation $A = \frac{1}{2} \int_{\alpha}^{\beta} [f(\theta)]^2 \, d\theta$ yields $A = \frac{1}{2} \int_{0}^{\pi/4} \cos^2 \theta \, d\theta$.

Thus, $A = \frac{1}{2} \int_{0}^{\pi/4} \cos^2 \theta \, d\theta = \frac{1}{2} \int_{0}^{\pi/4} \frac{1 + \cos 2\theta}{2} \, d\theta$

$= \frac{1}{4} [\theta \, \Big|_{0}^{\pi/4} + \frac{1}{2} \sin 2\theta \, \Big|_{0}^{\pi/4}] = \frac{1}{4} [\frac{\pi}{4} + \frac{1}{2}] = \frac{1}{8} [\frac{\pi}{2} + 1]$ square units.

3. Since $A = \frac{1}{2} \int_{\alpha}^{\beta} [f(\theta)]^2 \, d\theta$, we have $A = \frac{1}{2} \int_{0}^{\pi/6} e^{6\theta} \, d\theta$

$= \frac{1}{12} \int_{0}^{\pi/2} e^{6\theta} \, d(6\theta) = \frac{1}{12} e^{6\theta} \, \Big|_{0}^{\pi/2} = \frac{1}{12} [e^{3\pi} - 1]$ square units.

5. A graph of the desired region is indicated below.

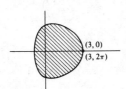

(3, 0)
(3, 2π)

Then on substitution into $A = \frac{1}{2} \int_{\alpha}^{\beta} [f(\theta)]^2 \, d\theta$ we obtain

$A = \frac{1}{2} \int_{0}^{2\pi} (2 + \cos \theta)^2 \, d\theta = \frac{1}{2} \int_{0}^{2\pi} (4 + 4 \cos \theta + \cos^2 \theta) \, d\theta$

$= \frac{1}{2} [4\theta \, \Big|_{0}^{2\pi} + 4 \sin \theta \, \Big|_{0}^{2\pi} + \int_{0}^{2\pi} \frac{1 + \cos^2 2\theta}{2} \, d\theta]$

$= \frac{1}{2} [8\pi + \frac{1}{2} (\theta \, \Big|_{0}^{2\pi} + \frac{1}{2} \sin 2\theta \, \Big|_{0}^{2\pi})] = \frac{9\pi}{2}$ square units.

7. Since $\sin^2 \alpha = \dfrac{1 - \cos 2\alpha}{2}$, we may rewrite $r = 2 \sin^2 \frac{1}{2} \theta$ as $r = 1 - \cos \theta$.

Thus, $A = \dfrac{1}{2} \displaystyle\int_0^{2\pi} (1 - \cos \theta)^2 \, d\theta = \dfrac{1}{2} \displaystyle\int_0^{2\pi} (1 - 2 \cos \theta + \cos^2 \theta) \, d\theta$

$= \dfrac{1}{2} \left[\theta \Big|_0^{2\pi} - 2 \sin \theta \Big|_0^{2\pi} + \displaystyle\int_0^{2\pi} \dfrac{1 + \cos 2\theta}{2} \, d\theta \right]$

$= \dfrac{1}{2} \left[2\pi + \dfrac{1}{2} \left(\theta \Big|_0^{2\pi} + \dfrac{1}{2} \sin 2\theta \Big|_0^{2\pi} \right) \right] = \dfrac{3\pi}{2}$ square units.

9. The graph of $r = 2 \sin 3\theta$ is sketched below.

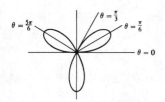

Since one loop of the graph is completed as θ ranges from 0 to $\pi/3$, we have

$A = \dfrac{1}{2} \displaystyle\int_0^{\pi/3} (2 \sin 3\theta)^2 \, d\theta = 2 \displaystyle\int_0^{\pi/3} \sin^2 3\theta \, d\theta = 2 \displaystyle\int_0^{\pi/3} \dfrac{1 - \cos 6\theta}{2} \, d\theta$

$= \left(\theta - \dfrac{1}{6} \sin 6\theta \right) \Big|_0^{\pi/3} = \dfrac{\pi}{3}$ square units.

11. The graph of $r = 1 + 2 \cos \theta$ is sketched below.

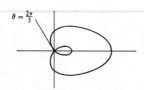

The top part of the larger loop is swept out as θ ranges from $\theta = 0$ to $\theta = 2\pi/3$. Because of symmetry about $\theta = 0$, the total area is given by

$A = 2 \left(\dfrac{1}{2} \displaystyle\int_0^{2\pi/3} (1 + 2 \cos \theta)^2 \, d\theta \right)$

$= \displaystyle\int_0^{2\pi/3} (1 + 4 \cos \theta + 4 \cos^2 \theta) \, d\theta = (\theta + 4 \sin \theta) \Big|_0^{2\pi/3}$

$+ 4 \displaystyle\int_0^{2\pi/3} \dfrac{1 + \cos 2\theta}{2} \, d\theta = \left(\dfrac{2\pi}{3} + 4 \dfrac{\sqrt{3}}{2} \right) + 2 \left(\theta + \dfrac{1}{2} \sin 2\theta \right) \Big|_0^{2\pi/3}$

$= \dfrac{2\pi}{3} + 2\sqrt{3} + \dfrac{4\pi}{3} - \dfrac{\sqrt{3}}{2} = 2\pi + \dfrac{3\sqrt{3}}{2}$ square units.

13. The region in question is indicated below.

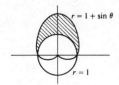

The area is given by $A = \dfrac{1}{2} \displaystyle\int_0^\pi (1 + \sin\theta)^2\, d\theta - \dfrac{1}{2}\displaystyle\int_0^\pi 1^2\, d\theta$

$= \dfrac{1}{2}\displaystyle\int_0^\pi (2\sin\theta + \sin^2\theta)\, d\theta = -\cos\theta \Big|_0^\pi + \dfrac{1}{2}\displaystyle\int_0^\pi \dfrac{1 - \cos 2\theta}{2}\, d\theta$

$= 1 + 1 + \dfrac{1}{4}\left(\theta - \dfrac{1}{2}\sin 2\theta\right)\Big|_0^\pi = 2 + \dfrac{\pi}{4}$ square units.

15.

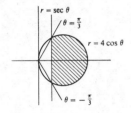

Then using the symmetry of the area about the line $\theta = 0$, we have

$A = \displaystyle\int_0^{\pi/3} (4\cos\theta)^2\, d\theta - \displaystyle\int_0^{\pi/3} \sec^2\theta\, d\theta = 16\displaystyle\int_0^{\pi/3} \dfrac{1 + \cos 2\theta}{2}\, d\theta - \tan\theta \Big|_0^{\pi/3}$

$= 8\left(\theta + \dfrac{1}{2}\sin 2\theta\right)\Big|_0^{\pi/3} - \tan\theta \Big|_0^{\pi/3} = 8\left(\dfrac{\pi}{3} - \dfrac{\sqrt{3}}{4}\right) - \sqrt{3} = \dfrac{8\pi}{3} + \sqrt{3}$ square units.

17. The graph of $r = 1 + 2\cos\theta$ along with the desired region is indicated below.

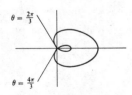

Since the small loop is swept out as θ ranges from $2\pi/3$ to $4\pi/3$, this area is

given by $A = \dfrac{1}{2}\displaystyle\int_{2\pi/3}^{4\pi/3} (1 + 2\cos\theta)^2\, d\theta = \dfrac{1}{2}\displaystyle\int_{2\pi/3}^{4\pi/3} (1 + 4\cos\theta + 4\cos^2\theta)\, d\theta$

$= \dfrac{1}{2}(\theta + 4\sin\theta)\Big|_{2\pi/3}^{4\pi/3} + 2\displaystyle\int_{2\pi/3}^{4\pi/3} \dfrac{1 + \cos 2\theta}{2}\, d\theta = \dfrac{1}{2}\left(\dfrac{2\pi}{3} - 4\sqrt{3}\right)$

$+ \left(\theta + \dfrac{1}{2}\sin\theta\right)\Big|_{2\pi/3}^{4\pi/3} = \dfrac{1}{2}\left(\dfrac{2\pi}{3} - 4\sqrt{3}\right) + \left(\dfrac{2\pi}{3} + \dfrac{\sqrt{3}}{2}\right) = \pi - \dfrac{3}{2}\sqrt{3}$ square units.

19.

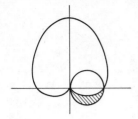

In this case $A = \frac{1}{2} \int_{\frac{3\pi}{2}}^{2\pi} \cos^2 \theta \, d\theta - \frac{1}{2} \int_{\frac{3\pi}{2}}^{2\pi} (1 + \sin \theta)^2 \, d\theta$. Since the integral

$\frac{1}{2} \int_{\frac{3\pi}{2}}^{2\pi} \cos^2 \theta \, d\theta$ gives the area of a semi-circle of radius $\frac{1}{2}$, $\frac{1}{2} \int_{\frac{3\pi}{2}}^{2\pi} \cos^2 \theta \, d\theta$

$= \frac{\pi}{8}$. Thus, $A = \frac{\pi}{8} - \frac{1}{2} \int_{\frac{3\pi}{2}}^{2\pi} (1 + 2 \sin \theta + \sin^2 \theta) \, d\theta = \frac{\pi}{8} - \frac{1}{2} ((\theta - 2 \cos \theta) \Big|_{\frac{3\pi}{2}}^{2\pi}$

$+ \int_{\frac{3\pi}{2}}^{2\pi} \frac{1 - \cos 2\theta}{2} \, d\theta) = \frac{\pi}{8} - \frac{1}{2} (\theta - 2 \cos \theta + \frac{1}{2} \theta - \frac{1}{4} \sin 2\theta) \Big|_{\frac{3\pi}{2}}^{2\pi}$

$= \frac{\pi}{8} - \frac{1}{2} (2\pi - 2 + \pi - \frac{3\pi}{2} - \frac{3\pi}{4}) = \frac{\pi}{8} - \frac{1}{2} (\frac{3\pi}{4} - 2) = 1 - \frac{\pi}{4}$ square units.

21. As illustrated below, the length of string that has been unwound while the end of the string has moved through an angle θ is $a\theta$.

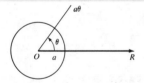

Thus, if r is the distance from the pole to the end of the string, we have $r = a + a\theta$. We are to find the area within the graph of $r = a + a\theta$ and outside the circle $r = a$ for $0 \leq \theta \leq \pi/2$. This area is given by

$A = \frac{1}{2} \int_0^{\pi/2} (a + a\theta)^2 \, d\theta - \frac{\pi a^2}{4}$

$= \frac{a^2}{a} \int_0^{\pi/2} (1 + 2\theta + \theta^2) \, d\theta - \frac{\pi a^2}{4} = \frac{a^2}{2} [\theta + \theta^2 + \frac{1}{3} \theta^3] \Big|_0^{\pi/2} - \frac{\pi a^2}{4}$

$= \frac{a^2}{2} [\frac{\pi}{2} + \frac{\pi^2}{4} + \frac{\pi^3}{24}] - \frac{\pi a^2}{4} = \frac{a\pi^2}{48} (6 + \pi)$ square units.

1. In this case $r = f(\theta) = 5$. Thus, substitution into

$$L = \int_{\alpha}^{\beta} \sqrt{[f'(\theta)]^2 + [f(\theta)]^2}\, d\theta \text{ gives } L = \int_0^{2\pi} \sqrt{25}\, d\theta = 10\pi \text{ units.}$$

3. The graph is a straight line segment with length 4.

5. In general, the arc length of $r = f(\theta)$, $\alpha \le \theta \le \beta$ is given by

$$L = \int_{\alpha}^{\beta} \sqrt{[f'(\theta)]^2 + [f(\theta)]^2}\, d\theta. \text{ In this case } f(\theta) = \theta^2 \text{ and } f'(\theta) = 2\theta, \text{ so}$$

$$L = \int_0^{\pi} \sqrt{4\theta^2 + \theta^4}\, d\theta \text{ and } L = \int_0^{\pi} \sqrt{4 + \theta^2}\; \theta\, d\theta = \frac{1}{3}(4 + \theta^2)^{3/2}\Big|_0^{\pi}$$

$$= \frac{1}{3}[(4 + \pi^2)^{3/2} - 8] \text{ units.}$$

7. In order to cover the entire curve, θ must go from 0 to π. Going from 0 to 2π covers it twice. Substitution of $f(\theta) = 9\cos\theta$ and $f'(\theta) = -9\sin\theta$ into

$$L = \int_0^{\pi} \sqrt{[f'(\theta)]^2 + [f(\theta)]^2}\, d\theta \text{ gives}$$

$$L = \int_0^{\pi} \sqrt{81\sin^2\theta + 81\cos^2\theta}\, d\theta = 9\int_0^{\pi} d\theta = 9\pi \text{ units.}$$

9. Substitution of $f(\theta) = e^{5\theta}$, $f'(\theta) = 5e^{5\theta}$ into

$$L = \int_{\alpha}^{\beta} \sqrt{[f'(\theta)]^2 + [f(\theta)]^2}\, d\theta \text{ gives } L = \int_0^{\ln 3} \sqrt{25e^{10\theta} + e^{10\theta}}\, d\theta$$

$$= \sqrt{26}\int_0^{\ln 3} e^{5\theta}\, d\theta = \frac{\sqrt{26}}{5} e^{5\theta}\Big|_0^{\ln 3} = \frac{\sqrt{26}}{5}(3^5 - 1) = \frac{242\sqrt{26}}{5} \text{ units.}$$

11. In order to cover the entire curve θ must go from 0 to π. Substitution of $f(\theta) = 12\sin\theta$ and $f'(\theta) = 12\cos\theta$ into $L = \int_{\alpha}^{\beta} \sqrt{[f'(\theta)]^2 + [f(\theta)]^2}\, d\theta$ gives

$$L = \int_0^{\pi} \sqrt{144\cos^2\theta + 144\sin^2\theta}\, d\theta = 12\int_0^{\pi} d\theta = 12\pi \text{ units.}$$

13. Substitution of $f(\theta) = 3\sec\theta$, $f'(\theta) = 3\sec\theta\tan\theta$ into

$$L = \int_{\alpha}^{\beta} \sqrt{[f(\theta)]^2 + [f'(\theta)]^2}\, d\theta \text{ gives } L = \int_0^{\pi/4} \sqrt{9\sec^2\theta + 9\sec^2\theta\tan^2\theta}\, d\theta$$

$$= 3\int_0^{\pi/4} \sec\theta\sqrt{1 + \tan^2\theta}\, d\theta. \text{ Then since } 1 + \tan^2\theta = \sec^2\theta, \text{ and } \sec\theta > 0$$

for $0 \le \theta \le \pi/4$, $L = 3\int_0^{\pi/4} \sec^2\theta\, d\theta = 3\tan\theta\Big|_0^{\pi/4} = 3$ units.

15. In order to cover the entire graph, θ must go from 0 to 2π. Substitution of this information along with $f(\theta) = 1 + \sin \theta$, $f'(\theta) = \cos \theta$ into

$$L = \int_\alpha^\beta \sqrt{[f(\theta)]^2 + [f'(\theta)]^2} \, d\theta \text{ gives } L = \int_0^{2\pi} \sqrt{1 + 2 \sin \theta + \sin^2 \theta + \cos^2 \theta} \, d\theta$$

$$= \sqrt{2} \int_0^{2\pi} \sqrt{1 + \sin \theta} \, d\theta. \text{ If we make the substitution } \theta = \alpha + \pi/2, \text{ we have}$$

$$L = \sqrt{2} \int_{-\pi/2}^{3\pi/2} \sqrt{1 + \sin (\alpha + \pi/2)} \, d\alpha. \text{ Then since } \sin (\alpha + \pi/2) = \cos \alpha,$$

$$L = \sqrt{2} \int_{-\pi/2}^{3\pi/2} \sqrt{1 + \cos \alpha} \, d\alpha. \text{ Now we can use the familiar identity}$$

$1 + \cos \alpha = 2 \cos^2 \alpha/2$ to get $L = 2 \int_{-\pi/2}^{3\pi/2} |\cos \alpha/2| \, d\alpha$. Since $\cos \alpha/2 \geq 0$ for

$-\pi \leq \alpha \leq \pi$ and $\cos \alpha/2 \leq 0$ for $\pi \leq \alpha \leq 3\pi$, we have

$$|\cos \alpha/2| = \begin{cases} \cos \alpha/2 \text{ for } -\pi/2 \leq \alpha \leq \pi \\ -\cos \alpha/2 \text{ for } \pi \leq \alpha \leq 3\pi/2. \end{cases} \text{ Thus,}$$

$$L = 2 \int_{-\pi/2}^{\pi} \cos \alpha/2 \, d\alpha - 2 \int_{\pi}^{3\pi/2} \cos \alpha/2 \, d\alpha = 4 \sin \alpha/2 \Big|_{-\pi/2}^{\pi} - 4 \sin \alpha/2 \Big|_{\pi}^{3\pi/2}$$

$$= 4 \left(1 + \frac{\sqrt{2}}{2}\right) - 4 \left(\frac{\sqrt{2}}{2} - 1\right) = 8 \text{ units.}$$

17. The full curve is traversed as θ increases from 0 to 3π. Then since

$f(\theta) = \sin^3 \theta/3$ and $f'(\theta) = \sin^2 \theta/3 \cos \theta/3$, we have

$$L = \int_0^{3\pi} \sqrt{\sin^6 \theta/3 + \sin^4 \theta/3 \cos^2 \theta/3} \, d\theta$$

$$= \int_0^{3\pi} \sin^2 \left(\frac{\theta}{3}\right) \sqrt{\sin^2 \theta/3 + \cos^2 \theta/3} \, d\theta$$

$$= \int_0^{3\pi} \sin^2 \left(\frac{\theta}{3}\right) d\theta = \int_0^{3\pi} \frac{1 - \cos (2\theta/3)}{2} \, d\theta = \frac{1}{2} \left(\theta - \frac{3}{2} \sin \left(\frac{2\theta}{3}\right)\right) \Big|_0^{3\pi} = \frac{3\pi}{2} \text{ units.}$$

Technique Review Exercises, Chapter 12 (page 535)

1. The graph of $r = f(\theta)$ will be symmetric with respect to the line $\theta = \pi/2$ if $f(\pi - \theta) = f(\theta)$. Since $\sin (\pi - \theta) = \sin \theta$, we have $3 + 5 \sin \theta = 3 + 5 \sin (\pi - \theta)$. Consequently, the graph is symmetric with respect to the line $\theta = \pi/2$.

2. Since $r = \sqrt{x^2 + y^2}$ and $y = r \sin \theta$, we have $\sin \theta = y/r = y/\sqrt{x^2 + y^2}$. Thus, on substitution we have $\sqrt{x^2 + y^2} = 3 + 5y/\sqrt{x^2 + y^2}$.

3. The following table which gives the change in r as θ varies is useful.

θ	sin θ	r
$0 \to \pi$	$0 \to 1$	$3 \to 8$
$\pi/2 \to \pi$	$1 \to 0$	$8 \to 3$
$\pi \to 3\pi/2$	$0 \to -1$	$3 \to -2$
$3\pi/2 \to 2\pi$	$-1 \to 0$	$-2 \to 3$

The graph can then be sketched as indicated below.

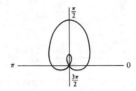

4. In this case $f(\theta) = e^{2\theta}$ and $f'(\theta) = 2e^{2\theta}$. Thus, substitution into

$$L = \int_0^{2\pi} \sqrt{[f'(\theta)]^2 + [f(\theta)]^2} \, d\theta \text{ gives } L = \int_0^{2\pi} \sqrt{4e^{4\theta} + e^{4\theta}} \, d\theta$$

$$= \sqrt{5} \int_0^{2\pi} e^{2\theta} \, d\theta = \frac{\sqrt{5}}{2} e^{2\theta} \Big|_0^{2\pi} = \frac{\sqrt{5}}{2} (e^{4\pi} - 1) \text{ units.}$$

5. We must find the length of $r = 1 + \cos\theta$ for $0 \le \theta \le 2\pi$. Since the graph is symmetric about the line $\theta = 0$, the total length is twice that for $0 \le \theta \le \pi$. Then since $f(\theta) = 1 + \cos\theta$ and $f'(\theta) = -\sin\theta$, substitution into

$$L = 2 \int_0^{\pi} \sqrt{[f'(\theta)]^2 + [f(\theta)]^2} \, d\theta \text{ gives } L = 2 \int_0^{\pi} \sqrt{\sin^2\theta + 1 + 2\cos\theta + \cos^2\theta}$$

$$d\theta = 2\sqrt{2} \int_0^{\pi} \sqrt{1 + \cos\theta} \, d\theta. \text{ Then since } \cos^2\theta = \frac{1 + \cos\theta}{2}, \text{ we have}$$

$$\sqrt{1 + \cos\theta} = \sqrt{2} \, |\cos\theta|. \text{ Consequently, } L = 4 \int_0^{\pi} |\cos\theta| \, d\theta$$

$$= 4(\int_0^{\pi/2} |\cos\theta| \, d\theta + \int_{\pi/2}^{\pi} |\cos\theta| \, d\theta). \text{ Then since } \cos\theta \ge 0 \text{ for } 0 \le \theta \le \pi/2$$

and $\cos \le 0$ for $\pi/2 \le \theta \le \pi$, we have $L = 4(\int_0^{\pi/2} \cos\theta \, d\theta - \int_{\pi/2}^{\pi} \cos\theta \, d\theta)$

$$= 4[\sin\theta \Big|_0^{\pi/2} - \sin\theta \Big|_{\pi/2}^{\pi}] = 4(1 + 1) = 8 \text{ units.}$$

6. A sketch of this graph is given below.

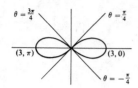

By symmetry the total area is $A = \int_{-\pi/4}^{\pi/4} 9 \cos 2\theta \; d\theta = \frac{9}{2} \sin 2\theta \Big|_{-\pi/4}^{\pi/4} = 9$

square units.

Additional Exercises, Chapter 12 (page 535)

17. Since $x^2 + y^2 = r^2$ and $x = r \cos \theta$, this equation becomes $r^2 = 2r \cos \theta$.

19. First we write this equation as $x^2 - 4x + 4 + y^2 = 1$ or, $x^2 + y^2 - 4x + 3 = 0$.
    Then using $x^2 + y^2 = r^2$ and $x = r \cos \theta$, we have $r^2 - 4r \cos \theta + 3 = 0$.

21. Since $\theta = \arctan y/x$, this equation becomes $\arctan y/x = \pi/6$. Thus,
    $y/x = \tan \pi/6 = \dfrac{1}{\sqrt{3}}$ or, $y = \dfrac{1}{\sqrt{3}} x$.

23. Multiplication by $\cos \theta - \sin \theta$ gives $r \cos \theta - r \sin \theta = 1$. Then since
    $x = r \cos \theta$ and $y = r \sin \theta$, this equation becomes $x - y = 1$ in rectangular
    form.

25. Substitution into $A = \dfrac{1}{2} \int_{\alpha}^{\beta} [f(\theta)]^2 \; d\theta$ gives $A = \dfrac{1}{2} \int_{\pi/2}^{3\pi/4} (e^{2\theta})^2 \; d\theta$. Thus,

    $A = \dfrac{1}{2} \int_{\pi/2}^{3\pi/4} e^{4\theta} \; d\theta = \dfrac{1}{8} e^{4\theta} \Big|_{\pi/2}^{3\pi/4} = \dfrac{1}{8} (e^{3\pi} - e^{2\pi})$ square units.

27. In this case $A = \dfrac{1}{2} \int_{0}^{2\pi} (2 - \sin \theta)^2 \; d\theta = \dfrac{1}{2} \int_{0}^{2\pi} (4 - 4 \sin \theta + \sin^2 \theta) \; d\theta$

    $= 2 \int_{0}^{2\pi} d\theta - 2 \int_{0}^{2\pi} \sin \theta \; d\theta + \dfrac{1}{2} \int_{0}^{2\pi} \dfrac{1 - \cos 2\theta}{2} \; d\theta = 4\pi + \dfrac{1}{4} \left(\theta - \dfrac{1}{2} \sin 2\theta\right) \Big|_{0}^{2\pi}$

    $= 4\pi + \pi/2 = 9\pi/2$ square units.

29. A sketch of the two graphs involved is indicated below.

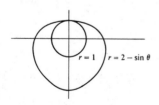

$r = 1$ $r = 2 - \sin \theta$

The area between the two curves is given by

$A = \dfrac{1}{2} \displaystyle\int_0^{2\pi} (2 - \sin\theta)^2 \, d\theta - \dfrac{1}{2} \displaystyle\int_0^{2\pi} d\theta$. Then using the result of Exercise 27,

$A = 9\pi/2 - \pi = 7\pi/2$ square units.

31. A sketch of the graph of $r = 1 - 2\sin\theta$ is shown below.

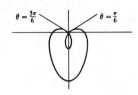

Since the smaller loop is covered as θ moves from $\pi/6$ to $5\pi/6$, the desired

area is $A = \dfrac{1}{2} \displaystyle\int_{\pi/6}^{5\pi/6} (1 - 2\sin\theta)^2 \, d\theta$

$= \dfrac{1}{2} \displaystyle\int_{\pi/6}^{5\pi/6} (1 - 4\sin\theta + 4\sin^2\theta) \, d\theta = \dfrac{1}{2} \displaystyle\int_{\pi/6}^{5\pi/6} d\theta - 2 \displaystyle\int_{\pi/6}^{5\pi/6}$

$\sin\theta \, d\theta + \displaystyle\int_{\pi/6}^{5\pi/6} (1 - \cos 2\theta) \, d\theta = \dfrac{1}{2} \, \theta \, \Big|_{\pi/6}^{5\pi/6} + 2\cos\theta \, \Big|_{\pi/6}^{5\pi/6}$

$+ \, \theta \, \Big|_{\pi/6}^{5\pi/6} - \dfrac{1}{2} \sin 2\theta \, \Big|_{\pi/6}^{5\pi/6} = \pi - 3\sqrt{3}/2$ square units.

33. $r = a\theta^2$, $0 \le \theta \le 2\pi$. Substitution into the arc length formula

$L = \displaystyle\int_\alpha^\beta \sqrt{[f'(\theta)]^2 + [f(\theta)]^2} \, d\theta$ gives

$L = \displaystyle\int_0^{2\pi} \sqrt{(2a\theta)^2 + (a\theta^2)^2} \, d\theta = \displaystyle\int_0^{2\pi} a\theta \sqrt{4 + a^2\theta^2} \, d\theta$

$= \dfrac{1}{2a} \displaystyle\int_0^{2\pi} \sqrt{4 + a^2\theta^2} \; 2a^2\theta \, d\theta = \dfrac{1}{2a} \left(\dfrac{2}{3}\right)(4 + a^2\theta^2)^{3/2} \, \Big|_0^{2\pi}$

$= \dfrac{1}{3a} [(4 + 4a^2\pi^2)^{3/2} - 4^{3/2}] = \dfrac{8}{3a} [(1 + a^2\pi^2)^{3/2} - 1]$ units.

35. In this case the entire graph is traversed as θ ranges from 0 to 2π. Thus,

$L = \displaystyle\int_0^{2\pi} \sqrt{(-\cos\theta)^2 + (1 - \sin\theta)^2} \, d\theta$

$= \displaystyle\int_0^{2\pi} \sqrt{\cos^2\theta + \sin^2\theta - 2\sin\theta + 1} \, d\theta = \sqrt{2} \displaystyle\int_0^{2\pi} \sqrt{1 - \sin\theta} \, d\theta$. If we let

$\theta = \alpha + \pi/2$ this integral becomes $L = \sqrt{2} \displaystyle\int_{-\pi/2}^{3\pi/2} \sqrt{1 - \cos\alpha} \, d\alpha$. Then since

$\sqrt{1 - \cos\alpha} = \sqrt{2} \, |\sin\alpha/2|$, $L = 2 \displaystyle\int_{-\pi/2}^{3\pi/2} \sin\alpha/2 \, d\alpha = 2\left[-\displaystyle\int_{-\pi/2}^{0} \sin\alpha/2 \, d\alpha\right.$

$$+ \int_0^{3\pi/2} \sin \alpha/2 \, d\alpha] = 2[2 \cos \alpha/2 \Big|_{-\pi/2}^{0} - 2 \cos \alpha/2 \Big|_0^{3\pi/2}]$$

$$= 4[1 - \frac{1}{\sqrt{2}} + \frac{1}{\sqrt{2}} + 1] = 8 \text{ units.}$$

37. Since $f(\theta) = \cos^3 (\theta/3)$ and $f'(\theta) = -\cos^2 (\theta/3) \sin (\theta/3)$,

$$L = \int_0^{\pi} \sqrt{\cos^4 (\theta/3) \sin^2 (\theta/3) + \cos^6 (\theta/3)} \, d\theta$$

$$= \int_0^{\pi} \cos^2 (\theta/3) \sqrt{\sin^2 (\theta/3) + \cos^2 (\theta/3)} \, d\theta$$

$$= \int_0^{\pi} \cos^2 (\theta/3) \, d\theta = \int_0^{\pi} \frac{1 + \cos (2\theta/3)}{2} \, d\theta = \frac{1}{2} (\theta + \frac{3}{2} \sin (2\theta/3)) \Big|_0^{\pi}$$

$$= \frac{1}{2} (\pi + \frac{3\sqrt{3}}{4}) \text{ units.}$$

39. In this case $f(\theta) = 2 \sec \theta$ and $f'(\theta) = 2 \sec \theta \tan \theta$. Thus,

$$L = \int_0^{\pi/4} \sqrt{4 \sec^2 \theta + 4 \sec^2 \theta \tan^2 \theta} \, d\theta$$

$$L = \int_0^{\pi/4} 2 \sec \theta \sqrt{1 + \tan^2 \theta} \, d\theta = 2 \int_0^{\pi/4} \sec^2 \theta \, d\theta$$

$$= 2 \tan \theta \Big|_0^{\pi/4} = 2 \text{ units.}$$

CHAPTER 13: DIFFERENTIAL EQUATIONS

Exercises 13.1: Differential Equations (page 541)

1. Since $y = Ce^{-kx}$, $\frac{dy}{dx} = -kCe^{-kx}$. Substitution then gives

$$\frac{dy}{dx} + ky = -kCe^{-kx} + kCe^{-kx} = 0 \text{ as was to be shown.}$$

3. Since $y = e^{2x} + C_1 e^x + C_2 e^{-3x}$, we have $y' = 2e^{2x} + C_1 e^x - 3C_2 e^{-3x}$ and

$y'' = 4e^{2x} + C_1 e^x + 9C_2 e^{-3x}$. Then substitution gives $y'' + 2y' - 3y$

$= 4e^{2x} + C_1 e^x + 9C_2 e^{-3x} + 2(2e^{2x} + C_1 e^x - 3C_2 e^{-3x}) - 3(e^{2x} + C_1 e^x + C_2 e^{-3x})$.

Thus, $y'' + 2y' - 3y = 5e^{2x}$ if $y = e^{2x} + C_1 e^x + C_2 e^{-3x}$ as was to be shown.

5. Since $x = C_1 \sin t + C_2 \cos t + \frac{1}{5} e^{2t} + t^2 - 2$, we have

$$\frac{dx}{dt} = C_1 \cos t - C_2 \sin t + \frac{2}{5} e^{2t} + 2t \text{ and } \frac{d^2x}{dt^2} = -C_1 \cos t - C_2 \sin t + \frac{4}{5} e^{2t}$$

+ 2. Then substitution gives $\dfrac{d^2x}{dt^2} + x = -C_1 \cos t - C_2 \sin t + \dfrac{4}{5} e^{2t} + 2$

$+ C_1 \sin t + C_2 \cos t + \dfrac{1}{5} e^{2t} + t^2 - 2$. Thus, $\dfrac{d^2x}{dt^2} = e^{2t} + t^2$, as was to be

shown.

7. Since $y = C_1 e^{-2x} + C_2 e^{-4x}$, $y' = -2C_1 e^{-2x} - 4C_2 e^{-4x}$, and $y'' = 4C_1 e^{-2x}$

$+ 16C_2 e^{-4x}$. Substitution yields $y'' + 6y' + 8y = 4C_1 e^{-2x} + 16C_2 e^{-4x}$

$- 12C_1 e^{-2x} - 24C_2 e^{-4x} + 8C_1 e^{-2x} + 8C_2 e^{-4x} = 0$ as was to be shown.

9. Since $y = e^{-3x} (C_1 \cos x + C_2 \sin x)$, we have

$y' = -3e^{-3x} (C_1 \cos x + C_2 \sin x) + e^{-3x} (-C_1 \sin x + C_2 \cos x)$

$= (C_2 - 3C_1) e^{-3x} \cos x - (C_1 + 3C_2) e^{-3x} \sin x$ and,

$y'' = -3(C_2 - 3C_1) e^{-3x} \cos x - (C_2 - 3C_1) e^{-3x} \sin x$

$+ 3(C_1 + 3C_2) e^{-3x} \sin x - (C_1 + 3C_2) e^{-3x} \cos x$

$y'' = (8C_1 - 6C_2) e^{-3x} \cos x + (6C_1 + 8C_2) e^{-3x} \sin x$. Substitution then yields

$y'' + 6y' + 10y = (8C_1 - 6C_2) e^{-3x} \cos x + (6C_1 + 8C_2) e^{-3x} \sin x$

$+ 6(C_2 - 3C_1) e^{-3x} \cos x - 6(C_1 + 3C_2) e^{-3x} \sin x + 10e^{-3x} (C_1 \cos x$

$+ C_2 \sin x) = 0$, as was to be shown.

11. Since the solution given is $y = Ce^{-kx}$, we are to find C such that $Ce^{-k \cdot 0} = 2$,

that is, $C = 2$. The desired particular solution is then $y = 2e^{-kx}$.

13. In Exercise 3 we have $y = e^{2x} + C_1 e^{x} + C_2 e^{-3x}$ and so,

$y' = 2e^{2x} + C_1 e^{x} - 3C_2 e^{-3x}$. Then since both y and y' are to be 0 when x = 0,

we must choose C_1 and C_2 such that $0 = e^{2 \cdot 0} + C_1 e^{0} + C_2 e^{-3 \cdot 0}$ and,

$0 = 2e^{2 \cdot 0} + C_1 e^{0} - 3C_2 e^{-3 \cdot 0}$. That is, such that $0 = 1 + C_1 + C_2$ and

$0 = 2 + C_1 - 3C_2$. Solution of these equations for C_1 and C_2 gives $C_1 = -5/4$

and $C_2 = 1/4$. Thus, the desired solution is $y = e^{2x} - \dfrac{5}{4} e^{x} + \dfrac{1}{4} e^{-3x}$.

15. In Exercise 7 we have $y = C_1 e^{-2x} + C_2 e^{-4x}$ and $y' = -2C_1 e^{-2x} - 4C_2 e^{-4x}$. Thus,

we must choose C_1 and C_2 so that $C_1 e^{0} + C_2 e^{0} = 0$ and $-2C_1 e^{0} - 4C_2 e^{0} = 2$. That is, $C_1 + C_2 = 0$, $-2C_1 - 4C_2 = 2$. Solution of these equations for C_1 and C_2 gives $C_1 = 1$ and $C_2 = -1$. Thus, the desired solution is $y = e^{-2x} - e^{-4x}$.

17. Let $y = e^{\lambda x}$, then $y' = \lambda e^{\lambda x}$ and $y'' = \lambda^2 e^{\lambda x}$. Then substitution gives $y'' - cy = \lambda^2 e^{\lambda x} - ce^{\lambda x}$. Consequently, we seek λ such that $\lambda^2 e^{\lambda x} - ce^{\lambda x} = 0$, or $e^{\lambda x}(\lambda^2 - c) = 0$. The desired value of λ are thus $\lambda = \pm\sqrt{c}$.

<u>Exercises 13.2</u>: Separable Differential Equations (page 545)

1. On division by y^2 we have $\frac{1}{y^2} \frac{dy}{dx} = 2x$. Then integrating with respect to x gives $\int \frac{1}{y^2} \frac{dy}{dx} dx = \int 2x \, dx$ or, $\int \frac{dy}{y^2} = \int 2x \, dx$. Thus, $-y^{-1} = x^2 + C$ or,

$$y = -\frac{1}{x^2 + C}.$$

5. This equation can be written as $\frac{1}{y} \frac{dy}{dx} = \frac{1 - x^2}{x^3}$. Then on integration with respect to x, we have $\int \frac{1}{y} \frac{dy}{dx} dx = \int \frac{1 - x^2}{x^3} dx$ or, $\int \frac{dy}{y} = \int (\frac{1}{x^3} - \frac{1}{x}) \, dx$. Thus,

$\ln y = -\frac{1}{2} x^{-2} - \ln x + C_1$ or, $y = e^{-x^{-2}/2 - \ln x + C_1}$. Thus,

$y = e^{-x^{-2}/2} e^{-\ln x} e^{C_1}$. Then on letting $C = e^{-C_1}$, we have $y = \frac{C}{x} e^{-x^{-2}/2}$.

7. Separation of the variables gives $\frac{dx}{x^2 + 1} = \frac{\cos y}{\sin y} dy$. On integration, we have

$\int \frac{dx}{x^2 + 1} = \int \frac{\cos y}{\sin y} dy$ or, $\arctan x = \ln (\sin y) + C$.

9. Division by $y + 3$ gives $\frac{1}{y + 3} \frac{dy}{dx} = -\frac{1}{x - 2}$. Then integration with respect to x yields $\int \frac{1}{y + 3} \frac{dy}{dx} dx = \int \frac{dx}{x - 2}$ or, $\ln |y + 3| = -\ln |x - 2| + C_1$. Consequently, $e^{\ln|y+3|} = e^{-\ln|x-2|} e^{C_1}$. Then if we let $C = e^{C_1}$, we have

$|y + 3| = \frac{C}{|x - 2|}$. Since the particular solution is to pass through the point (3,4), we must have $y = 4$ when $x = 3$. That is, $7 = \frac{C}{1}$. Thus, the desired particular solution is $|y + 3| = \frac{7}{|x + 1|}$.

11. If we write this equation as $\frac{dy}{dx} = 10x^4$ we can integrate to obtain $y = 2x^5 + C$. Since $y = 8$ when $x = 1$, we must have $8 = 2 + C$ or, $C = 6$. Consequently, $y = 2x^5 + 6$ is the desired solution.

13. Division by $1 + y^2$ gives $\dfrac{1}{1 + y^2} \dfrac{dy}{dx} = e^x$. Then $\int \dfrac{1}{1 + y^2} \dfrac{dy}{dx} dx = \int e^x dx$ or,

arctan $y = e^x + C$. Consequently, $y = \tan(e^x + C)$ is the desired solution.

15. Since $\dfrac{dN}{dt} = kN$, we have $\dfrac{1}{N} \dfrac{dN}{dt} = k$. Then on integration we have

$\int \dfrac{1}{N} \dfrac{dN}{dt} dt = k \int dt$ or, $\ln N = kt + C_1$. Thus, $e^{\ln N} = e^{kt} e^{C_1}$ or, on letting

$e^{C_1} = C$, $N = Ce^{kt}$. Note that in this case the population grows without bound

as t increases. In Example 2, the population had a limiting value.

17. If x denotes the weight of the radioactive substance present at time t, then

$\dfrac{dx}{dt}$ is the rate of change of the weight. Consequently, $\dfrac{dx}{dt} = -kx$. Thus,

$\dfrac{1}{x} \dfrac{dx}{dt} = -k$ and on integration we have $\int \dfrac{1}{x} \dfrac{dx}{dt} dt = -k \int dt$ or, $\ln x = -kt + C_1$.

Solution for x gives $x = Ce^{-kt}$ where $C = e^{C_1}$. Note that $x = C$ when $t = 0$.

Thus, C is the initial amount of the substance present. We know that $x = C/2$

when $t = 90$. Thus, $\dfrac{C}{2} = Ce^{-90k}$ or, $e^{-90k} = \dfrac{1}{2}$. Thus, $e^{-k} = 2^{-1/90}$ and we have

$x = C \cdot 2^{-t/90}$ where C is the initial amount present, x is measured in pounds

and t in years.

19. Let x denote the temperature of the body and T denote the temperature of the

surroundings. Then, the rate at which the body cools is $\dfrac{dx}{dt}$. Thus,

$\dfrac{dx}{dt} = k(T - x)$. On integration we find $\ln(T - x) = -kt + C_1$ or,

$T - x = e^{-kt} e^{C_1}$. Thus, if we set $C = e^{C_1}$, we have $x = T - Ce^{-kt}$. In the

specific case given in the problem, $T = 20°$, and $x = 600°$ when $t = 0$. Thus,

$600 = 20 - C$ or, $C = -580$ and we have $x = 20 + 580e^{-kt}$. In order to find k,

note that $x = 300°$ when $t = 30$ min. Thus, $300 = 20 + 580e^{-30k}$. Consequently,

$e^{-30k} = \dfrac{280}{580}$ so $e^{-k} = (\dfrac{280}{580})^{1/30} = (\dfrac{14}{19})^{1/30}$. Thus, $x = 20 + 580 (\dfrac{14}{29})^{t/30}$.

21. If we let x denote the number of people holding the given idea at time t, and

let N denote the size of the total population, then $\dfrac{dx}{dt} = kx(N - x)$. Thus,

$\int \dfrac{1}{x(N - x)} dx = k \int t \, dt$. Use of partial fractions on the first integral gives

$\int \dfrac{1}{x(N - x)} dx = \dfrac{1}{N} \int (\dfrac{1}{x} + \dfrac{1}{N - x}) dx = \dfrac{1}{N} (\ln(x) - \ln(N - x)) + C_1$. Thus,

$\dfrac{1}{N} \ln \dfrac{x}{N - x} = kt + C_2$. Hence, $\dfrac{x}{N - x} = e^{Nkt} e^{NC_2}$ and on letting $C = e^{NC_2}$, we have

$\dfrac{x}{N - x} = Ce^{Nkt}$. We need now only solve for x. Then, $x = NCe^{Nkt} - xCe^{Nkt}$ so

$x(1 + Ce^{Nkt}) = NCe^{Nkt}$ or finally, $x = \dfrac{NCe^{Nkt}}{1 + Ce^{Nkt}}$.

23. Since $f'(t)/f(t)$ is inversely proportional to t, we have $\dfrac{f'(t)}{f(t)} = \dfrac{k}{t}$. Thus,

$\int \dfrac{f'(t)\, dt}{f(t)} = k \int \dfrac{dt}{t}$ or, $\ln f(t) = k \ln t + C_1$. On solving for $f(t)$, we have

$f(t) = e^{k \ln t} e^{C_1}$ or, $f(t) = Ct^k$ where $C = e^{C_1}$. Since $f(1) = 1$, we must have

$C = 1$. Thus, $f(t) = t^k$. To find k we use the information that $f(2) = 3$.

Thus, we have $3 = 2^k$ or, $k = \dfrac{\ln 3}{\ln 2}$, and so $f(t) = t^{(\ln 3)/(\ln 2)}$.

25. Let x denote the weight of glucose in grams in the bloodstream. Then $\dfrac{dx}{dt} = kx$

or on integration, $\ln x = kt + C_1$. Solving for x and setting $C = e^{C_1}$ gives

$x = Ce^{kt}$.

Exercises 13.3: First Order Linear Differential Equations (page 550)

1. Since $P(x) = \dfrac{5}{x}$, $\rho(x) = e^{\int P(x)\, dx} = e^{\int \frac{5}{x}\, dx} = e^{5 \ln x} = x^5$ is an integrating

factor. On multiplication by x^5, we have $x^5 \dfrac{dy}{dx} + 5x^4 y = x^7 + 2x^6$ or,

$\dfrac{d(x^5 y)}{dx} = x^7 + 2x^6$. On integrating we obtain $x^5 y = \dfrac{1}{8} x^8 + \dfrac{2}{7} x^7 + C$ or,

$y = \dfrac{1}{8} x^3 + \dfrac{2}{7} x^2 + Cx^{-5}$. The particular solution we seek must have $y = 0$ when

$x = 1$. Thus, we choose C such that $0 = \dfrac{1}{8} + \dfrac{2}{7} + C$. That is, we select

$C = -23/56$. Consequently, $y = \dfrac{1}{8} x^3 + \dfrac{2}{7} x^2 - \dfrac{23}{56} x^{-5}$ is the desired particular

solution.

3. If we write this equation as $\dfrac{ds}{dt} - s \cos t = e^{\sin t}$ we see that $P(t) = -\cos t$.

Thus, $\rho(t) = e^{-\int \cos t} = e^{-\sin t}$ is an integrating factor. Multiplication by

$e^{-\sin t}$ gives $e^{-\sin t} \dfrac{ds}{dt} - s \cos t (e^{-\sin t}) = 1$ or, $\dfrac{d}{dt} (se^{-\sin t}) = 1$. Con-

sequently, on integration we have $se^{-\sin t} = t + C$ or, $s = e^{\sin t} (t + C)$.

5. On division by x^2 we have $y' - \dfrac{2y}{x} = \dfrac{2}{x^2}$. Thus, $p(x) = -2/x$ and an integrating

factor is given by $\rho(x) = e^{\int p(x)\, dx} = e^{-2 \int \frac{dx}{x}} = e^{-2 \ln x} = x^{-2}$. Then multipli-

cation by the integrating factor gives $x^{-2} y' - 2yx^{-3} = 2x^{-4}$ or, $\dfrac{d(x^{-2} y)}{dx}$

$= 2x^{-4} + C$. Then on integrating, we have $x^{-2} y = -\dfrac{2}{3} x^{-3}$ or,

$y = x^2 (-\dfrac{2}{3} x^{-3} + C)$.

9. Since $p(x) = 1$, an integrating factor is given by $\rho(x) = e^{\int p(x)\, dx} = e^{\int dx}$
$= e^x$. Multiplication by the integrating factor yields $y'e^x + ye^x = x^2 e^x$ or,
$\frac{d(ye^x)}{dx} = x^2 e^x$. Then we have $ye^x = \int x^2 e^x\, dx$. Two applications of integration
by parts then gives $ye^x = x^2 e^x - 2x\, e^x + 2e^x + C$ or, $y = x^2 - 2x + 2 + Ce^{-x}$.
Then since we must have $y = 5$ when $x = 0$, we have $5 = 2 + C$ or, $C = 3$. The
desired particular solution is then $y = x^2 - 2x + 2 + 3e^{-x}$.

11. Since $p(x) = -\tan x$, an integrating factor is given by
$\rho(x) = e^{\int p(x)\, dx} = e^{-\int \tan x\, dx} = e^{\ln |\cos x|} = \cos x$. Multiplication by this
integrating factor gives $(\cos x)\, y' - (\sin x)\, y = x \cos x$ or,
$\frac{d}{dx}((\cos x)\, y) = x \cos x$. Consequently, $y \cos x = \int x \cos x\, dx$. Evaluation
of this integral by integration by parts then gives $y \cos x = x \sin x + \cos x$
$+ C$. Then since we are to have $y = 1$ when $x = \pi/4$, we must choose C such that
$1/\sqrt{2} = \pi/4\sqrt{2} + 1/\sqrt{2} + C$ or, $C = -\pi/4\sqrt{2}$. The desired particular solution is
thus $y \cos x = x \sin x + \cos x - \pi/4\sqrt{2}$.

15. Since the roles of x and y are interchanged $\rho(y) = e^{\int -\frac{2}{y}\, dy}$ will be an integrating factor. Then $\rho(y) = e^{-2 \ln y} = y^{-2}$ and on multiplication by y^{-2} we have
$y^{-2} \frac{dx}{dy} - 2xy^{-3} = 2y$ or, $\frac{d(y^{-2}x)}{dy} = 2y$. Integration then gives $y^{-2}x = y^2 + C$ or,
$x = y^2(y^2 + C)$.

17. in Exercise 16 it was shown that a substitution of the form $y^{1-n} = z$ will
change the equation $\frac{dy}{dx} + P(x)y = Q(x)y^n$ into an equation that is linear in
$\frac{dz}{dx}$ and z. Thus, to solve $y' - y = xy^{1/2}$ we substitute $z = y^{1-1/2} = y^{1/2}$.
Then $y = z^2$, so $\frac{dy}{dx} = 2z \frac{dz}{dx}$. Substitution then yields $2z \frac{dz}{dx} - z^2 = xz$ or,
$\frac{dz}{dx} - \frac{z}{2} = \frac{x}{2}$, a linear equation with integrating factor $\rho(x) = e^{-\int \frac{dx}{2}} = e^{-x/2}$.
Multiplication by the integrating factor gives $e^{-x/2} \frac{dz}{dx} - \frac{z}{2} e^{-x/2} = \frac{x}{2} e^{-x/2}$ or,
$\frac{d(ze^{-x/2})}{dx} = \frac{x}{2} e^{-x/2}$. On integration we then have $ze^{-x/2} = -e^{-x/2}(x + 2) + C$.
(Note that one may use integration by parts to evaluate $\int \frac{x}{2} e^{-x/2}\, dx$.) Thus,
$z = -x - 2 + Ce^{x/2}$, and since $z = y^{1/2}$, $y^{1/2} = -x - 2 + Ce^{x/2}$ or,
$y = [Ce^{x/2} - x - 2]^2$.

21. Here $\frac{dr}{d\omega} = -\frac{r}{2}$ so $\frac{1}{r}\frac{dr}{d\omega} = -\frac{1}{2}$. On integration we have $\ln r = -\frac{1}{2}\omega + C_1$ or,

$\omega = -2\ln r + 2C_1$. On letting $C = 2C_1$, we have the simpler form

$\omega = C - 2\ln r$.

Exercises 13.4: Two Special Types of Second Order Differential Equations (page 555)

1. Since the dependent variable, x, is missing, we make the substitution $\frac{dx}{dt} = w$,

$\frac{d^2x}{dt^2} = \frac{dw}{dt}$. Then we have $\frac{dw}{dt} + \frac{w}{t} = 1$, a first order linear equation with inte-

grating factor $\rho(t) = e^{\int dt/t} = t$. Then on multiplication by t, we have

$t\frac{dw}{dt} + w = t$ or, $\frac{d(wt)}{dt} = t$. Thus, $wt = \frac{1}{2}t^2 + C_1$ or, $w = \frac{1}{2}t + C_1 t^{-1}$. Then

since $w = \frac{dx}{dt}$, $\frac{dx}{dt} = \frac{1}{2}t + C_1 t^{-1}$ or on integration $x = \frac{1}{4}t^2 + C_1\ln|t| + C_2$,

the desired solution.

3. Since the independent variable x is missing, we make the substitution

$y' = \frac{dy}{dx} = w$ and $\frac{d^2y}{dx^2} = \frac{dw}{dx} = \frac{dw}{dy}\frac{dy}{dx} = w\frac{dw}{dy}$. Then we have $w^2 + 2yw\frac{dw}{dy} = 0$. Con-

sequently, $2y\frac{dw}{dy} = -w$ or, $\frac{2}{w}\frac{dw}{dy} = -\frac{1}{2y}$. Then on integration we have $\ln|w|$

$= -\frac{1}{2}\ln|y| + C_1$ or, $|w| = \frac{C}{\sqrt{y}}$ where $C_2 = e^{C_1}$. Thus, $w = \frac{C_2}{\pm\sqrt{y}}$ or since $w = \frac{dy}{dx}$,

$\frac{dy}{dx} = \frac{C_2}{\pm\sqrt{y}}$. If we again separate variables we have $\sqrt{y}\frac{dy}{dx} = \pm C_2$ and on integrat-

ing with respect to x, we get $\frac{2}{3}y^{3/2} = \pm C_2 x + C_3$. If we let $C_4 = \pm C_2$, C_4 is an

arbitrary constant and we can write $\frac{2}{3}y^{3/2} = C_4 x + C_3$. We must now evaluate the

constants in such a way that $y'(0) = y(0) = 1$. Use of $y(0) = 1$ gives $\frac{2}{3} = C_3$.

Moreover, since $\frac{dy}{dx} = \frac{C_4}{\sqrt{y}}$ and $y = 1$ and $y' = 1$ when $x = 0$, we have $1 = \frac{C_4}{1}$ or,

$C_4 = 1$. Substitution then gives $\frac{2}{3}y^{3/2} = x + \frac{2}{3}$.

5. Since the dependent variable y is missing, we use the substitution $\frac{dy}{dx} = w$ and

$\frac{d^2y}{dx^2} = \frac{dw}{dx}$. Then we have $x^2\frac{dw}{dx} + x^2 = 2xw$. On division by x^2 we get $\frac{dw}{dx} + 1 = \frac{2w}{x}$

or, $\frac{dw}{dx} - \frac{2}{x}w = -1$, a first order linear equation. Since $p(x) = -\frac{2}{x}$, this equa-

tion has $\rho(x) = e^{-2\int\frac{dx}{x}} = e^{-2\ln x} = x^{-2}$ as an integrating factor.

Multiplication by the integrating factor gives $x^{-2} \frac{dw}{dx} - 2wx^{-3} = -x^{-2}$ or,

$\frac{dx^{-2}w}{dx} = -x^{-2}$. Then integration gives $x^{-2}w = x^{-1} + C_1$ or, $w = x^2(x^{-1} + C_1)$.

Since $w = \frac{dy}{dx}$, we have $\frac{dy}{dx} = x^2(x^{-1} + C_1) = x + C_1 x^2$. Integration yields

$y = \frac{1}{2} x^2 + \frac{C_1}{3} x^3 + C_2$. If we then let $C_3 = \frac{C_1}{3}$, we have $y = \frac{1}{2} x^2 + C_3 x^3 + C_2$

where C_2 and C_3 are arbitrary constants.

7. The dependent variable y is again missing. Thus, we use the substitution

$\frac{dy}{dx} = w$ and $\frac{d^2y}{dx^2} = \frac{dw}{dx}$. Then we have $x^2 \frac{dw}{dx} - 2xw = 2$. On writing this as

$\frac{dw}{dx} - \frac{2}{x} w = \frac{2}{x^2}$, we recognize the equation as a first order linear equation with

integrating factor $\rho(x) = e^{-2\int \frac{dx}{x}} = e^{-2 \ln x} = x^{-2}$. Multiplication by this

integrating factor yields $x^{-2} \frac{dw}{dx} - 2wx^{-3} = 2x^{-4}$ or, $\frac{d(x^{-2}w)}{dx} = 2x^{-4}$. Thus, on

integrating we get $x^{-2}w = -\frac{2}{3} x^{-3} + C_1$ or, $w = -\frac{2}{3} x^{-1} + C_1 x^2$. Since $w = \frac{dy}{dx}$,

we have $\frac{dy}{dx} = -\frac{2}{3} x^{-1} + C_1 x^2$. Integrating again gives $y = -\frac{2}{3} \ln |x| + \frac{C_1}{3} x^3$

$+ C_2$. If we then let $C_3 = C_1/3$, we can write the solution in the simpler form

$y = -\frac{2}{3} \ln |x| + C_3 x^3 + C_2$. We must now determine the constants C_2 and C_3 in

such a way that $y'(1) = y(1) = 0$. Since $y(1) = 0$, we must have

$0 = -\frac{2}{3} \ln 1 + C_3 + C_2$ or, (*) $C_2 + C_3 = 0$. Since $y' = -\frac{2}{3} x^{-1} + 3C_3 x^2$ and

$y'(1) = 0$, we must have $0 = -\frac{2}{3} + 3C_3$ or, $C_3 = 2/9$. Substitution into (*)

above gives $C_2 = -2/9$. Thus, the desired particular solution is

$y = -\frac{2}{3} \ln |x| + \frac{2}{9} x^3 - \frac{2}{9}$.

9. Since the independent variable x is missing, we use the substitution $\frac{dy}{dx} = w$

and $\frac{d^2y}{dx^2} = \frac{dw}{dx} = \frac{dw}{dy} \frac{dy}{dx} = w \frac{dw}{dy}$. Then the equation becomes $w \frac{dw}{dy} + y = 0$. If we

then separate variables and integrate, we have $\frac{1}{2} w^2 = -\frac{1}{2} y^2 + C_1$ or,

$(\frac{dy}{dx})^2 = -y^2 + 2C_1$. Since $y(0) = 0$ and $y'(0) = 1$, we must have $1 = 2C_1$ or,

$C_1 = 1/2$. Thus, $(\frac{dy}{dx})^2 = 1 - y^2$ or, $\frac{dy}{dx} = \sqrt{1 - y^2}$. (Note that since $y'(0) = 1$,

we use the positive square root.) If we write $\frac{1}{\sqrt{1 - y^2}} \frac{dy}{dx} = 1$, we may integrate to get arcsin $y = x + C_2$. Then since $y(0) = 0$, we have arcsin $0 = C_2$ or, $C_2 = 0$. Consequently, we have arcsin $y = x$ or, $y = \sin x$.

11. Since the dependent variable s is missing, we will use the substitution $\frac{ds}{dt} = w$ and $\frac{d^2s}{dt^2} = \frac{dw}{dt}$. On making these substitutions, we have $t \frac{dw}{dt} = w$ or, $\frac{1}{w} \frac{dw}{dt} = \frac{1}{t}$. Integration then gives $\ln |w| = \ln |t| + C_1$. If $t = 1$, $w = s' = 1$. Thus, we must have $\ln |1| = \ln |1| + C_1$ or, $C_1 = 0$. Consequently, $\ln |w| = \ln |t|$ or, $w = t$. Since $w = \frac{ds}{dt}$, we have $\frac{ds}{dt} = t$ or on integration $s = \frac{1}{2} t^2 + C_2$. Then using the condition $s(1) = 0$, we must have $0 = \frac{1}{2} + C_2$ or, $C_2 = -\frac{1}{2}$. Consequently, $s = \frac{1}{2} t^2 - \frac{1}{2}$.

13. Since the dependent variable y is missing, we will use the substitution $y' = w$ and $y'' = w'$. On making these substitutions, we have $xw' - w^2 = -4$ or, $\frac{1}{w^2 - 4} \frac{dw}{dx} = \frac{1}{x}$. Integration yields $\frac{1}{4} \ln \left|\frac{w - 2}{w + 2}\right| = \ln |x| + C$. Thus, $\ln \left|\frac{w - 2}{w + 2}\right| = \ln |x|^4 + 4C$, or $\frac{w - 2}{w + 2} = C'x^4$ where $C' = \pm e^{4C}$. On solving for w, we obtain $w = \frac{2C'x^4 + 2}{1 - C'x^4}$, or $\frac{dy}{dx} = \frac{2C'x^4 + 2}{1 - C'x^4}$. Thus, $y = 2 \int \frac{C'x^4 + 1}{1 - C'x^4} dx$. On dividing we have $y = -2 \int dx + 4 \int \frac{dx}{1 - C'x^4}$. In order to evaluate the last integral we can use partial fractions. Then we have

$$\int \frac{dx}{1 - C'x^4} = \frac{1}{4} \int \frac{dx}{1 - \sqrt[4]{C'}\, x} + \frac{1}{4} \int \frac{dx}{1 + \sqrt[4]{C'}\, x} + \frac{1}{2} \int \frac{dx}{1 + \sqrt[4]{C'}\, x^2}$$

so, $\int \frac{dx}{1 - C'x^4} = \frac{1}{4\sqrt[4]{C'}} \ln \left|\frac{1 + \sqrt[4]{C'}\, x}{1 - \sqrt[4]{C'}\, x}\right| + \frac{1}{2\sqrt[4]{C'}} \arctan \sqrt[4]{C'}\, x + C_2$.

Finally, then $y = -2x + \frac{1}{\sqrt[4]{C'}} \ln \left|\frac{1 + \sqrt[4]{C'}\, x}{1 + \sqrt[4]{C'}\, x}\right| + \frac{2}{\sqrt[4]{C'}} \arctan \sqrt[4]{C'}\, x + C_2$. If we let $\sqrt[4]{C} = C_1$, this result takes on the form

$$y = -2x + \frac{1}{C_1} \ln \left|\frac{1 + C_1 x}{1 - C_1 x}\right| + \frac{2}{C_1} \arctan C_1 x + C_2.$$

15. Since the dependent variable s is missing, we use the substitution $w = \frac{ds}{dt}$, $\frac{dw}{dt} = \frac{d^2s}{dt^2}$. On making this substitution, the equation becomes $t \frac{dw}{dt} - w = 2/t - \ln t$ or, $\frac{d}{dt} (tw) = 2/t - \ln t$. Then on integrating we have

$tw = 2 \ln |t| - t \ln |t| + t + C_1$ or, $w = 2 \dfrac{\ln |t|}{t} - \ln |t| + 1 + C_1/t$. Conse-

$\dfrac{ds}{dt} = 2 \dfrac{\ln |t|}{t} - \ln |t| + 1 + C_1/t$. Integration then gives

$s = (\ln |t|)^2 - t \ln |t| + t + t + C_1 \ln |t| + C_2$ or,

$s = (\ln |t|)^2 + (C_1 - t) \ln |t| + 2t + C_2$.

17. In Exercise 3 we found that $\dfrac{2}{3} y^{3/2} = C_4 x + C_3$ where C_3 and C_4 are arbitrary

constants. Then since $y(0) = 0$, we must have $C_3 = 0$. Since $\sqrt{y} \dfrac{dy}{dx} = C_4$ and

$y'(0) = 0$, we must have $C_4 = 0$. Consequently, the desired particular solution

is $y = 0$.

19. We let x denote the displacement from the equilibrium position in centimeters.
Then the spring exerts a force of 2x in the direction of decreasing x. Thus,

$F = -2x$. Substitution into $F = ma$ then gives $-2x = 50 \dfrac{d^2 x}{dt^2}$, since $m = 50$ kilo-

grams and $a = \dfrac{d^2 x}{dt^2}$. To solve the differential equation $25 \dfrac{d^2 x}{dt^2} = -x$ we will let

$w = \dfrac{dx}{dt}$, then $\dfrac{d^2 x}{dt^2} = \dfrac{dw}{dt} = \dfrac{dw}{dx} \dfrac{dx}{dt} = w \dfrac{dw}{dx}$. On substitution we have $25w \dfrac{dw}{dx} = -x$ or

on integration, $\dfrac{25}{2} w^2 = -\dfrac{x^2}{2} + C_1$. Thus, $\dfrac{25}{2} (\dfrac{dx}{dt})^2 = -\dfrac{x^2}{2} + C_1$. Since the body

is initially displaced 5 cm and released from rest, we must have $x = 5$ and

$\dfrac{dx}{dt} = 0$ when $t = 0$. Substitution of these conditions gives $0 = -\dfrac{25}{2} + C_1$ or,

$C_1 = \dfrac{25}{2}$. Consequently, $\dfrac{25}{2} (\dfrac{dx}{dt})^2 = -\dfrac{x^2}{2} + \dfrac{25}{2}$ or, $(\dfrac{dx}{dt})^2 = -\dfrac{x^2}{25} + 1$. Thus,

$\dfrac{1}{\sqrt{1 - (\frac{x}{5})^2}} \dfrac{dx}{dt} = \pm 1$. Integration then gives $5 \arcsin \dfrac{x}{5} = \pm t + C_2$ or,

(*) $x = 5 \sin (\pm t/5 + C_2/5)$. Since $x = 5$ when $t = 0$, we must have

$5 = 5 \sin C_2/5$ so we must have $\sin C_2/5 = 1$. Thus, we may select $C_2/5 = \pi/2$.
Substitution into (*) then gives $x = 5 \sin (\pm t/5 + \pi/2)$. Since
$\sin (\pi/2 - \alpha) = \sin (\pi/2 + \alpha)$, we can write this result as $x = 5 \sin (t/5 + \pi/2)$.

Exercises 13.5: Series Solutions (page 560)

1. We attempt to find a solution of the form

$y = \displaystyle\sum_{k=0}^{\infty} a_k x^k = a_0 + a_1 x + a_2 x^2 + \ldots + a_n x^n + \ldots$ In this case

$y' = \sum_{k=0}^{\infty} k a_k x^{k-1} = a_1 + 2a_2 x + \ldots + n a_n x^{n-1} + \ldots$ Then on substitution of

these results into the differential equation, we have

$a_1 + 2a_2 x + \ldots + n a_n x^{n-1} + \ldots + 3x^2 (a_0 + a_1 x + \ldots + a_n x^n + \ldots) = 1$ or,

$a_1 + 2a_2 x + \ldots + n a_n x^{n-1} + \ldots + 3a_0 x^2 + 3a_1 x^3 + \ldots + 3a_n x^{n+2} + \ldots = 1.$

Then on gathering like terms, we have $a_1 + 2a_2 x + (3a_3 + 3a_0) x^2 + (4a_4 + 3a_1)$

$x^3 + \ldots + ((n + 3) a_{n+3} + 3a_n) x^{n+2} + \ldots = 1.$ Then from the uniqueness of

power series expansions, we have $a_1 = 1$

$$2a_1 = 0$$

$$3a_3 + 3a_0 = 0$$

$$4a_4 + 3a_1 = 0$$

and, in general, $(n + 3) a_{n+3} + 3a_n = 0.$ In succession these equations yield

$$a_1 = 1$$

$$a_2 = 0$$

$$a_3 = -a_0$$

$$a_4 = -\frac{3}{4} a_1 = -\frac{3}{4}$$

and, in general, (*) $a_{n+3} = -\frac{3}{n + 3} a_n.$ Thus, continuing we have if $n = 2$,

$a_5 = -\frac{3}{5} a_2 = 0$; if $n = 3$, $a_6 = -\frac{1}{2} a_3 = \frac{a_0}{2}$; if $n = 4$, $a_7 = -\frac{3}{7} a_4 - \frac{3}{7} (-\frac{3}{4})$

$= \frac{9}{28}.$ The desired solution is then $y = a_0 + x - a_0 x^3 - \frac{3}{4} x^4 + \frac{a_0}{2} x^5 + \frac{9}{28} x^7$

$+ \ldots = a_0 (1 - x^3 + \frac{1}{2} x^5 - \ldots) + (x - \frac{3}{4} x^4 + \frac{9}{28} x^7 + \ldots)$ where a_0 is an

arbitrary constant. Further coefficients can be found by use of (*).

5. Again we seek a solution of the form

$y = \sum_{k=0}^{\infty} a_k x^k = a_0 + a_1 x + a_2 x^2 + \ldots + a_n x^n + \ldots$ Then,

$y' = \sum_{k=0}^{\infty} k a_k x^{k-1} = a_1 + 2a_2 x + \ldots + n a_n x^{n-1} + \ldots$ and

$y'' = \sum_{k=0}^{\infty} k(k - 1) a_k x^{k-2} = 2a_2 + 3 \cdot 2a_3 x + \ldots + n(n - 1) a_n x^{n-2} + \ldots$ Then

on substitution of these results into the differential equation we have

$(x^2 - 1)(2a_2 + 3 \cdot 2a_3 x + \ldots + n(n - 1) a_n x^{n-2} + \ldots) + 6x(a_1 + 2a_2 x + \ldots$

$+ na_n x^{n-1} + \ldots) + 3(a_0 + a_1 x + a_2 x^2 + \ldots + a_n x^n + \ldots) = 0$ or,

$2a_2 x^2 + 3 \cdot 2a_3 x^3 + \ldots + n(n - 1) a_n x^n + \ldots - 2a_2 - 3 \cdot 2a_3 x - \ldots$

$- n(n - 1) a_n x^{n-2} - \ldots + 6a_1 x + 6 \cdot 2a_2 x^2 + \ldots + 6na_n x^n + \ldots) + 3a_0 + 3a_1 x$

$+ 3a_2 x^2 + \ldots + 3a_n x^n + \ldots = 0$. On gathering like terms we have

$(-2a_2 + 3a_0) + (-6a_3 + 6a_1 + 3a_1) x + (2a_2 - 12a_4 + 12a_2 + 3a_2) x^2 + \ldots$

$+ (n(n - 1) a_n - (n + 2)(n + 1) a_{n+2} + 6na_n + 3a_n) x^n + \ldots = 0$ or,

$(-2a_2 + 3a_0) + (9a_1 - 6a_3) x + (17a_2 - 12a_4) x^2 + \ldots + ((n(n - 1) + 6n + 3) a_n$

$- (n + 2)(n + 1) a_{n+2}) x^n + \ldots = 0$. By the uniqueness of power series expan-

sions we have
$$-2a_2 + 3a_0 = 0$$
$$9a_1 - 6a_3 = 0$$
$$17a_2 - 12a_4 = 0$$

and, in general, $(n(n - 1) + 6n + 3) a_n - (n + 2)(n + 1) a_{n+2} = 0$. In succes-

sion these equations yield $a_2 = \frac{3}{2} a_0$,

$$a_3 = \frac{3}{2} a_1,$$

$$a_4 = \frac{17}{12} a_2 = \frac{17}{12} (\frac{3}{2}) a_0 = \frac{51}{24} a_0$$

and, in general, (*) $a_{n+2} = \frac{n(n - 1) + 6n + 3}{(n + 2)(n + 1)} a_{n+2}$. Consequently, the desired

solution is $y = a_0 + a_1 x + \frac{3}{2} a_0 x^2 + \frac{3}{2} a_1 x^3 + \ldots$ where a_0 and a_1 are arbitrary

constants. Further coefficients can be found by use of (*).

7. Let $y = \sum_{k=0}^{\infty} a_k x^k = a_0 + a_1 x + a_2 x^2 + \ldots + a_n x^n + \ldots$ Then,

$y' = \sum_{k=0}^{\infty} ka_k x^{k-1} = a_1 + 2a_2 x + \ldots + na_n x^{n-1} + \ldots$ and,

$y'' = \sum_{k=0}^{\infty} k(k - 1) a_k x^{k-2} = 2a_2 + 3 \cdot 2a_3 x + \ldots + n(n - 1) a_n x^{n-2} + \ldots$

Substitution of these results into the differential equation gives

$2a_2 + 3 \cdot 2a_3 x + \ldots + n(n - 1) a_n x^{n-2} + \ldots - a_1 x - 2a_2 x^2 - \ldots - na_n x^n$

$- \ldots - a_0 - a_1 x - a_2 x^2 - \ldots - a_n x^n - \ldots = 0$. On gathering like terms we

have $(2a_2 - a_0) + (3 \cdot 2a_3 - 2a_1) x + (4 \cdot 3a_4 - 3a_2) x^2 + \ldots$

$+ ((n + 2)(n + 1) a_{n+2} - na_n - a_n) x^n + \ldots = 0.$ By the uniqueness of power series expansions we have

$$2a_2 - a_0 = 0$$
$$3 \cdot 2a_3 - 2a_1 = 0$$
$$4 \cdot 3a_4 - 3a_2 = 0$$

and, in general, $(n + 2)(n + 1) a_{n+2} - (n + 1) a_n = 0.$ These equations yield

$$a_2 = a_0/2, \quad a_3 = \frac{a_1}{3},$$

$$a_4 = \frac{a_2}{4} = \frac{a_0}{8}$$

and, in general, (*) $a_{n+2} = (n + 1) a_n/(n + 2)(n + 1) = \frac{a_n}{n + 2}.$ Consequently, the desired solution is $y = a_0 + a_1 x + \frac{a_0}{2} x^2 + \frac{a_1}{3} x^3 + \ldots$ where a_0 and a_1 are arbitrary constants. Further coefficients can be found by use of (*).

11. If we let $y = \sum_{k=0}^{\infty} a_k(x - 1)^k = a_0 + a_1(x - 1) + a_2(x - 1)^2 + \ldots + a_n(x - 1)^n$

$+ \ldots,$ we have $y' = \sum_{k=0}^{\infty} ka_k(x - 1)^{k-1} = a_1 + 2a_2 (x - 1) + \ldots + na_n(x - 1)^{n-1}$

$+ \ldots$ Substitution into the differential then gives (*) $a_1 + 2a_2(x - 1) + \ldots$

$+ na_n(x - 1)^{n-1} + \ldots = 2x(a_0 + a_1(x - 1) + a_2(x - 1)^2 + \ldots + a_n(x - 1)^n$

$+ \ldots) + x^2.$ The computation is simpler if we use the facts that $x = 1 + (x - 1)$ and, $x^2 = 1 + 2(x - 1) + (x - 1)^2.$ Then we can write (*) as

$a_1 + 2a_2(x - 1) + \ldots + na_n(x - 1)^{n-1} + \ldots = 2a_0 + 2a_1(x - 1) + 2a_2(x - 1)^2$

$+ \ldots + 2a_n(x - 1)^n + \ldots + 2a_0(x - 1) + 2a_1(x - 1)^2 + 2a_2(x - 1)^3 + \ldots$

$+ 2a_n(x - 1)^{n+1} + \ldots + 1 + 2(x - 1) + (x - 1)^2.$ On gathering like terms we

have $a_1 + 2a_2(x - 1) + \ldots + na_n(x - 1)^{n-1} + \ldots = (2a_0 + 1) + (2a_1 + 2a_0 + 2)$

$(x - 1) + (2a_2 + 2a_1 + 1)(x - 1)^2 + (2a_3 + 2a_2)(x - 1)^3 + \ldots + (2a_n + 2a_{n-1})$

$(x - 1)^n + \ldots$ Then from the uniqueness of power series expansions we have

$$a_1 = 2a_0 + 1$$
$$2a_2 = 2a_1 + 2a_0 + 2$$
$$3a_3 = 2a_2 + 2a_1 + 1$$

and, in general, $na_n = 2a_{n-1} + 2a_{n-2}$ for $n \geq 3.$ Since $y = a_0 + a_1(x - 1)$

$+ a_2(x - 1)^2 + \ldots$ and we are to have $y = 0$ when $x = 1$, we must have $a_0 = 0$.

Thus, from the above equations we have

$$a_1 = 1$$
$$a_2 = 2$$
$$a_3 = 7/3$$
$$a_4 = \frac{1}{4}(2a_3 + 2a_2) = \frac{1}{4}(\frac{14}{3} + 4)$$
$$= 13/6$$

Thus, the desired solution is $y = (x - 1) + 2(x - 1)^2 + \frac{7}{3}(x - 1)^3 + \frac{13}{6}$

$(x - 1)^4 + \ldots$ Further coefficients may be found from the equation

$na_n = 2a_{n-1} + 2a_{n-2}$ for $n \geq 3$.

13. Since $y = \sum\limits_{k=0}^{\infty} a_k(x - 1)^k$, we have $y' = \sum\limits_{k=0}^{\infty} ka_k(x - 1)^{k-1} = a_1 + 2a_2(x - 1)$

$+ 3a_3(x - 1)^2 + \ldots$ and, $y'' = \sum\limits_{k=0}^{\infty} k(k - 1) a_k(x - 1)^{k-2} = 2a_2 + 3 \cdot 2a_3(x - 1)$

$+ 4 \cdot 3a_4(x - 2)^2 + \ldots$ On substitution into the differential equation we ob-

tain $2a_2 + 3 \cdot 2a_3(x - 1) + 4 \cdot 3a_3(x - 1)^2 + \ldots + n(n - 1) a_n(x - 1)^{n-2}$

$+ \ldots + a_1(x - 1) + 2a_2(x - 1)^2 + 3a_3(x - 1)^3 + \ldots + na_n(x - 1)^n + \ldots + a_0$

$+ a_1(x - 1) + a_2(x - 1)^2 + \ldots + a_n(x - 1)^n + \ldots = 0$. On gathering like terms

we have $2a_2 + a_0 + (3 \cdot 2a_3 + 2a_1)(x - 1) + (4 \cdot 3a_4 + 3a_2)(x - 1)^2 + \ldots$

$+ ((n + 2)(n + 1) a_{n+2} + (n + 1) a_n)(x - 1)^n + \ldots = 0$. Thus,

$$2a_2 + a_0 = 0$$
$$3 \cdot 2a_3 + 2a_1 = 0$$
$$4 \cdot 3a_4 + 3a_2 = 0$$

and, in general, $(n + 2)(n + 1) a_{n+2} + (n + 1) a_n = 0$. Solution of these equa-

tions then gives $a_2 = -a_0/2$

$$a_3 = -a_1/3$$
$$a_4 = -a_2/4 = a_0/8$$

and, in general, (*) $a_{n+2} = -a_n/(n + 2)$. Thus, the solution we seek is

$y = a_0 + a_1(x - 1) - \frac{a_0}{2}(x - 1)^2 - \frac{a_1}{3}(x - 1)^3 + \frac{a_0}{8}(x - 1)^4 + \ldots$ where a_0

and a_1 are arbitrary constants. Further coefficients may be found from (*).

1. $y' = \dfrac{x + y}{y}$, $(0,1)$ $0 \le x \le 1$. The interval $0 \le x \le 1$ is subdivided into three
equal parts by introduction of the points $x = .\overline{3}$, $x = .\overline{6}$. Then since the solution is to pass through the point $(0,1)$, it will have slope $y' = \dfrac{x + y}{y} = \dfrac{0 + 1}{1}$
$= 1$ at that point. Consequently, we take the line through $(0,1)$ with slope 1
as an approximate solution. This line has the equation $y = x + 1$ and thus
passes through the point $(.\overline{3},1.\overline{3})$. At this point the solution must have slope
$y' = \dfrac{x + y}{y} = \dfrac{.\overline{3} + 1.\overline{3}}{1.\overline{3}} = 1.25$. The equation of the line through the point
$(.\overline{3},1.\overline{3})$ with slope 1.25 is $y - 1.\overline{3} = 1.25(x - .\overline{3})$ or, $y = 1.25x + .91\overline{6}$. This
line passes through the point $(.\overline{6},1.75)$. At this point, that is, when $x = .\overline{6}$
and $y = 1.75$, the solution must have slope $\dfrac{dy}{dx} = \dfrac{x + y}{y} = \dfrac{.\overline{6} + 1.75}{1.75} \approx 1.381$.
Finally, the line through the point $(.\overline{6},1.75)$ with slope 1.381 has equation
$y - 1.75 = 1.381 (x - .\overline{6})$ or, $y = 1.381x + .829$. Thus, the desired polygonal
approximation is $y = x + 1$, $0 \le x \le 1/3$
$$y = 1.25x + .916, \quad 1/3 \le x \le 2/3$$
$$y = 1.381x + .829, \quad 2/3 \le x \le 1.$$
Since the equation $y' = \dfrac{x + 1}{y}$ is separable, we can find the precise solution
easily. Then since $y \dfrac{dy}{dx} = x + 1$, integration gives $\dfrac{1}{2} y^2 = \dfrac{1}{2} x^2 + x + C$. Since
the solution is to pass through the point $(0,1)$, we must have $C = \dfrac{1}{2}$. Thus, we
have $\dfrac{1}{2} y^2 = \dfrac{1}{2} x^2 + x + \dfrac{1}{2}$ or, $y = \sqrt{x^2 + 2x + 1} = |x + 1|$. (Note that the posi-
tive square root must be selected since $y > 0$ when $x = 0$.) Thus, the exact
solution passes through the points $(1/3,4/3)$, $(2/3,5/3)$, and $(1,2)$. The approximate solution passes through the points $(1/3,4/3)$, $(2/3,7/4)$, and $(1,2.210)$.

3. The interval $0 \le x \le 1$ is subdivided into three equal parts by the introduction
of the points $x = 1/3$ and $x = 2/3$. Then since the solution is to pass through
the point $(0,1)$, it will have slope $y' = 0$ at $(0,1)$. This line has equation
$y = 1$ and passes through the point $(1/3,1)$. At this point the solution has
slope $y' = (1/3)^2 = 1/9$. The equation of the line through $(1/3,1)$ with slope
$1/9$ is $y = x/9 + 26/27$. This line passes through the point $(2/3,28/27)$. At
this point the solution must have slope $y' = (2/3)^2/(28/27) = 3/7$. The equation of the line through $(2/3,28/27)$ with slope $3/7$ is $y = 3x/7 + 142/189$.
Thus, the desired polynomial approximation is

$$y = 1, \ 0 \le x \le 1/3,$$
$$y = x/9 + 26/27, \ 1/3 \le x \le 2/3,$$
$$y = 3x/7 + 142/189, \ 2/3 \le x \le 1.$$

5. The interval $0 \le x \le 0.3$ is subdivided into three equal parts by the introduction of the points $x = .1$ and $x = .2$. Then since the solution is to pass through the point $(0,0)$, it will have slope $y' = 1$ at $(0,0)$. The line through $(0,0)$ with slope 1 has equation $y = x$. This line passes through the point $(.1,.1)$. At this point the solution must have slope $y' = \sqrt{1 - (.1)^2} = \sqrt{.99}$. The equation of the line through $(.1,.1)$ with slope $\sqrt{.99}$ is $y = \sqrt{.99}x + (.1 - \sqrt{.99}/10)$ or, $y = \sqrt{.99}x + (.1 - \sqrt{.0099})$. This line passes through the point $(.2,.1 + \sqrt{.0099})$. At this point the solution must slope $y' = \sqrt{1 - (.1 + \sqrt{.0099}}^2 = \sqrt{.9801 - .2\sqrt{.0099}}$. The equation of the line through $(.2,1 + \sqrt{.0099})$ with slope $\sqrt{.9801 - .2\sqrt{.0099}}$ is

$$y = \sqrt{.9801 - .2\sqrt{.0099}} \ x + (1 + \sqrt{.0099} - .2\sqrt{.9801 - .2\sqrt{.0099}}) \text{ or,}$$

$$y = \sqrt{.9999 - \sqrt{.000396}} \ x + (1 + \sqrt{.0099} - \sqrt{.03920 - \sqrt{.0000006}}).$$

Thus, the desired polygonal approximation is

$$y = x, \ 0 \le x \le .1,$$

$$y = \sqrt{.99}x + (.1 - \sqrt{.0099}), \ .1 \le x \le .2,$$

$$y = \sqrt{.9999 - \sqrt{.000396}} \ x + (1 + \sqrt{.0099} - \sqrt{.03920 - \sqrt{.0000006}}), \ .2 \le x \le .3.$$

<u>Technique Review Exercises, Chapter 13</u> (page 564)

1. If we divide by x^2, this equation becomes $y' + \frac{2}{x} y = \frac{1}{x^2}$, a first order linear equation with $P(x) = \frac{2}{x}$. Thus, this equation has the integrating factor $\rho(x) = e^{\int P(x)dx} = e^{2\int \frac{1}{x} dx} = e^{2 \ln x} = x^2$. On multiplication by the integrating factor we have $x^2 y' + 2xy = 1$ or, $\frac{d(x^2 y)}{dx} = 1$. Integration then yields $x^2 y = x + C$ or, $y = x^{-2}(x + C)$. Since $y = 0$ when $x = 1$, we must have $0 = 1 + C$ or, $C = -1$. Substitution of $C = -1$ then gives the desired particular solution, $y = x^{-2}(x - 1)$.

2. We can separate variables as follows. $\frac{dx}{dt} = x^2 \sin t + 2x^2$ so if $x \ne 0$,

$\frac{1}{x^2} \frac{dx}{dt} = \sin t + 2$. Thus, $\int \frac{1}{x^2} \frac{dx}{dt} \ dt = \int (\sin t + 2) \ dt$ so $-x^{-1} = -\cos t + 2t$

$+ \ C$ or, $x = \dfrac{1}{\cos t - 2t - C}$ where C is an arbitrary constant. We have assumed above that $x \neq 0$. In fact, $x(t) = 0$ is a solution to the given equation.

3. Since the dependent variable y is missing, we will use the substitution

$y' = \frac{dy}{dx} = w$ and $y'' = \frac{dw}{dx}$. Then the equation becomes $\frac{dw}{dx} + \frac{1}{x} w = x$, a first order

linear equation with integrating factor $\rho(x) = e^{\int P(x) \, dx} = e^{\int \frac{1}{x} \, dx} = e^{\ln x} = x$.

Multiplication by the integrating factor gives $x \frac{dw}{dx} + w = x^2$ or, $\frac{d(xw)}{dx} = x^2$.

Then on integrating we obtain $xw = \frac{1}{3} x^3 + C_1$ or, $w = \frac{1}{3} x^2 + C_1 x^{-1}$. Since

$w = \frac{dy}{dx}$, we have $\frac{dy}{dx} = \frac{1}{3} x^2 + C_1 x^{-1}$. Integrating again then gives $y = f(x)$

$= \frac{1}{9} x^3 + C_1 \ln |x| + C_2$. Since we want $f(1) = 0$, we must have $0 = \frac{1}{9} + C_2$ or,

$C_2 = -1/9$. Since $f'(x) = \frac{1}{3} x^2 + C_1 x^{-1}$ and we want $f'(1) = 0$, we must have

$0 = \frac{1}{3} + C_1$ or, $C_1 = -1/3$. Consequently, the desired particular solution is

$y = \frac{1}{9} x^3 - \frac{1}{3} \ln |x| - \frac{1}{9}$.

4. Since the independent variable t is missing, we use the substitution $\frac{dx}{dt} = w$ and

$\frac{d^2 x}{dt^2} = \frac{dw}{dt} = \frac{dw}{dx} \frac{dx}{dt} = w \frac{dw}{dx}$. Then the equation becomes $2xw \frac{dw}{dx} + w^2 = 0$. We can

separate the variables as $\frac{2}{w} \frac{dw}{dx} = -\frac{1}{x}$. Then integration yields $2 \ln w = -\ln x$

$+ C_1$ or, $w^2 = C_2 x^{-1}$ where $C_2 = e^{C_1}$. On taking the square root of both sides,

we have $w = \pm \sqrt{C_2} \, x^{-1/2}$. Then if we let $C_3 = \pm \sqrt{C_2}$, we have $w = C_3 x^{-1/2}$. Since

$w = \frac{dx}{dt}$, we have $\frac{dx}{dt} = C_3 x^{-1/2}$ or, $x^{1/2} \frac{dx}{dt} = C_3$. Integration then gives

$\frac{2}{3} x^{3/2} = C_3 t + C_4$ or on letting $C_5 = \frac{3}{2} C_3$ and $C_6 = \frac{3}{2} C_4$, we have $x^{3/2} = C_5 t$

$+ C_6$ or, $x = (C_5 t + C_6)^{2/3}$ where C_5 and C_6 are arbitrary constants.

5. We attempt to find a solution of the form

$y = \sum_{k=0}^{\infty} a_k x^k = a_0 + a_1 x + a_2 x^2 + \ldots + a_n x^n + \ldots$ Then,

$y' = \sum_{k=0}^{\infty} k a_k x^{k-1} = a_1 + 2a_2 x + 3a_3 x^2 + \ldots + n a_n x^{n-1} + \ldots$ and

$y'' = \sum_{k=0}^{\infty} (k-1) k a_k x^{k-2} = 2a_2 + 2 \cdot 3a_3 x + 3 \cdot 4a_4 x^2 + \ldots + (n-1) n a_n x^{n-2}$

$+ \ldots$ Substitution into the differential equation then gives $(2a_2 + 2 \cdot 3a_3 x$

$+ 3 \cdot 4a_4 x^2 + \ldots + (n - 1) na_n x^{n-2} + \ldots) - 2x(a_1 + 2a_2 x + 3a_3 x^2 + \ldots$

$+ na_n x^{n-1} + \ldots) - 4(a_0 + a_1 x + a_2 x^2 + \ldots + a_n x^n + \ldots) = 0.$ On gathering

like terms, we have $(2a_2 - 4a_0) + (6a_3 - 2a_1 - 4a_1) x + (12a_4 - 4a_2 - 4a_2) x^2$

$+ \ldots + ((n + 1)(n + 2) a_{n+2} - 2na_n - 4a_n) x^n + \ldots = 0.$ Thus, for the unique-

ness of power series, we have $2a_2 - 4a_0 = 0$

$$6a_3 - 6a_1 = 0$$

$$12a_4 - 8a_2 = 0$$

and, in general, $(n + 1)(n + 2) a_{n+2} - (2n + 4) a_n = 0.$ These equations yield

$$a_2 = 2a_0$$

$$a_3 = a_1$$

$$a_4 = \frac{2}{3} a_2 = \frac{4}{3} a_0$$

and, in general, $a_n = \frac{(n + 1)(n + 2)}{2(n + 2)} a_{n+2} = \frac{n + 1}{2} a_{n+2}$ or,

(*) $a_{n+2} = \frac{2}{n + 1} a_n.$ Thus, the solution we seek is

$y = a_0 + a_1 x + 2a_0 x^2 + a_1 x^3 + \frac{4}{3} a_0 x^4 + \ldots$ where a_0 and a_1 are arbitrary con-

stants. Further coefficients may be computed from (*) above.

6. The solution passes through the point $(0,1)$ and has slope $\frac{dy}{dx} = x + y = 0 + 1$

= 1 at that point. Consequently, we take the line through $(0,1)$ with slope 1
as an approximate solution. Since this line has equation $y = x + 1$, it passes
through the point $(.1,1.1)$. At this point the solution must have slope

$\frac{dy}{dx} = x + y = .1 + 1.1 = 1.2.$ The straight line through the point $(.1,1.1)$

with slope 1.2 has equation $y - 1.1 = 1.2 (x - .1)$ or, $y = 1.2x + .98.$ This
straight line passes through the point $(.2,1.22)$. When $x = .2$ and $y = 1.22$,

the solution must have slope $\frac{dy}{dx} = x + y = .2 + 1.22 = 1.42.$ Finally, the equa-

tion of the line through the point $(.2,1.22)$ with slope 1.42 is $y - 1.22 = 1.42$

$(x - .2)$ or, $y = 1.42x + .936.$ The desired polygonal approximation is thus

$$y = x + 1 \text{ for } 0 \le x \le .1$$
$$y = 1.2x + .98 \text{ for } .1 \le x \le .2$$
$$y = 1.42x + .936 \text{ for } .2 \le x \le .3.$$

1. $y'' + 4y = 0$, $y = C_1 \cos 2x + C_2 \sin 2x$. Since $y = C_1 \cos 2x + C_2 \sin 2x$, we have $y' = -2C_1 \sin 2x + 2C_2 \cos 3x$ and, $y'' = -4C_1 \cos 2x - 4C_2 \sin 2x$. Thus, on substitution we have $y'' + 4y = -4C_1 \cos 2x - 4C_2 \sin 2x + 4(C_1 \cos 2x + C_2 \sin 2x) = 0$ as was to be shown.

3. Since $y = Ce^{-\arcsin x}$, we have $y' = -Ce^{-\arcsin x} \dfrac{1}{\sqrt{1 - x^2}}$. Then on substitution we find $\dfrac{1}{\sqrt{1 - x^2}} + \dfrac{y'}{y} = \dfrac{1}{\sqrt{1 - x^2}} - Ce^{-\arcsin x} \dfrac{1}{\sqrt{1 - x^2}} \dfrac{1}{Ce^{-\arcsin x}} = 0$ as was to be shown.

5. Since $y = C_1 e^{3x} + C_2 e^{-5x}$, we have $y' = 3C_1 e^{3x} - 5C_2 e^{-5x}$ and $y'' = 9C_1 e^{3x} + 25C_2 e^{-5x}$. Thus, $y'' + 2y' - 15y = 9C_1 e^{3x} + 25C_2 e^{-5x} + 6C_1 e^{3x} - 10C_2 e^{-5x} - 15C_1 e^{3x} - 15C_2 e^{-5x} = 0$ as was to be shown.

7. In Exercise 4 we found that any expression of the form $y = 1 + C(x - 1)$ is a solution to the given differential equation. Since we require $y = 0$ when $x = 0$, we must have $0 = 1 + C(0 - 1)$ or, $C = 1$. Thus, the desired particular solution is $y = x$.

9. In Exercise 4 we found that any expression of the form $y = 1 + C(x - 1)$ is a solution to the given differential equation. Since we require $y = 1$ when $x = 2$, we must have $1 = 1 + C(2 - 1)$ or, $C = 0$. Thus, the desired particular solution is $y = 1$.

11. Here $\dfrac{1}{y - 1} \dfrac{dy}{dx} = x - 1$ so on integrating both sides with respect to x we have $\int \dfrac{1}{y - 1} \dfrac{dy}{dx}\, dx = \int (x - 1)\, dx$ or, $\ln |y - 1| = \dfrac{x^2}{2} - x + C$. Thus, $|y - 1| = e^{\frac{1}{2} x^2 - x + C}$ or, $y - 1 = \pm e^{\frac{1}{2} x^2 - x + C}$. Thus, $y = \pm e^C e^{\frac{1}{2}x^2 - x} + 1$. On letting $C' = \pm e^C$, we have $y = C' e^{\frac{1}{2} x^2 - x} + 1$.

13. Since $\dfrac{dy}{dx} - y = x^2 y$, we have $\dfrac{dy}{dx} = y(x^2 + 1)$ or, $\dfrac{1}{y} \dfrac{dy}{dx} = (x^2 + 1)$. Then on integrating both sides with respect to x we obtain $\ln |y| = \dfrac{1}{3} x^3 + x + C$ or, $y = \pm e^{x^3/3 + x + C} = \pm e^C e^{x^3/3 + x}$. On letting $C' = \pm e^C$, we have $y = C' e^{x^3/3 + x}$

15. Separation of variables gives $\dfrac{1}{1 - x^2} \dfrac{dx}{dt} = t$. Thus,

$\int \frac{1}{1 - x^2} \frac{dx}{dt} dt = \int t \, dt$ or, $\frac{1}{2} \ln \left| \frac{x + 1}{x - 1} \right| = \frac{1}{2} t^2 + C$. Thus,

$\frac{x + 1}{x - 1} = \pm e^{t^2 + 2C}$. On solving for x, we have $x = -\frac{\pm e^{t^2 + 2C} + 1}{1 - \pm e^{t^2 + 2C}}$. Finally, if we

let $C' = \pm e^{2C}$ the solution takes on the simpler form $x = -\frac{C'e^{t^2} + 1}{1 - C'e^{t^2}}$

$= \frac{C'e^{t^2} + 1}{c'e^{t^2} - 1}$. We are to find that particular solution with the property that

$x = 0$ when $t = 0$. Thus, we must have $0 = -\frac{C' + 1}{1 - C'}$ so $C' = -1$. The desired

particular solution is thus $x = \frac{e^{t^2} - 1}{1 + e^{t^2}}$.

19. This equation is of the form $y' + P(x)y = Q(x)$ and so is linear. Since $P(x)$

$= \frac{3}{x}$, $\rho(x) = e^{\int P(x) \, dx} = e^{3 \int dx/x} = e^{3 \ln x} = x^3$ is an integrating factor. On

multiplication by $\rho(x) = x^3$ we have $x^3 y' + 3yx^2 = x^6 + 3x^4$ or, $\frac{d(x^3 y)}{dx} = x^6$

$+ 3x^4$. Integration then gives $x^3 y = \frac{1}{7} x^7 + \frac{3}{5} x^5 + C$ or,

$y = x^{-3} (\frac{1}{7} x^7 + \frac{3}{5} x^5 + C)$.

21. This equation is of the form $\frac{ds}{dt} + P(t) s = Q(t)$ and so it is linear. An inte-

grating factor is then given by $\rho(t) = e^{\int P(t) \, dt} = e^{\int \tan t \, dt} = e^{\ln (\sec t)}$

$= \sec t$. Multiplication by the integrating factor yields $\sec t \frac{ds}{dt} + s \tan t$

$\sec t = \sec^2 t$ or, $\frac{d}{dt} (s \sec t) = \sec^2 t$. On integrating we obtain $s \sec t$

$= \tan t + C$ or, $s = \cos t (\tan t + C) = \sin t + C \cos t$.

23. Division by t parts puts the differential equation in the form $\frac{ds}{dt} + 3 \frac{s}{t^3} = t^{-3}$,

a linear differential equation with $P(t) = 3/t^3$. Thus, $\rho(t) = e^{\int P(t) \, dt}$

$= e^{3 \int dt/t^3} = e^{-\frac{3}{2} t^{-2}}$ is an integrating factor. Multiplication by $\rho(t) = e^{-\frac{3}{2} t^{-2}}$

gives $e^{-\frac{3}{2} t^{-2}} \frac{ds}{dt} + 3 \frac{s}{t^3} e^{-\frac{3}{2} t^{-2}} = t^{-3} e^{-\frac{3}{2} t^{-2}}$ or, $\frac{d}{dt} (se^{-\frac{3}{2} t^{-2}}) = t^{-3} e^{-\frac{3}{2} t^{-2}}$.

On integrating we obtain $se^{-\frac{3}{2} t^{-2}} = \frac{1}{3} e^{-\frac{3}{2} t^{-2}} + C$ or, $s = \frac{1}{3} + Ce^{\frac{3}{2} t^{-2}}$.

25. If we let $u = y^{1/2}$, then $y = u^2$ and so $\frac{dy}{dx} = 2u \frac{du}{dx}$. Substitution then gives

$2u \frac{du}{dx} - u^2 = xu$. Division by 2u then puts the equation into the form of a linear equation. That is, $\frac{du}{dx} - \frac{u}{2} = \frac{x}{2}$. An integrating factor is given by

$\rho(x) = e^{-\frac{1}{2}\int dx} = e^{-x/2}$. Multiplication by the integrating factor gives

$e^{-x/2} \frac{du}{dx} - \frac{u}{2} e^{-x/2} = \frac{x}{2} e^{-x/2}$ or, $\frac{d}{dx} (ue^{-x/2}) = \frac{x}{2} e^{-x/2}$. Integration (note that

integration by parts may be used to evaluate the integral on the right) then

gives $ue^{-x/2} = -e^{-x/2} (x + 2) + C$ or, $y = -(x + 2) + Ce^{x/2}$. Then since

$u = y^{1/2}$, $y^{1/2} = (-x - 2 + Ce^{x/2})$ or, $y = (-x - 2 + Ce^{x/2})^2$.

27. $\frac{d^2y}{dx^2} + 2 \frac{dy}{dx} = 0$, $y'(0) = y(0) = 1$. Since both x and y are missing from this

equation, we may use either method to solve the equation. If we let $y' = w$

then $y'' = w'$ so the equation becomes $\frac{dw}{dx} + 2w = 0$, a separable differential

equation. On separating the variables and integrating we have $w = Ce^{-2x}$.

Thus, $\frac{dy}{dx} = Ce^{-2x}$ or, $y = -\frac{1}{2} Ce^{-2x} + C_2$. On letting $C_1 = -\frac{1}{2} C$, the solution

assumes the simpler form $y = C_1e^{-2x} + C_2$. Since we want $y'(0) = y(0) = 1$, we

must select C_1 and C_2 so that $C_1 + C_2 = 1$ and $-2C_1 = 0$. Solution of these

equations gives $C_1 = 0$ and $C_2 = 1$. Thus, the desired particular solution is

$y = 1$.

29. Since the independent variable x is missing, we make the substitutions $\frac{dy}{dx} = w$

$\frac{d^2y}{dx^2} = w \frac{dw}{dy}$. Then the equation becomes $yw = \frac{dw}{dy} = 1 + w^2$. On separating vari-

ables, we have $\frac{w}{1 + w^2} \frac{dw}{dy} = \frac{1}{y}$. Thus, $\frac{1}{2} \ln (1 + w^2) = \ln y + C_1'$ or,

$w^2 = C_1y^2 - 1$, where $C_1 = e^{2C_1'}$. Thus, since $w = \frac{dy}{dx}$, $\frac{dy}{dx} = \pm\sqrt{C_1y^2 - 1}$, or,

$\frac{1}{\sqrt{C_1y^2 - 1}} \frac{dy}{dx} = \pm 1$. On using the trigonometric substitution $y = \frac{\sec \theta}{\sqrt{C_1}}$ to evalu-

ate the integral we obtain $\frac{1}{\sqrt{C_1}} \ln |\sqrt{C_1} y + \sqrt{C_1y^2 - 1}| = \pm x + C_2$.

31. Since the dependent variable y is missing, we use the substitutions $\frac{dy}{dx} = w$ and

$\frac{d^2y}{dx^2} = \frac{dw}{dx}$. Then the equation becomes $\frac{dw}{dx} - x^3 = -\frac{w}{x}$ or, $\frac{dw}{dx} + \frac{w}{x} = x^3$, a first

order linear differential equation with integrating factor $\rho(x) = e^{\int \frac{dx}{x}} = e^{\ln x}$

$= x$. Multiplication by this integrating factor gives $x \frac{dw}{dx} + w = x^4$ or,

$\frac{d(xw)}{dx} = x^4$. Then integration gives $xw = \frac{1}{5} x^5 + C_1$ or, $w = \frac{1}{5} x^4 + \frac{C_1}{x}$. Since

$w = \frac{dy}{dx}$, we have $\frac{dy}{dx} = \frac{1}{5} x^4 + \frac{C_1}{x}$. Thus, $y = \frac{1}{25} x^5 + C_1 \ln |x| + C_2$.

33. Since the dependent variable s is missing, we use the substitution $\frac{ds}{dt} = w$ and

$\frac{d^2 s}{dt^2} = \frac{dw}{dt}$. Then the equation becomes $t^2 \frac{dw}{dt} = w^2$ or, $w^{-2} \frac{dw}{dt} = t^{-2}$. Integration

then yields $-w^{-1} = -t^{-1} + C_1$ or, $w = \frac{1}{t^{-1} - C_1} = \frac{t}{1 - tC_1}$. Since $w = \frac{ds}{dt}$, we have

$\frac{ds}{dt} = \frac{t}{1 - tC_1} = -\frac{1}{C_1} (1 + \frac{1}{tC_1 - 1})$ or, on integration $s = -\frac{1}{C_1} (t + \frac{1}{C_1} \ln$

$|tC_1 - 1|) + C_2$.

35. If we let $\sum\limits_{k=0}^{\infty} a_k x^k = a_0 + a_1 x + a_2 x^2 + \ldots + a_n x^n + \ldots$, we have

$y' = \sum\limits_{k=0}^{\infty} k a_k x^{k-1} = a_1 + 2a_2 x + \ldots + n a_n x^{n-1} + \ldots$ Then on substitution of

these results into the differential equation we have $a_1 + 2a_2 x + \ldots + n a_n x^{n-1}$

$+ \ldots - x(a_0 + a_1 x + \ldots + a_n x^n + \ldots) = x$ or, $a_1 + 2a_2 x + \ldots + n a_n x^{n-1} + \ldots$

$- a_0 x - a_1 x^2 - \ldots - a_n x^{n+1} - \ldots = x$. Then on gathering like terms, we have

$a_1 + (2a_2 - a_0) x + (3a_3 - a_1) x^2 + \ldots + ((n + 2) a_{n+2} - a_n) x^{n+1} + \ldots = x$.

Then from the uniqueness of the power series expansion, we must have

$$a_1 = 0$$

$$2a_2 - a_0 = 1$$

$$3a_3 - a_1 = 0$$

and, in general, $(n + 2) a_{n+2} - a_n = 0$, $n > 0$. In succession these equations

yield
$$a_1 = 0$$

$$a_2 = (1 + a_0)/2$$

$$a_3 = a_1/3 = 0$$

and, in general, (*) $a_{n+2} = a_n/(n + 2)$. Thus, continuing we have if $n = 2$,

$a_4 = a_2/4 = (1 + a_0)/8$; if $n = 3$, $a_5 = a_3/5 = 0$. Thus, the desired series

solution is $y = a_0 + \frac{1 + a_0}{2} + \frac{1 + a_0}{8} x^4 + \ldots$, where a_0 is an arbitrary con-

stant. Further coefficients may be found by use of (*). Note that if we let

$1 + a_0 = C$ the solution can be written as $y = C(1 + \frac{x^2}{2} + \frac{x^4}{8} + \ldots) - 1$.

37. We seek a solution of the form $y = \sum\limits_{k=0}^{\infty} a_k x^k = a_0 + a_1 x + a_2 x^2 + \ldots + a_n x^n$

$+ \ldots$ Thus, $y' = \sum\limits_{k=0}^{\infty} k a_k x^{k-1} = a_1 + 2a_2 x + \ldots + n a_n x^{n-1} + \ldots$ Substitution

then yields $2(a_1 + 2a_2 x + \ldots + n a_n x^{n-1} + \ldots) + (x^2 - 1)(a_0 + a_1 x + a_2 x^2$

$+ \ldots + a_n x^n + \ldots) = x$ or, $2a_1 + 4a_2 x + \ldots + 2n a_n x^{n-1} + a_0 x^2 + a_1 x^3 + a_2 x^4$

$+ \ldots + a_n x^{n+2} + \ldots - a_0 - a_1 x - a_2 x^2 - \ldots - a_n x^n - \ldots = x$. On collecting

like terms we have $(2a_1 - a_0) + (4a_2 - a_1) x + (6a_3 + a_0 - a_2) x^2 + \ldots$

$+ (2(n + 3) a_{n+3} + a_n - a_{n+2}) x^{n+2} + \ldots = x$. Thus,

$$2a_1 - a_0 = 0$$

$$4a_2 - a_1 = 1$$

$$6a_3 + a_0 - a_2 = 0$$

and, in general, $2(n + 3) a_{n+3} + a_n - a_{n+2} = 0$. Solution of these equations

gives $\quad\quad\quad\quad\quad a_1 = a_0/2,$

$$a_2 = (1 + a_1)/4 = (2 + a_0)/8$$

$$a_3 = (a_2 - a_0)/6 = (2 - 7a_0)/48$$

and, in general, (*) $a_{n+3} = (a_{n+2} - a_n)/2(n + 3)$. Thus, the desired solution

is $y = a_0 + \frac{a_0}{2} x + \frac{(2 + a_0)}{8} x^2 + \frac{2 - 7a_0}{48} x^3 + \ldots$ where a_0 is an arbitrary con-

stant. Further coefficients are obtained by use of (*).

39. Let $y = \sum\limits_{k=0}^{\infty} a_k x^k = a_0 + a_1 x + a_2 x^2 + \ldots + a_n x^n + \ldots$ Then,

$y' = \sum\limits_{k=0}^{\infty} k a_k x^{k-1} = a_1 + 2a_2 x + \ldots + n a_n x^{n-1} + \ldots$ and

$y'' = \sum\limits_{k=0}^{\infty} k(k - 1) a_k x^{k-2} = 2a_2 + 6a_3 x + \ldots + n(n - 1) a_n x^{n-2} + \ldots$ Substitu-

tion then gives $2a_2 + 6a_3 x + \ldots + n(n - 1) a_n x^{n-2} + \ldots + a_1 x + 2a_2 x^2 + \ldots$

$+ n a_n x^n + \ldots = 1$. On gathering like terms we have $2a_2 + (6a_3 + a_1) x + \ldots$

$+ ((n + 2)(n + 1) a_{n+2} + n a_n) x^n + \ldots = 1$. Thus, $2a_2 = 1$

$$6a_3 + a_1 = 0$$

and, in general, $(n + 2)(n + 1) a_{n+2} + na_n = 0$. Solution of these equations gives

$$a_2 = 1/2$$
$$a_3 = -a_1/6$$

and, in general, (*) $a_{n+2} = - \dfrac{n}{(n + 2)(n + 1)} a_n$. The solution we seek is then

$y = a_0 + a_1 x + \dfrac{1}{2} x^2 - \dfrac{a_1}{6} x^3 + \ldots$ Further coefficients can be found by use of (*).

41. Let $y = \displaystyle\sum_{k=0}^{\infty} a_k x^k = a_0 + a_1 x + a_2 x^2 + \ldots + a_n x^n + \ldots$ Then,

$$y' = \sum_{k=0}^{\infty} k a_k x^{k-1} = a_1 + 2a_2 x + 3a_3 x^2 + \ldots + n a_n x^{n-1} + \ldots \text{ and}$$

$y'' = \displaystyle\sum_{k=0}^{\infty} k(k - 1) a_k x^{k-1} = 2a_2 + 6a_3 x + \ldots + n(n - 1) a_n x^{n-2} + \ldots$ Substitution then gives $(1 - x)(2a_2 + 6a_3 x + \ldots + n(n - 1) a_n x^{n-2} + \ldots)$

$+ x(a_0 + a_1 x + a_2 x^2 + \ldots + a_n x^n + \ldots) - x = 0$ or, $2a_2 + 6a_3 x + \ldots$

$+ n(n - 1) a_n x^{n-2} + \ldots - 2a_2 x - 6a_3 x^2 - \ldots - n(n - 1) a_n x^{n-1} - \ldots$

$+ a_0 x + a_1 x^2 + a_2 x^3 + \ldots + a_n x^{n+1} - x = 0$. On gathering like terms we obtain

$2a_2 + (6a_3 - 2a_2 + a_0 - 1) x + (12a_4 - 6a_3 + a_1) x^2 + \ldots + ((n + 3)(n + 2)$

$a_{n+3} - (n + a)(n + 1) a_{n+2} + a_n) x^{n+1} + \ldots = 0$. Thus,

$$2a_2 = 0$$
$$6a_3 - 2a_2 + a_0 - 1 = 0$$
$$12a_4 - 6a_3 + a_1 = 0$$

and, in general, $(n + 3)(n + 2) a_{n+3} - (n + 2)(n + 1) a_{n+2} + a_n = 0$. Solution of these equations yields

$$a_2 = 0$$
$$a_3 = (2a_2 - a_0 + 1)/6 = (1 - a_0)/6$$
$$a_4 = (6a_3 - a_1)/12 = (1 - a_0 - a_1)/12$$

and, in general, (*) $a_{n+3} = ((n + 2)(n + 1) a_{n+2} - a_n)/(n + 3)(n + 2)$. Thus, the solution we seek is $y = a_0 + a_1 x + \dfrac{1 - a_0}{6} x^3 + \dfrac{1 - a_0 - a_1}{12} x^4 + \ldots$ where a_0 and a_1 are arbitrary constants. Further coefficients may be found by use of (*).

43. If we let $y = \sum\limits_{k=0}^{\infty} a_k (x - 1)^k = a_0 + a_1 (x - 1) + \ldots + a_n (x - 1)^n + \ldots$ then

$y' = \sum\limits_{k=0}^{\infty} k a_k (x - 1)^{k-1} = a_1 + 2a_2 (x - 1) + \ldots + n a_n (x - 1)^{n-1} + \ldots$ In

order to facilitate solution, we will write the differential equation as

$y' - (x - 1) y - y = (x - 1) + 1$. Then on substitution we obtain

$a_1 + 2a_2 (x - 1) + \ldots + n a_n (x - 1)^{n-1} + \ldots - a_0 (x - 1) - a_1 (x - 1)^2 - \ldots$

$- a_n (x - 1)^{n+1} - \ldots - a_0 - a_1 (x - 1) - \ldots - a_n (n - 1)^n - \ldots = (x - 1) + 1$.

On gathering like terms on the left hand side we have $(a_1 - a_0) + (2a_2 - a_0$

$- a_1)(x - 1) + (3a_3 - a_1 - a_2)(x - 1)^2 + \ldots + ((n + 2) a_{n+2} - a_n - a_{n+1})$

$(x - 1)^{n+1} + \ldots = (x - 1) + 1$. Then since power series expansions are unique,

we must have
$$a_1 - a_0 = 1$$
$$2a_2 - a_0 - a_1 = 1$$
$$3a_3 - a_1 - a_2 = 0$$

and, in general, $(n + 2) a_{n+2} - a_n - a_{n+1} = 0$. Solution of these equations

then yields
$$a_1 = 1 + a_0$$
$$a_2 = (1 + a_0 + a_1)/2 = 1 + a_0$$
$$a_3 = (a_1 + a_2)/3 = 2(1 + a_0)/3$$

and, in general, (*) $a_{n+2} = (a_n + a_{n+1})/(n + 2)$. Thus, the solution we seek is

$y = a_0 + (1 + a_0) x + (1 + a_0) x^2 + \frac{2}{3} (1 + a_0) x^3 + \ldots$ where a_0 is an

arbitrary constant. Further coefficients can be found by use of (*).

45. The interval $0 \leq x \leq \pi/2$ is subdivided into three parts by introduction of the
points $x = \pi/6$ and $x = \pi/3$. Since the solution is to pass through the point
$(0,1)$, it will have slope $y' = \sin (0) = 0$ at that point. Consequently, we
take the line with equation $y = 1$ as an approximate solution. This line passes
through the point $(\pi/6,1)$. At this point the solution must have slope

$y' = \sin \pi/6 = \frac{1}{2}$. The equation of the line through $(\pi/6,1)$ with slope 1/2 is

$y - 1 = \frac{1}{2} (x - \pi/6)$ or, $y = \frac{1}{2} x + 1 - \pi/12$. This line passes through the point

$(\pi/3, \pi/12 + 1)$ and has slope $y' = \sin (\frac{\pi/3}{\pi/12 + 1}) = \sin (\frac{4\pi}{\pi + 12})$ there. The

equation of the line through the point $(\pi/3, \pi/12 + 1)$ with slope $\sin (\frac{4\pi}{\pi + 12})$

is $y - (\pi/12 + 1) = \sin (\frac{4\pi}{\pi + 12})(x - \pi/3)$. Thus, the desired polygonal approximation to the solution is

$$y = 1 \text{ for } 0 \leq x \leq \pi/6$$

$$y = \frac{1}{2} x + 1 - \pi/12 \text{ for } \pi/6 \leq x \leq \pi/3$$

$$y = \sin \left(\frac{4\pi}{\pi + 12}\right)(x - \pi/3) + \pi/12 + 1 \text{ for } \pi/3 \leq x \leq \pi/2.$$

CHAPTER 14: VECTORS AND 3-SPACE

Exercises 14.1: The Three Dimensional Coordinate System (page 571)

17. Since x and y must both be 0 and z is unrestricted, this is the z-axis.

21. Since x - 1, y + 2, and z must all be 0, this is just the point $(1, -2, 0)$.

23. This is equivalent to $y = \pm 2$, so it is two planes parallel to the xz-plane, each being 2 units away from the xz-plane.

Exercises 14.2: Distance (page 573)

1. $\sqrt{(-6 + 4)^2 + (3 - 5)^2 + (2 + 1)^2} = \sqrt{4 + 4 + 9} = \sqrt{17}$

3. $\sqrt{(-2 - 0)^2 + (5 + 2)^2 + (1 - 3)^2} = \sqrt{4 + 49 + 4} = \sqrt{57}$

7. $r^2 = (-2 + 4)^2 + (5 - 2)^2 + (1 - 3)^2 = 4 + 9 + 4 = 17$ so an equation of the sphere is $(x + 4)^2 + (y - 2)^2 + (z - 3)^2 = 17$.

9. The center is at the midpoint $\left(\frac{-2 + 4}{2}, \frac{3 + 1}{2}, \frac{5 + 3}{2}\right)$ of the diameter. Thus, the center is at $(1, 2, 4)$. Thus, $r^2 = (1 + 2)^2 + (2 - 3)^2 + (4 - 5)^2$ $= 9 + 1 + 1 = 11$ and an equation of the sphere is $(x - 1)^2 + (y - 2)^2 + (z - 4)^2 = 11$.

13. We found that an equation of the sphere is $(x + 4)^2 + (y - 2)^2 + (z - 3)^2 = 17$. So the set of all points outside this sphere has inequality $(x + 4)^2 + (y - 2)^2 + (z - 3)^2 > 17$.

15. We are to find inequalities for the set of points inside the sphere $(x - 1)^2 + (y + 2)^2 + (z - 3)^2 = 49$ and outside the sphere $(x + 4)^2 + (y - 2)^2 + (z - 3)^2 = 17$. The radii of these spheres are 7 and $\sqrt{17}$ and the distance between their centers is $\sqrt{(1 + 4)^2 + (-2 - 2)^2 + (3 - 3)^2} = \sqrt{25 + 16 + 0} = 6$ so the spheres overlap. The set of points in question is described by the two inequalities: $(x - 1)^2 + (y + 2)^2 + (z - 3)^2 < 49$ and $(x + 4)^2 + (y - 2)^2 + (z - 3)^2 > 17$.

17. The distance between a point (x,y,z) and the origin is $\sqrt{x^2 + y^2 + z^2}$. The distance between (x,y,z) and the plane y = 4 is $|y - 4|$. So an equation of the set is $\sqrt{x^2 + y^2 + z^2} = |y - 4|$, or $x^2 + z^2 = 16 - 8y$.

19. The distances from a point (x,y,z) to these points are

$\sqrt{(x + 1)^2 + (y - 2)^2 + (z - 3)^2}$ and $\sqrt{(x + 2)^2 + (y + 3)^2 + (z - 5)^2}$. The distances are equal if and only if their squares are equal so an equation of the set is $(x + 1)^2 + (y - 2)^2 + (z - 3)^2 = (x + 2)^2 + (y + 3)^2 + (z - 5)^2$. Simplifying we obtain 2x + 10y - 4z = 24 or, x + 5y - 2z = 12.

Exercises 14.3: Sketching Graphs (page 580)

9. A plane of the form z = t intersects the graph in $x^2 + y^2 = z^2 - 1$. Thus, there is no intersection of such a plane for -1 < z < 1 and a circle if z < -1 z > 1. In the xz-plane for the yz-plane we get the hyperbolas $z^2 - x^2 = 1$ and $z^2 - y^2 = 1$. The graph is indicated in the answer section of the text.

13. In the xy-plane we get the parabola $x = 4y^2$. In the xz-plane we get the parabola $z^2 = x$ and a plane x = r, r > 0 produces an ellipse $4y^2 + z^2 = r$, but no graph of r < 0. The graph is shown in the answer section of the text.

19. This is a "sphere of radius zero" centered at the point (3,-2,4). That is, it is the point (3,-2,4).

23. This is a cylinder whose generating line moves parallel to the y-axis and around the circle $x^2 + z^2 = 16$ in the xz-plane. The graph is shown in the answer section of the text.

25. This is a cylinder whose generating line moves parallel to the z-axis and around the parabola $y = x^2$ in the xy-plane. The graph is shown in the text.

27. This is a cylinder whose generating line is parallel to the z-axis and moves around the ellipse $4x^2 + y^2 = 16$ in the xy-plane. The graph is shown in the text.

29. This is a plane generated by a line moving parallel to the z-axis along the line y = 2x in the xy-plane. The graph is shown in the text.

33. This is a cylinder whose generating line moves parallel to the z-axis along the graph of $y^2 - x^2 = 0$ in the xy-plane. That is, it moves along the lines y = ±x in the xy-plane. We get two intersecting planes. The graph is shown in the text.

7. $6[(1,-2) + (4,5)] = 6(5,3) = (30,18)$

9. $3[(1,-2) - (4,5)] = 3(-3,-7) = (-9,-21)$

11. $(2 + 3)(-4,2,1) = 5(-4,2,1) = (-20,10,5)$

13. $X = (2,-3,4) - (5,1,3) = (-3,-4,1)$

15. $2X = (4,2) - (1,5) = (3,-3)$ so $X = (3/2,-3/2)$

17. $x = 6$, $2y = 10$ so $x = 6$, $y = 5$.

19. $x + 2y = 5$ and $7x + y = 4$ so $x = 3/13$ and $y = 31/13$.

21. $2x + 5 = 8$ and $7 = y - 1$ so $y = 8$, $x = 3/2$.

23. $2x = 6$, $x = 3$. Also $xy = 7$ so $3y = 7$, $y = 7/3$.

25. $\sqrt{4^2 + 6^2} = 2\sqrt{13}$ 27. $\sqrt{1^2 + 2^2 + (-2)^2} = 3$

29. $\sqrt{4^2 + 0^2 + 2^2 + 1^2 + (-3)^2} = \sqrt{30}$

33. $\dfrac{1}{\sqrt{9 + 36 + 36}} (3,-6,6) = \dfrac{1}{9} (3,-6,6) = (1/3,-2/3,2/3)$

35. $\dfrac{1}{\sqrt{36 + 16 + 144}} (6,4,-12) = \dfrac{1}{14} (6,4,-12) = (3/7,2/7,-6/7)$

37. $A - B = (-2,1,-3)$ and $B - C = (6,-3,9)$. Thus, $A - B = -3(B - C)$, so $A - B$ and $B - C$ are parallel. Therefore, the points lie on the same line.

39. $A - B = (-6,21,-6)$ and $B - C = (14,-49,14)$. Thus, $-\dfrac{3}{7} (B - C) = (-6,21,-6)$

    $= A - B$, so $A - B$ and $B - C$ are parallel. Therefore, the points lie on the same side.

41. $A - B = (-2,1,-3)$ and $A - C = (4,-2,6)$. Thus, $A - C = -2(A - B)$, so $A - B$ and $A - C$ are parallel. Hence, the points lie on the same line.

43. Let $A = (2,-3,5)$, $B = (6,-1,4)$, $C = (-3,7,1)$, and $D = (1,9,0)$. Then $A - C = (5,-10,4)$ and $B - D = (5,-10,4)$, so $A - C = B - D$. Therefore, the points are the vertices of a parallelogram.

Exercises 14.5: Dot and Cross Products (page 596)

1. (a) $A \cdot B = 0 + 0 + 0 = 0$

    (b) $A \times B = \begin{vmatrix} i & j & k \\ 1 & 0 & 0 \\ 0 & 0 & 1 \end{vmatrix} = i(0) - j(1) + k(0) = (0,-1,0)$

    (c) $B \times A = -(A \times B) = (0,1,0)$

    (d) Since $A - B = (1,0,-1)$, $|A - B| = \sqrt{1 + 0 + 1} = \sqrt{2}$

    (e) $|B - A| = |A - B| = \sqrt{2}$

    (f) $\cos \theta = \dfrac{A \cdot B}{|A||B|} = \dfrac{0}{1 \cdot 1} = \dfrac{0}{1} = 0$ so $\theta = \pi/2$

    (g) $\dfrac{A \cdot B}{|B|^2} B = \dfrac{0}{|B|^2} = (0,0,0)$

3. (a) $A \cdot B = 10 - 2 - 8 = 0$

(b) $\begin{vmatrix} i & j & k \\ 2 & -1 & 2 \\ 5 & 2 & -4 \end{vmatrix} = i(4 - 4) - j(-8 - 10) + k(4 + 5) = (0,18,9)$

(c) $B \times A = -(A \times B) = (0,-18,-9)$

(d) $A - B = (-3,-3,6)$ so $|A - B| = \sqrt{9 + 9 + 36} = 3\sqrt{6}$

(f) $\cos\theta = \dfrac{A \cdot B}{|A||B|} = \dfrac{0}{|A||B|} = 0$ so $\theta = \pi/2$

(g) $\dfrac{A \cdot B}{|B|^2} B = \dfrac{0}{|B|^2} B = (0,0,0)$

5. (a) $A \cdot B = -1 - 1 + 6 = 4$

(b) $\begin{vmatrix} i & j & k \\ 1 & 1 & \sqrt{6} \\ -1 & -1 & \sqrt{6} \end{vmatrix} = i(\sqrt{6} + \sqrt{6}) - j(\sqrt{6} + \sqrt{6}) + k(-1 + 1) = (2\sqrt{6},-2\sqrt{6},0)$

(d) $A - B = (2,2,0)$ so $|A - B| = \sqrt{4 + 4 + 0} = 2\sqrt{2}$

(f) $\cos\theta = \dfrac{A \cdot B}{|A||B|} = \dfrac{4}{\sqrt{8}\sqrt{8}} = \dfrac{4}{8} = \dfrac{1}{2}$, so $\theta = \pi/3$.

(g) $\dfrac{A \cdot B}{|B|^2} B = \dfrac{4}{1 + 1 + 6} (-1,-1,\sqrt{6}) = \dfrac{1}{2} (-1,-1,\sqrt{6})$

9. $\begin{vmatrix} i & j & k \\ 1 & -2 & 3 \\ 2 & 1 & 4 \end{vmatrix} = i(-8 - 3) - j(4 - 6) + k(1 + 4) = (-11,2,5)$

11. $\begin{vmatrix} i & j & k \\ -1 & 5 & 2 \\ 4 & 1 & -1 \end{vmatrix} = i(-5 - 2) - j(1 - 8) + k(-1 - 20) = (-7,7,-21)$

27. Let the vectors A and B be as follows:
$A = (-1,4,2) - (3,2,1) = (-4,2,1)$,
$B = (0,3,5) - (3,2,1) = (-3,1,4)$.

The area of the triangle with the given vertices is half of the area of the parallelogram having A and B as adjacent sides. Therefore, the area of the triangle in question is $\dfrac{1}{2} |A \times B|$. Now

$A \times B = \begin{vmatrix} i & j & k \\ -4 & 2 & 1 \\ -3 & 1 & 4 \end{vmatrix} = (7,13,2)$. So $|A \times B| = \sqrt{49 + 169 + 4} = \sqrt{222}$. Thus, the

area is $\dfrac{1}{2} \sqrt{222}$.

31. Three adjacent edges of the parallelopiped are replicas of the vectors
$(5,2,-1) - (2,-1,4) = (3,3,-5)$, $(-2,5,1) - (2,-1,4) = (-4,6,-3)$,
$(1,0,-2) - (2,-1,4) = (-1,1,-6)$. Thus, the volume equals the volume of the parallelopiped having the vectors $(3,3,-5)$, $(-4,6,-3)$, and $(-1,1,-6)$ as adjacent edges. Thus, the volume is the absolute value of

$$\begin{vmatrix} 3 & 3 & -5 \\ -4 & 6 & -3 \\ -1 & 1 & -6 \end{vmatrix} = -172.$$ So the volume is 172 units^2.

Exercises 14.6: Equations of Lines and Planes (page 604)

5. (b) Solving for t in (a), we get $t = \dfrac{8 - x}{3}$, $t = \dfrac{y + 2}{6}$, $t = \dfrac{4 - z}{5}$. Thus, the symmetric form is $\dfrac{8 - x}{3} = \dfrac{y + 2}{6} = \dfrac{4 - z}{5}$.

11. The line is parallel to the vector $(5 - 1, 2 - 3, 1 + 2) = (4, -1, 3)$ so has equations $x = 2 + 4t$, $y = -1 - t$, $z = 4 + 3t$.

13. The line is parallel to the vector $(5, 2, -1)$ so has equations $x = 8 + 5t$, $y = -2 + 2t$, $z = 3 - t$.

17. A vector perpendicular to the plane is the cross product of the vectors $(3, 1, 4) - (2, 0, -1) = (1, 1, 5)$ and $(0, -2, 5) - (2, 0, -1) = (-2, -2, 6)$. Thus,
$$\begin{vmatrix} i & j & k \\ 1 & 1 & 5 \\ -2 & -2 & 6 \end{vmatrix} = (16, -16, 0)$$ is perpendicular to the plane. Since the plane passes through the point $(2, 0, -1)$, an equation of the plane is $16(x - 2) - 16(x - 0) + 0(z + 1) = 0$, or $x - y = 2$.

19. Substituting we get $(t + 2) - (2t - 1) + 3(t) = 9$ so $2t = 6$, $t = 3$ and the intersection point has coordinates $x = 3 + 2 = 5$, $y = 2(3) - 1 = 5$, $z = 3$. Thus, it is the point $(5, 5, 3)$.

21. Solving for t we get $3(2t + 1) - 2(5t) - (1 - t) = 8$, $-3t = 6$, $t = -2$. Thus, the point has coordinates $x = 2(-2) + 1 = -3$, $y = 5(-2) = -10$, $z = 1 + 2 = 3$. Thus, it is the point $(-3, -10, 3)$.

23. We can show that it is parallel to the plane by showing it does not intersect the plane. Substituting we get $2t - 3 - 6(t + 2) + 4(t - 5) = 7$, or $-35 = 7$, an impossibility. Thus, there is no solution to the equation for t and so the line does not intersect the plane.

25. We obtain $5(x - 3) + 6(y + 2) + 1(z - 4) = 0$, or $5x + 6y + z = 7$.

27. We get $1(x + 3) + 0(y - 2) + 6(z - 0) = 0$, or $x + 6z + 3 = 0$.

29. The plane must be perpendicular to the vector $(2 - 1, 4 + 2, 5 - 3) = (1, 6, 2)$. Thus, an equation of the plane is $1(x + 2) + 6(y - 6) + 2(z - 1) = 0$, or $x + 6y + 2z = 36$.

31. The vector $(3, -5, 7)$ is perpendicular to the given plane and the one whose equation we are finding, so we get $3(x - 1) - 5(y - 5) + 7(z + 3) = 0$, or $3x - 5y + 7z + 43 = 0$.

33. Every plane has an equation of the form $ax + by + cz = d$. Since the given points are in the plane, we get

$a(0) + b(0) + c(3) = d$, or $3c = d$,

$a(0) + b(-2) + c(0) = d$, or $-2b = d$,

$a(1) + b(0) + c(0) = d$, or $a = d$.

For any non-zero d we get an equation of the plane so take d to be the convenient number 6. Then $c = 2$, $b = -3$, and $a = 6$ and we get $6x - 3y + 2z = 6$.

37. The vector $(1,1,\sqrt{6})$ is perpendicular to the first plane and the vector $(1,1,-\sqrt{6})$ is perpendicular to the second plane. The angle θ between these two vectors satisfies $\cos\theta = \dfrac{1 + 1 - 6}{\sqrt{1 + 1 + 6}\,\sqrt{1 + 1 + 6}} = \dfrac{-4}{8} = -\dfrac{1}{2}$. So $\theta = \dfrac{2\pi}{3}$. The acute angle between the planes is the supplement of this angle, so it is $\pi/3$.

$\underline{\text{Exercises 14.7:}}$ Functions from R^n to R^m (page 610)

9. Since $\sqrt{9 - x^2 - y^2} = \sqrt{9 - (x^2 + y^2)}$, we get Range: $0 \le f(x,y) \le 3$ and Domain: all points (x,y) with $x^2 + y^2 \le 9$.

11. Range: all non-negative real numbers. To find the domain note that $\dfrac{x + y}{x - y}$ is positive if and only if $(x + y)(x - y) = x^2 - y^2$ is positive. Thus, the domain is all points (x,y) with $|x| \ge |y|$ but $x \ne y$.

15. Since $Y(t) = (r\cos\omega t, r\sin\omega t, 0)$ and $V(t) = (-r\omega\sin\omega t, r\omega\sin\omega t, 0)$,

$V(t) \cdot Y(t) = -r^2\omega\sin\omega t\cos\omega t + r^2\omega^2\sin\omega t\cos\omega t = 0$. Thus, $V(t)$ is perpendicular to $Y(t)$. We conclude that the velocity is always directed tangentially to the path of the point.

17. (c) Since $(2\cos t)^2 + (2\sin t)^2 = 4(\sin^2 t + \cos^2 t) = 4$, the path of P is on the cylinder. As t increases, z increases so the path looks like a vine winding around the cylinder getting higher as it winds.

19. Since the derivative with respect to t of the position vector $X(t)$ is $(9t^2, 12t^2, 6t + 5)$, $X(t) = (3t^3 + C_1, 4t^3 + C_2, 3t^2 + 5t + C_3)$. But $X(2) = (4,-1,2) = (24 + C_1, 32 + C_2, 22 + C_3)$, so $C_1 = -20$, $C_2 = -33$, $C_3 = -20$. Thus, $X(t) = (3t^3 - 20, 4t^3 - 33, 3t^2 + 5t - 20)$.

21. (c) Since $x = y$, path is in the plane $x = y$. Also, since $x^2 + y^2 + z^2 = \cos^2\omega t + \cos^2\omega t + 2\sin^2\omega t = 2$, the path is also on the sphere of radius $\sqrt{2}$ centered at the origin. Thus, the path is a circle of radius $\sqrt{2}$ centered at the origin and in the plane $x = y$.

1. (a) $2(-2,-3,6,-1) + 3(4,1,5,4) = (8,-3,27,10)$

   (b) $6(-2,-3,6,-1) \cdot (4,1,5,4) = 6(-8 - 3 + 30 - 4) = 90$

3. (a) $\displaystyle \lim_{(x,y)\to(3,5)} f(x,y) = \frac{3}{3 + 5} = \frac{3}{8}$

   (b) For $y = kx$, $\dfrac{|x|}{|x| + |y|} = \dfrac{|k||x|}{|x| + |k||x|} = \dfrac{|k|}{1 + |k|}$. So in any small disk cent-

   ered at $(0,0)$ the function assumes the value $\dfrac{|k|}{1 + |k|}$ for every number k. Thus,

   the function does not get arbitrarily close to a unique number and the limit

   does not exist.

5. (b) In any small disk centered at $(1,3)$ there are points for which $y = 3x$ and

   $y \neq 3x$ so the function gets arbitrarily close to both 4 and 5. Since it does

   not get arbitrarily close to a unique number, the limit does not exist.

2. $\sqrt{(-5 + 2)^2 + (2 + 3)^2 + (1 - 3)^2} = \sqrt{9 + 25 + 4} = \sqrt{38}$

3. The radius is $\sqrt{(2 - 4)^2 + (-3 - 1)^2 + (1 + 2)^2} = \sqrt{4 + 16 + 9} = \sqrt{29}$. Thus, an

   equation of the sphere is $(x - 2)^2 + (y + 3)^2 + (z - 1)^2 = 29$.

4. The graph cuts the xy-plane in $x^2 = 4y^2$. That is, in the lines $x = \pm 2y$. It

   cuts the yz-plane in the two lines $z = \pm y$. A plane of the form $y = s$ cuts the

   graph in the ellipse $x^2 + 4z^2 = 4s^2$. We get the graph shown in the answer sec-

   tion of the text.

5. $[(1,2,-3,4) - 2(-2,1,2,5)] \cdot (1,2,-3,4) = (5,0,-7,-6) \cdot (1,2,-3,4) = 5 + 0$

   $+ 21 - 24 = 2$.

6. $(2,-5,8) \cdot (-2,4,3) = -4 - 20 + 24 = 0$ so they are perpendicular.

7. A vector perpendicular to both vectors is their cross product

   $\begin{vmatrix} i & j & k \\ 2 & -1 & -3 \\ 3 & 2 & 4 \end{vmatrix} = i(-4 + 6) - j(8 + 9) + k(4 + 3) = (2,17,7)$. Thus, a unit vector

   perpendicular to them is $\dfrac{1}{\sqrt{4 + 289 + 49}} (2,17,7) = \dfrac{1}{3\sqrt{38}} (2,17,7)$.

8. $\dfrac{A \cdot B}{|B|^2} B = \dfrac{-2 - 6 + 15}{1 + 4 + 9} (1,-2,3) = \dfrac{1}{2} (1,-2,3)$

9. $\cos \theta = \dfrac{16 - 36 - 12}{\sqrt{16 + 36 + 12} \ \sqrt{16 + 36 + 12}} = -\dfrac{1}{2}$, so $\theta = \dfrac{2\pi}{3}$

10. $x = 2 + t(-3 - 2)$, $y = 5 + t(5 - 5)$, $z = -1 + t(2 + 1)$, so parametric equations

    are $x = 2 - 5t$, $y = 5$, $z = -1 + 3t$.

11. The line must be parallel to $(4,-3,1)$, so it has symmetric equations

$$\frac{x-5}{4} = \frac{y-8}{-3} = \frac{z+6}{1}, \text{ or } \frac{x-5}{4} = \frac{8-y}{3} = z+6.$$

12. The plane must be perpendicular to the vector $(6-7, 4+1, -1+2) = (-1,5,-3)$. Thus, an equation for the line is $-1(x-5) + 5(y+4) - 3(z+6) = 0$, or $x - 5y + 3z = 7$.

13. The velocity vector is $(10t, 3t^2, 2e^{2t})$. When $t = 1$ we get $(10, 3, 2e^2)$.

14. (b) In any small disk centered at $(-2,-2)$ there are points for which $x \neq y$ and for which $x = y$ where the function is arbitrarily close to both 0 and 4. Since the function does not get arbitrarily close to a unique number, the limit does not exist.

15. When the point (x,y) is not on the line $x = y$ it is continuous. The only point on the line $x = y$ for which it is continuous is $(0,0)$. Thus, the function is continuous for all points (x,y) not on the line $x = y$ and also continuous for the point $(0,0)$.

Additional Exercises, Chapter 14 (page 621)

5. $\sqrt{(8+3)^2 + (3-5)^2 + (1+2)^2} = \sqrt{134}$

9. The center is at the midpoint $(\frac{5+3}{2}, \frac{-3+1}{2}, \frac{8-2}{2})$ of the ends of the diameter. That is, the center is at $(4,-1,3)$. Thus, the radius is

$\sqrt{(5-4)^2 + (-3+1)^2 + (8-3)^2} = \sqrt{30}$. An equation of the sphere is $(x-4)^2 + (y+1)^2 + (z-3)^2 = 30$.

11. The radius is $\sqrt{(2+1)^2 + (6-7)^2 + (-1-9)^2} = \sqrt{110}$. Thus, an inequality for all points outside the sphere is $(x+1)^2 + (y\bullet-7)^2 + (z-9)^2 > 110$.

23. $3X = (1,-2,3) - (-3,5,7) = (4,-7,-4)$, so $X = (4/3, -7/3, -4/3)$.

25. We get $2\sqrt{9+4+16} = 2\sqrt{29}$.

27. Let A, B, and C be the position vectors of these points respectively. We can show the points are on the same line by showing that the vectors $A - B$ and $B - C$ are parallel. Now $A - B = (3,-2,5)$ and $B - C = (-15,10,-25)$, so $B - C = -5(A - B)$ and hence $A - B$ and $B - C$ are parallel and the points must lie on the same line.

29. $A \cdot B = 2(5) - 1(6) + 3(-4) + 8(1) = 10 - 6 - 12 + 8 = 0$

31. $\dfrac{A \cdot B}{|B|^2} B = \dfrac{6+2-4}{4+1+16} (2,-1,4) = \dfrac{4}{21}(2,-1,4)$

33. $\begin{vmatrix} i & j & k \\ 1 & 2 & -3 \\ -2 & -1 & 1 \end{vmatrix} = i(2-3) - j(1-6) + k(-1+4) = (-1,5,3)$. So a unit vector

perpendicular to the given vector is $\dfrac{1}{\sqrt{1 + 25 + 9}}\,(-1,5,3) = \dfrac{1}{\sqrt{35}}\,(-1,5,3)$.

35. We get $x = 5 + t(1 - 5)$, $y = -7 + t(2 + 7)$, $z = 2 + t(-3 - 2)$, or $x = 5 - 4t$, $y = -7 + 9t$, $z = 2 - 5t$.

37. The line must be parallel to the vector $(4,-7,3)$. Thus, we get $x = 8 + 4t$, $y = 5 - 7t$, $z = 1 + 3t$.

39. We get $-2(x - 9) + 3(y - 1) + 5(z - 3) = 0$, or $2x - 3y - 5z = 0$.

41. The plane must be perpendicular to the vector $(5,2,-1)$ so we get $5(x - 5) + 2(y + 3) - 1(z - 2) = 0$, or $5x + 2y - z = 17$.

45. The position vector must be of the form $(e^t + C_1, 2t^2 + C_2, t^3 + C_3)$. When $t = 2$ we must have $e^2 + C_1 = 4$, $2(2)^2 + C_2 = -1$, $2^3 + C_3 = 3$, so $C_1 = 4 - e^2$, $C_2 = -9$, $C_3 = -5$ and the position vector is $(e^t + 4 - e^2, 2t^2 - 9, t^3 - 5)$.

CHAPTER 15: DIFFERENTIAL CALCULUS OF FUNCTIONS OF SEVERAL VARIABLES

Exercises 15.1: Partial Derivatives (page 628)

1. (a) $\dfrac{\partial f}{\partial x} = 2$, (b) $\dfrac{\partial f}{\partial y} = 3$, (c) $\dfrac{\partial^2 f}{\partial x^2} = \dfrac{\partial 2}{\partial x} = 0$, (d) $\dfrac{\partial^2 f}{\partial y^2} = \dfrac{\partial 3}{\partial y} = 0$, (e) $\dfrac{\partial^2 f}{\partial x \partial y} = \dfrac{\partial 3}{\partial x} = 0$

3. (a) $\dfrac{\partial f}{\partial x} = 2xy + 2y$, (b) $\dfrac{\partial f}{\partial y} = x^2 + 2x$, (c) $\dfrac{\partial^2 f}{\partial x^2} = \dfrac{\partial}{\partial x}(2xy + 2y) = 2y$,

(d) $\dfrac{\partial^2 f}{\partial y^2} = \dfrac{\partial}{\partial y}(x^2 + 2x) = 0$, (e) $\dfrac{\partial^2 f}{\partial x \partial y} = \dfrac{\partial}{\partial x}(x^2 + 2x) = 2x + 2$

5. (a) $\dfrac{\partial f}{\partial x} = x(x^2 + y^2)^{-1/2}$, (b) $\dfrac{\partial f}{\partial y} = y(x^2 + y^2)^{-1/2}$,

(c) $\dfrac{\partial^2 f}{\partial x^2} = (x^2 + y^2)^{-1/2} - x^2(x^2 + y^2)^{-3/2} = y^2(x^2 + y^2)^{-3/2}$,

(d) $\dfrac{\partial^2 f}{\partial y^2} = (x^2 + y^2)^{-1/2} - y^2(x^2 + y^2)^{-3/2} = x^2(x^2 + y^2)^{-3/2}$,

(e) $\dfrac{\partial^2 f}{\partial x \partial y} = -xy(x^2 + y^2)^{-3/2}$

7. (a) $\dfrac{\partial f}{\partial x} = \cos(x + y) + 2$, (b) $\dfrac{\partial f}{\partial y} = \cos(x + y)$,

(c) $\dfrac{\partial^2 f}{\partial x^2} = \dfrac{\partial}{\partial x}(\cos(x + y) + 2) = -\sin(x + y)$,

(d) $\dfrac{\partial^2 f}{\partial y^2} = \dfrac{\partial}{\partial y}\cos(x + y) = -\sin(x + y)$,

(e) $\dfrac{\partial^2 f}{\partial x \partial y} = \dfrac{\partial}{\partial x}\cos(x + y) = -\sin(x + y)$

9. (a) $\dfrac{\partial f}{\partial x} = \dfrac{1}{x} + \cos x$, (b) $\dfrac{\partial f}{\partial y} = y^{-1}$, (c) $\dfrac{\partial^2 f}{\partial x^2} = \dfrac{\partial}{\partial x}\left(\dfrac{1}{x} + \cos x\right) = -x^{-2} - \sin x$,

(d) $\dfrac{\partial^2 f}{\partial y^2} = \dfrac{\partial}{\partial y}\left(\dfrac{1}{y}\right) = -y^2$, (e) $\dfrac{\partial^2 f}{\partial x \partial y} = \dfrac{\partial}{\partial x}\left(\dfrac{1}{y}\right) = 0$

11. (a) $\dfrac{\partial f}{\partial x} = \tan(xy) + xy \sec^2(xy)$, (b) $\dfrac{\partial f}{\partial y} = x^2 \sec^2(xy)$,

(c) $\dfrac{\partial^2 f}{\partial x^2} = \dfrac{\partial}{\partial x}(\tan(xy) + xy \sec^2(xy)) = 2y \sec^2(xy) + 2y^2 x \sec^2(xy) \tan(xy)$,

(d) $\dfrac{\partial^2 f}{\partial y^2} = \dfrac{\partial}{\partial y}(x^2 \sec^2(xy)) = 2x^3 \sec^2(xy) \tan(xy)$,

(e) $\dfrac{\partial^2 f}{\partial x \partial y} = \dfrac{\partial}{\partial x}(x^2 \sec^2(xy)) = 2x \sec^2(xy) + 2x^2 y \sec^2(xy) \tan(xy)$

13. (a) $\dfrac{\partial f}{\partial x} = \dfrac{-y/x^2}{1 + (y/x)^2} = \dfrac{-y}{x^2 + y^2}$, (b) $\dfrac{\partial f}{\partial y} = \dfrac{1/x}{1 + (y/x)^2} = \dfrac{x}{x^2 + y^2}$,

(c) $\dfrac{\partial^2 f}{\partial x^2} = \dfrac{2xy}{(x^2 + y^2)^2}$, (d) $\dfrac{\partial^2 f}{\partial y^2} = \dfrac{-2xy}{(x^2 + y^2)^2}$,

(e) $\dfrac{\partial^2 f}{\partial x \partial y} = \dfrac{\partial}{\partial x}\left(\dfrac{x}{x^2 + y^2}\right) = \dfrac{(x^2 + y^2) - 2x^2}{(x^2 + y^2)^2} = \dfrac{-x^2 + y^2}{(x^2 + y^2)^2}$

17. (a) $\dfrac{\partial f}{\partial x} = 4(x + yz)^3 + yz \cos(xyz)$, (b) $\dfrac{\partial f}{\partial y} = 4z(x + yz)^3 + xz \cos(xyz)$,

(c) $\dfrac{\partial^2 f}{\partial x^2} = \dfrac{\partial}{\partial x}(4(x + yz)^3 + yz \cos(xyz) = 12(x + yz)^2 - (yz)^2 \sin(xyz)$,

(d) $\dfrac{\partial^2 f}{\partial y^2} = \dfrac{\partial}{\partial y}(4z(x + yz)^3 + xz \cos(xyz) = 12z^2(x + yz)^2 - (xz)^2 \sin(xyz)$,

(e) $\dfrac{\partial^2 f}{\partial x \partial y} = \dfrac{\partial}{\partial x}(4z(x + yz)^3 + xz \cos(xyz)) = 12z(x + yz)^2 + z \cos(xyz)$

$- xyz^2 \sin(xyz)$

19. $\dfrac{\partial z}{\partial y} = xe^y$. Thus, $m = \dfrac{\partial z}{\partial y}\Big|_{\substack{x=1 \\ y=1}} = e$.

21. $\dfrac{\partial z}{\partial y} = \dfrac{x}{1 + x^2 y^2}$. Thus, $m = \dfrac{\partial z}{\partial y}\Big|_{\substack{x=1 \\ y=1}} = \dfrac{1}{2}$.

23. $f_{x_i}(x_1, \ldots, x_n) = \dfrac{\partial}{\partial x_i}(x_1 + x_2^2 + x_3^3 + \ldots + x_i^i + \ldots + x_n^n) = ix_i^{i-1}$.

25. Since $f(X) = X \cdot X$, $f(x_1, \ldots, x_i, \ldots, x_j, \ldots, x_n) = x_1^2 + \ldots + x_i^2 + \ldots + x_j^2 + \ldots + x_n^2$. Thus, $f_{x_i} = \dfrac{\partial f}{\partial x_i} = 2x_i$ and $f_{x_i x_j} = \dfrac{\partial}{\partial x_j}\dfrac{\partial f}{\partial x_i} = \dfrac{\partial}{\partial x_j}(2x_i) = 0$.

27. (a) $f_x(x,y) = \dfrac{\partial}{\partial x}(x^y) = yx^{y-1}$

(b) $f_y(x,y) = \dfrac{\partial}{\partial y}(x^y)$. Let $z = x^y$. Then $\ln z = y \ln x$ and so, $\dfrac{1}{z}\dfrac{\partial z}{\partial y} = \ln x$.

Thus, $\dfrac{\partial z}{\partial y} = z \ln x = x^y \ln x$.

Exercises 15.2: Chain Rules (page 635)

1. Substitution into $\dfrac{df}{dt} = \dfrac{\partial f}{\partial x}\dfrac{dx}{dt} + \dfrac{\partial f}{\partial y}\dfrac{dy}{dt}$ gives $\dfrac{df}{dt} = (1 + 4xy)(3) + 2x^2(2t)$. Then

since $x = 8$ and $y = 1$ when $t = 1$, we have $\dfrac{df}{dt} = (1 + 32)(3) + 2(64)(2) = 355$

when $t = 1$.

3. Substitution into $\dfrac{df}{dt} = \dfrac{\partial f}{\partial x}\dfrac{dx}{dt} + \dfrac{\partial f}{\partial y}\dfrac{dy}{dt}$ gives $\dfrac{df}{dt} = \dfrac{y}{x}(-\sin t) + (\ln|x|)\cdot 1$.

Then since $x = \cos \pi = -1$, and $y = \pi$ when $t = \pi$, we have

$\dfrac{df}{dt} = \dfrac{\pi}{-1}(-\sin \pi) + \ln|-1| = 0$ if $t = \pi$.

5. Substitution into $\dfrac{df}{dt} = \dfrac{\partial f}{\partial x}\dfrac{dx}{dt} + \dfrac{\partial f}{\partial y}\dfrac{dy}{dt}$ gives $\dfrac{df}{dt} = (\arctan y)\dfrac{1}{t} + \dfrac{x}{1 + y^2}\sec t$

$\tan t$.

7. Substitution into $\dfrac{df}{dt} = \dfrac{\partial f}{\partial x}\dfrac{dx}{dt} + \dfrac{\partial f}{\partial y}\dfrac{dy}{dt}$ gives $\dfrac{df}{dt} = e^x(1) + e^y(1) = e^x + e^y$

$= f(x,y)$.

9. Substitution into $\dfrac{df}{dt} = \dfrac{\partial f}{\partial x}\dfrac{dx}{dt} + \dfrac{\partial f}{\partial y}\dfrac{dy}{dt} + \dfrac{\partial f}{\partial z}\dfrac{dz}{dt}$ gives $\dfrac{df}{dt} = (2x + y^2)\cos t$

$- 2xy \sin t + a$.

11. Substitution into $\dfrac{df}{dt} = \dfrac{\partial f}{\partial x_1}\dfrac{\partial x_1}{\partial t} + \dfrac{\partial f}{\partial x_2}\dfrac{\partial x_2}{\partial t} + \ldots + \dfrac{\partial f}{\partial x_n}\dfrac{\partial x_n}{\partial t}$ gives

$\dfrac{df}{dt} = 2x_1 + 2x_2 + \ldots + 2x_n$.

13. Since $\dfrac{\partial z}{\partial x} = \dfrac{\frac{1}{y}}{1 + \frac{x^2}{y^2}} = \dfrac{y}{y^2 + x^2}$ and $\dfrac{\partial x}{\partial y} = \dfrac{-\frac{x}{y^2}}{1 + \frac{x^2}{y^2}} = -\dfrac{x}{y^2 + x^2}$, substitution into

$\dfrac{dz}{dt} = \dfrac{\partial z}{\partial x}\dfrac{dx}{dt} + \dfrac{\partial z}{\partial y}\dfrac{dy}{dt}$ gives $\dfrac{dz}{dt} = \dfrac{y}{y^2 + x^2}\cos t + (-\dfrac{x}{y^2 + x^2})(-\sin t)$. Then since

$x = \sin t$ and $y = \cos t$, we have $\dfrac{dz}{dt} = \dfrac{\cos t}{\cos^2 t + \sin^2 t}\cos t + \dfrac{\sin t}{\cos^2 t + \sin^2 t}\sin t$

$= \cos^2 t + \sin^2 t = 1$. Note that since $z = \arctan\dfrac{x}{y}$ and $x = \sin t$ and $y = \cos t$,

$z = \arctan\dfrac{\sin t}{\cos t} = \arctan(\tan t) = t$. Consequently, the result $\dfrac{dz}{dt} = 1$ is no

surprise.

15. Substitution into $\frac{\partial f}{\partial u} = \frac{\partial f}{\partial x}\frac{\partial x}{\partial u} + \frac{\partial f}{\partial y}\frac{\partial y}{\partial u}$ gives $\frac{\partial f}{\partial u} = 3x^2y^5$ (1) $+ 5x^3y^4$ (1)

$= 3x^2y^5 + 5x^3y^4$. Then since $x = -1$ and $y = 1$ when $u = 0$ and $v = 1$, we have

$\frac{\partial f}{\partial u} = 3 - 5 = -2$ when $u = 0$ and $v = 1$. Substitution into

$\frac{\partial f}{\partial v} = \frac{\partial f}{\partial x}\frac{\partial x}{\partial v} + \frac{\partial f}{\partial y}\frac{\partial y}{\partial v}$ gives $\frac{\partial f}{\partial v} = 3x^2y^5$ (-1) $+ 5x^3y^4$ (1) $= -3x^2y^5 + 5x^3y^4$. Then

since $x = -1$ and $y = 1$ when $u = 0$ and $v = 1$, we have $\frac{\partial f}{\partial v} = -3 - 5 = -8$ when

$u = 0$ and $v = 1$.

17. Substitution into $\frac{\partial f}{\partial u} = \frac{\partial f}{\partial x}\frac{\partial x}{\partial u} + \frac{\partial f}{\partial y}\frac{\partial y}{\partial u}$ gives $\frac{\partial f}{\partial u} = yv \cos(xy) + x \cos(xy)$.

Substitution into $\frac{\partial f}{\partial v} = \frac{\partial f}{\partial x}\frac{\partial x}{\partial v} + \frac{\partial f}{\partial y}\frac{\partial y}{\partial v}$ gives $\frac{\partial f}{\partial v} = yu \cos(xy)$.

19. Substitution into $\frac{\partial f}{\partial u} = \frac{\partial f}{\partial x}\frac{\partial x}{\partial u} + \frac{\partial f}{\partial y}\frac{\partial y}{\partial u} + \frac{\partial f}{\partial z}\frac{\partial z}{\partial u}$ gives $\frac{\partial f}{\partial u} = yz \cos u + xz$ (0)

$+ xy$ (1) $= yz \cos u + xy$. Then since $x = \sin \pi/2 = 1$, $y = 0$, and $z = \pi/2$ when

$u = \pi/2$ and $v = 0$, we have $\frac{\partial f}{\partial u} = 0$ when $u = \pi/2$ and $v = 0$. Substitution into

$\frac{\partial f}{\partial v} = \frac{\partial f}{\partial x}\frac{\partial x}{\partial v} + \frac{\partial f}{\partial y}\frac{\partial y}{\partial v} + \frac{\partial f}{\partial z}\frac{\partial z}{\partial v}$ gives $\frac{\partial f}{\partial v} = yz$ (0) $+ xz$ (1) $+ xy$ (1) $= xz + xy$. Then

since $x = 1$, $y = 0$, and $z = \pi/2$ when $u = \pi/2$ and $v = 0$, we have $\frac{\partial f}{\partial v} = \pi/2$ when

$u = \pi/2$ and $v = 0$.

21. Substitution into $\frac{\partial f}{\partial u} = \frac{\partial f}{\partial x}\frac{\partial x}{\partial u} + \frac{\partial f}{\partial y}\frac{\partial y}{\partial u} + \frac{\partial f}{\partial z}\frac{\partial z}{\partial u}$ gives $\frac{\partial f}{\partial u} = -2x \sin u + 2y \cos u$

$+ 2zv$. Substitution into $\frac{\partial f}{\partial v} = \frac{\partial f}{\partial x}\frac{\partial x}{\partial v} + \frac{\partial f}{\partial y}\frac{\partial y}{\partial v} + \frac{\partial f}{\partial z}\frac{\partial z}{\partial v}$ gives $\frac{\partial f}{\partial v} = 2x$ (0) $+ 2y$ (0)

$+ 2zu = 2zu$.

23. Since $\frac{\partial f}{\partial \rho} = \frac{\partial f}{\partial x}\frac{\partial x}{\partial \rho} + \frac{\partial f}{\partial y}\frac{\partial y}{\partial \rho} + \frac{\partial f}{\partial z}\frac{\partial z}{\partial \rho}$, we have

$\frac{\partial f}{\partial \rho} = 2x \sin \phi \cos \theta + 2y \sin \phi \sin \theta + 2z \cos \phi$. Then on substitution for x,

y, and z, we get $\frac{\partial f}{\partial \rho} = 2\rho \sin^2 \phi \cos^2 \theta + 2\rho \sin^2 \phi \sin^2 \theta + 2\rho \cos^2 \phi$

$= 2\rho (\sin^2 \phi + \cos^2 \phi) = 2\rho$. Similarly $\frac{\partial f}{\partial \phi} = 2x\rho \cos \phi \cos \theta + 2y\rho \cos \phi \sin \theta$

$= 2z\rho \sin \phi = 2\rho^2 \sin \phi \cos \phi \cos^2 \theta + 2\rho^2 \cos \phi \sin \phi \sin^2 \theta - 2\rho^2 \sin \phi \cos \phi$

$= 2\rho^2 \sin \phi \cos \phi - 2\rho^2 \sin \phi \cos \phi = 0$. Finally

$\frac{\partial f}{\partial \theta} = -2x\rho \sin \phi \sin \theta + 2y\rho \sin \phi \cos \theta = 2\rho^2 \sin^2 \phi \sin \theta \cos \theta + 2\rho^2 \sin^2 \phi$

$\cos \theta \sin \theta = 0$.

25. Let $x = f(t)$ and $y = g(t)$ give the x and y coordinates of the particle at time

t. Then if $F(x,y)$ is the magnitude of the force, we can use the chain rule to

write $\frac{dF}{dt} = \frac{\partial F}{\partial x}\frac{dx}{dt} + \frac{\partial F}{\partial y}\frac{dy}{dt}$. Since $F(x,y) = (x^2 + y^2)^{-1}$, we have

$\frac{\partial F}{\partial x} = \frac{-2x}{(x^2 + y^2)^2}$ and $\frac{\partial F}{\partial y} = \frac{-2y}{(x^2 + y^2)^2}$. Thus, $\frac{dF}{dt} = \frac{-2x}{(x^2 + y^2)^2} \frac{dx}{dt} - \frac{2y}{(x^2 + y^2)^2} \frac{dy}{dt}$.

While we do not know the functions f and g, we are given that $\frac{dx}{dt} = 3$ and

$\frac{dy}{dt} = -1$ when $x = 1$ and $y = 2$. Thus, when $x = 1$ and $y = 2$ we have

$\frac{dF}{dt} = \frac{-2}{25} \cdot 3 - \frac{4}{25} (-1) = -\frac{2}{25}$.

Exercises 15.3: The Gradient (page 641)

1. Since $\nabla f(x,y) = (\frac{\partial f}{\partial x}, \frac{\partial f}{\partial y})$, we have $\nabla f(x,y) = (y + 2x, x + 2y)$. Thus,

$\nabla f(1,-1) = (1,-1)$.

3. Substitution into $\nabla f(x,y) = (\frac{\partial f}{\partial x}, \frac{\partial f}{\partial y})$ gives $\nabla f(x,y) = (1 + ye^x, e^x)$.

5. Substitution into $\nabla f(x,y) = (\frac{\partial f}{\partial x}, \frac{\partial f}{\partial y})$ gives $\nabla f(x,y) = (-\sin (x - y) + 2, \sin$

$(x - y))$.

7. Substitution into $\nabla f(x,y) = (\frac{\partial f}{\partial x}, \frac{\partial f}{\partial y})$ gives $\nabla f(x,y) = (\tan (xy) + xy \sec^2 (xy),$

$x^2 \sec^2 (xy))$.

9. Substitution into $\nabla f(x,y) = (\frac{\partial f}{\partial x}, \frac{\partial f}{\partial y}, \frac{\partial f}{\partial z})$ gives $\nabla f(x,y,z) = (2x, 2y, 2z)$.

11. Since $\nabla f(x,y,z) = (\frac{\partial f}{\partial x}, \frac{\partial f}{\partial y}, \frac{\partial f}{\partial z})$, we have $\nabla f(x,y,z) = (y, x + z, y)$. Thus,

$\nabla f(1,0,-1) = (0,0,0)$.

13. Since $f_x(x,y,z) = \cos xy - xy \sin xy$, $f_y(x,y,z) = -x^2 \sin xy + e^z$, and

$f_z(x,y,z) = ye^z$, we have $\nabla f(x,y,z) = (\cos xy - xy \sin xy, -x^2 \sin xy + e^z,$

$ye^z)$.

15. Substitution into $\nabla f(x_1,x_2,x_3,x_4) = (\frac{\partial f}{\partial x_1}, \frac{\partial f}{\partial x_2}, \frac{\partial f}{\partial x_3}, \frac{\partial f}{\partial x_4})$ gives

$\nabla f(x_1,x_2,x_3,x_4) = (x_2 x_3 x_4, x_1 x_3 x_4, x_1 x_2 x_4, x_1 x_2 x_3)$.

17. Since $f(x_1,x_2,x_3,x_4) = \ln (x_1 x_2 x_3 x_4) = \ln x_1 + \ln x_2 + \ln x_3 + \ln x_4$, substitu-

tion into $\nabla f(x_1,x_2,x_3,x_4) = (\frac{\partial f}{\partial x_1}, \frac{\partial f}{\partial x_2}, \frac{\partial f}{\partial x_3}, \frac{\partial f}{\partial x_4})$ gives $\nabla f(x_1,x_2,x_3,x_4)$

$= (1/x_1, 1/x_2, 1/x_3, 1/x_4)$.

19. In this case $f(X) = f(x_1, \ldots, x_n) = \sqrt{x_1^2 + x_2^2 + \ldots + x_n^2}$. Thus,

$\frac{\partial f}{\partial x_i} = x_i (x_1^2 + x_2^2 + \ldots + x_n^2)^{-1/2}$ for $i = 1, 2, \ldots, n$. Then since

$\nabla f(X) = (\frac{\partial f}{\partial x_1}, \frac{\partial f}{\partial x_2}, \ldots, \frac{\partial f}{\partial x_n})$, we have $\nabla f(X)$

$$= (\frac{x_1}{\sqrt{x_1^2 + \ldots + x_n^2}}, \frac{x_2}{\sqrt{x_1^2 + \ldots + x_n^2}}, \ldots, \frac{x_n}{\sqrt{x_1^2 + \ldots + x_n^2}})$$

$$= \frac{1}{\sqrt{x_1^2 + \ldots + x_n^2}} (x_1, x_2, \ldots, x_n).$$ Then in terms of X we have $\nabla f(X) = \frac{1}{|X|} X$.

21. The vectors $\pm \nabla g(a,b,c)$ are normal to the surface $g(x,y,z) = k$ at (a,b,c). Then if we let $g(x,y,z) = 2x^2 + y^2 + z^2$ we have $\nabla g(x,y,z) = (4x, 2y, 2z)$. Thus, $\pm \nabla g(1,1,1) = \pm(4,2,2)$ are normal to the surface at $(1,1,1)$. In order to obtain the desired unit vectors we need only divide by the length $\sqrt{16 + 4 + 4} = 2\sqrt{6}$. Thus, the desired vectors are $\pm \frac{\nabla g(1,1,1)}{|\nabla g(1,1,1)|} = \pm \frac{(4,2,2)}{2\sqrt{6}} = \pm \frac{(2,1,1)}{\sqrt{6}}$.

23. If we let $g(x,y,z) = x^2 + y^2 - z$, the vector $\nabla g(1,2,5)$ is normal to the surface (and so normal to the tangent plane) at $(1,2,5)$. Since $\nabla g(x,y,z) = (2x, 2y, -1)$, we have $\nabla g(1,2,5) = (2,4,-1)$. Thus, we seek an equation of the plane normal to the vector $(2,4,-1)$ and passing through the point $(1,2,5)$. An equation is $2(x - 1) + 4(y - 2) - 1(z - 5) = 0$, or $2x + 4y - z = 5$.

25. Since $f(X) = \frac{1}{|X|}$, we have $f(x_1, x_2, \ldots, x_n) = \frac{1}{\sqrt{x_1^2 + x_2^2 + \ldots + x_n^2}}$.

Thus, $\frac{\partial f}{\partial x_i} = -\frac{1}{2}(x_1^2 + \ldots + x_n^2)^{-3/2}(2x_i) = \frac{-x_i}{(x_1^2 + \ldots + x_n^2)^{3/2}} = -\frac{x_i}{|X|^3}$.

Substitution into $\nabla f(X) = (\frac{\partial f}{\partial x_i}, \ldots, \frac{\partial f}{\partial x_n})$ yields

$\nabla f(X) = (\frac{x_1}{|X|^3}, \frac{x_2}{|X|^3}, \ldots, \frac{x_n}{|X|^3}) = \frac{1}{|X|^3}(x_1, x_2, \ldots, x_n) = \frac{X}{|X|^3}$. Note in addition

$|\nabla f(X)| = |\frac{X}{|X|^3}| = \frac{|X|}{|X|^3} = \frac{1}{|X|^2}$.

Exercises 15.4: Directional Derivatives (page 646)

1. Since $\nabla f(x,y) = (\frac{\partial f}{\partial x}, \frac{\partial f}{\partial y})$, we have $\nabla f(x,y) = (2xy^4, 4x^2y^3)$. Thus, $\nabla f(1,1)$ $= (2,4)$. Then since $D_{(U)}f(1,1) = \nabla f(1,1) \cdot U$, we have $D_{(U)}f(1,1)$

$= (2,4) \cdot \frac{1}{\sqrt{5}}(1,2) = \frac{10}{\sqrt{5}} = 2\sqrt{5}$.

3. In this case $\nabla f(x,y) = (2xy^3, 3x^2y^2)$ so $\nabla f(2,1) = (4,12)$. Since

$U = (\cos \pi/4, \sin \pi/4) = (1/\sqrt{2}, 1/\sqrt{2})$ is a unit vector in the desired direc-

tion, $D_{(\pi/4)}f(2,1) = (4,12) \cdot (1/\sqrt{2}, 1/\sqrt{2}) = 8\sqrt{2}$.

5. Since $\nabla f(x,y) = (\frac{\partial f}{\partial x}, \frac{\partial f}{\partial y})$, we have $\nabla f(x,y) = (2x + 2y, 2x + 2y)$. Thus,

$f(2,1) = (6,6)$. Then since $D_{(U)}f(2,1) = \nabla f(2,1) \cdot U$, we have

$D_{(U)}f(2,1) = (6,6) \cdot \dfrac{1}{\sqrt{2}} (1,1) = \dfrac{12}{\sqrt{2}} = 6\sqrt{2}$.

7. Since $\nabla f(x,y) = (e^{xy} + xye^{xy}, x^2e^{xy})$, we have $\nabla f(1,2) = (3e^2, e^2)$. Then since

$U = (\cos (-\pi/4), \sin (-\pi/4)) = (1/\sqrt{2}, -1/\sqrt{2})$ is a unit vector making an angle

$-\pi/4$ with the positive x-axis, we have $D_{(-\pi/4)}f(1,2) = \nabla f(1,2) \cdot U$

$= (3e^2, e^2) \cdot (1/\sqrt{2}, -1/\sqrt{2}) = \sqrt{2}e^2$.

9. Since $\nabla f(x,y) = (e^x \cos y, -e^x \sin y)$, we have $\nabla f(0,0) = (1,0)$. Then since

$U = (\cos 2\pi/3, \sin 2\pi/3) = (-1/2, \sqrt{3}/2)$ is a unit vector making an angle

$2\pi/3$ with the positive x-axis, we have $D_{(2\pi/3)}f(0,0) = \nabla f(0,0) \cdot U$

$= (1,0) \cdot (-1/2, \sqrt{3}/2) = -1/2$.

11. Since $\nabla f(x,y) = (3x^2 + 3y, 3x)$, we have $\nabla f(0,1) = (3,0)$. Then since

$U = (\cos 3\pi/4, \sin 3\pi/4) = (-1/\sqrt{2}, 1/\sqrt{2})$ is a unit vector making an angle $3\pi/4$

with the positive x-axis, we have $D_{(3\pi/4)}f(0,1) = \nabla f(0,1) \cdot U$

$= (3,0) \cdot (-1/\sqrt{2}, 1/\sqrt{2}) = -3/\sqrt{2}$.

13. In this case $\nabla f(x,y,z) = (y + 2xy, x - z + x^2, -y)$. Thus, $\nabla f(-1,1,2)$
$= (-1,-2,-1)$. Then since $D_{(U)}f(-1,1,2) = \nabla f(-1,1,2) \cdot U$, we have $D_{(U)}f(-1,1,2)$

$= (-1,-2,1) \cdot \dfrac{1}{\sqrt{3}} (1,1,-1) = -4/\sqrt{3}$.

15. Since $f(x_1,x_2,x_3,x_4) = x_1^2 + x_2^2 + x_3^2 + x_4^2$, we have

$\nabla f(x_1,x_2,x_3,x_4) = (2x_1, 2x_2, 2x_3, 2x_4)$ and $\nabla f(1,1,1,1) = (2,2,2,2)$. Then since

$D_{(U)}f(1,1,1,1) = \nabla f(1,1,1,1) \cdot U = (2,2,2,2) \cdot \dfrac{1}{\sqrt{7}} (1,-1,2,-1) = 2/\sqrt{7}$.

17. Since $\nabla f(x,y) = (2x, \sqrt{3}y)$, $\nabla f(1/2,1) = (1, \sqrt{3})$. Consequently, the maximum value

of the directional derivative at $(1/2,1)$ is $|\nabla f(1/2,1)| = |(1,\sqrt{3})| = 2$ and this

maximum value is in the direction of the unit vector $U = \dfrac{\nabla f(1/2,1)}{|\nabla f(1/2,1)|} = \dfrac{1}{2} (1,\sqrt{3})$.

19. Since $\nabla f(x,y) = (2x \sin xy + x^2 y \cos xy, x^3 \cos xy)$, $\nabla f(0,1) = (0,0)$. Consequently, the maximum value of the directional derivative at $(0,1)$ is 0. The value remains the same in all directions.

21. (a) The gradient of $f(x,y)$ is a vector in the direction of the maximal rate of change of $f(x,y)$. Moreover, the absolute value of the gradient is the maximal rate of change. In this case $\nabla f(x,y) = (e^{xy} + xye^{xy}, x^2 e^{xy})$ so, $\nabla f(2,0)$

$= (1,4)$. Thus, the magnitude of the maximal rate of change is $|\nabla f(2,0)| = \sqrt{17}$ and this maximal rate of change is in the direction of the unit vector

$$U = \frac{\nabla f(2,0)}{|\nabla f(2,0)|} = \frac{1}{\sqrt{17}} (1,4).$$

(b) If we let $g(x,y,z) = xe^{xy} - z$, we know that $\pm \nabla g(2,0,2)$ is normal to the graph of $g(x,y,z) = 0$, or $xe^{xy} - z = 0$ at $(2,0,2)$.

$\nabla g(x,y,z) = (e^{xy} + xye^{xy}, x^2 e^{xy}, -1)$. Thus, the vectors $\pm \nabla g(2,0,2) = \pm(1,4,-1)$ are normal to the surface at $(2,0,2)$.

25. Since $f(x,y) = f(y,x)$, $f_x(a,a) = f_y(a,a)$. Thus, $\nabla f(x,y) = (f_x(x,y), f_y(x,y))$.

Then at (a,a), $\nabla f(a,a) = (f_y(a,a), f_y(a,a))$. Consequently, the gradient vector

must be in the direction of the vector $\dfrac{1}{\sqrt{2}\, f_y(a,a)} (f_y(a,a), f_y(a,a))$

$= (1/\sqrt{2}, 1/\sqrt{2})$. This vector makes an angle of $\pi/4$ with the positive x-axis.

Exercises 15.5: Local Extrema of Functions of Two Variables (page 655)

1. Since $f_x(x,y) = 4x - 1$ and $f_y(x,y) = 4y + 2$, $(1/4,-1/2)$ is the only critical point. Then since $f_{xx} = 4$, $f_{yy} = 4$, and $f_{xy} = 0$, we have

$f_{xy}^2 (1/4,-1/2) - f_{xx} (1/4,-1/2) f_{yy}(1/4,-1/2) = -16 < 0$. Thus, since $f_{xx} (1/4,-1/2) > 0$, $(1/4,-1/2)$ determines a local minimum.

3. Since $f_x(x,y) = y - 2$ and $f_y(x,y) = x + 3$, we know that $(-3,2)$ is the only critical point. Then since $f_{xx}(x,y) = 0$, $f_{yy}(x,y) = 0$, and $f_{xy}(x,y) = 1$, we have $f_{xy}^2(-3,2) - f_{xx}(-3,2) f_{yy}(-3,2) = 1 > 0$. Thus, $(-3,2)$ is a saddle point.

5. Since $f_x(x,y) = 2x + 4y$ and $f_y(x,y) = 4x - 2y$, we solve the equations

$$2x + 4y = 0$$
$$4x - 2y = 0$$

to find the critical points. In this case $(0,0)$ is the only critical point. Then since $f_{xx}(x,y) = 2$, $f_{yy}(x,y) = -2$, and $f_{xy}(x,y) = 4$, we have

$f_{xy}^2(0,0) - f_{xx}(0,0) f_{yy}(0,0) = 16 + 4 > 0$. Thus, $(0,0)$ is a saddle point.

7. Since $f_x(x,y) = 3x^2 - 3$ and $f_y(x,y) = 2y$, we find that $(1,0)$ and $(-1,0)$ are the only possible critical points. Since $f_{xx}(x,y) = 6x$, $f_{xy}(x,y) = 0$, and $f_{yy}(x,y) = 2$, we have $f_{xy}^2(x,y) - f_{xx}(x,y) f_{yy}(x,y) = -12x$. Thus,

$f_{xy}^2(1,0) - f_{xx}(1,0) f_{yy}(1,0) = -12 < 0$ and

$f_{xy}^2(-1,0) - f_{xx}(-1,0) f_{yy}(-1,0) = 12 > 0$. Thus, $(1,0)$ determines a local minimum while $(-1,0)$ is a saddle point.

9. Here $f_x(x,y) = 3x^2 - 12y^2$ and $f_y(x,y) = -24xy + 3y^2 + 45$. Thus, we want to find all points (x,y) such that $3x^2 - 12y^2 = 0$ and $-24xy + 3y^2 + 45 = 0$. From the first equation we have $x = \pm 2y$. If $x = 2y$ the second equation becomes $-48y^2 + 3y^2 + 45 = 0$, or $-45y^2 + 45 = 0$. Thus, $y = \pm 1$, and $(2,1)$ and $(-2,-1)$ are critical points. If $x = -2y$ the second equation becomes $48y^2 + 3y^2 + 45 = 0$, or $51y^2 + 45 = 0$. Since this equation has no real zeros, $(2,1)$ and $(-2,-1)$ are the only critical points. Since $f_{xx}(x,y) = 6x$, $f_{yy}(x,y) = -24x + 6y$, and $f_{xy}(x,y) = -24y$, we have $f_{xy}^2(2,1) - f_{xx}(2,1) f_{yy}(2,1)$

$= -24 - (12)(-42) > 0$ and $f_{xy}^2(-2,-1) - f_{xx}(-2,-1) f_{yy}(-2,-1) = 24 - (-12)(42)$

> 0. Consequently, both $(2,1)$ and $(-2,-1)$ are saddle points.

11. In this case $f_x(x,y) = 6x - 4y - 2$ and $f_y(x,y) = -4x + 10y + 3$. Simultaneous solution of the equations $6x - 4y - 2 = 0$
$$-4x + 10y + 3 = 0$$
gives $x = 2/11$, $y = -5/22$. Consequently, $(2/11, -5/22)$ is the only critical point. Since $f_{xx}(x,y) = 6$, $f_{yy}(x,y) = 10$, and $f_{xy}(x,y) = -4$, we have

$f_{xy}^2(2/11,-5/22) - f_{xx}(2/11,-5/22) f_{yy}(2/11,-5/22) = 16 - 60 < 0$. Then since

$f_{xx}(2/11,-5/22) = 6 > 0$, $(2/11,-5/22)$ determines a local minimum.

13. Here $f_x(x,y) = 2(x + y) + 4x^3$ and $f_y(x,y) = 2(x + y)$. Thus, we want to find those points (x,y) such that $2(x + y) + 4x^3 = 0$ and $2(x + y) = 0$. Substitution of the second equation into the first gives $4x^3 = 0$, or $x = 0$. Then substitution into the second equation gives $y = 0$. Consequently, the only critical point to be considered is $(0,0)$. Since $f_{xx}(x,y) = 2 + 12x^2$, $f_{yy}(x,y) = 2$,

$f_{xy}(x,y) = 2$, we have $f_{xy}^2(0,0) - f_{xx}(0,0) f_{yy}(0,0) = 4 - 2 \cdot 2 = 0$ and no conclusion can be drawn from the second derivative test. However, since $f(x,y)$

$= (x + y)^2 + x^4 \geq 0$ and $f(0,0) = 0$, we see that $(0,0)$ must determine a local minimum.

17. In this case $f_x(x,y) = -6x - 3y + 9$ and $f_y(x,y) = 6y^2 - 3x$. Then $f_x(x,y) = 0$ and $f_y(x,y) = 0$ implies $x = 2$ and $y = -1$ or $x = 9/8$ and $y = 3/4$. Since $f_{xx}(x,y) = -6$, $f_{xy}(x,y) = -3$, and $f_{yy}(x,y) = 12y$, we have

$f_{xy}^2(x,y) - f_{xx}(x,y) \, f_{yy}(x,y) = 9 + 72y$. Thus,

$f_{xy}^2(2,-1) - f_{xx}(2,-1) \, f_{yy}(2,-1) < 0$ and

$f_{xy}^2(9/8,3/4) - f_{xx}(9/8,3/4) \, f_{yy}(9/8,3/4) > 0$. Consequently, $(2,-1)$ determines a local maximum and $(9/8,3/4)$ is a saddle point.

19. Since $f_x(x,y) = -2yx$ and $f_y(x,y) = 9y^2 - x^2 + 1$. Thus, we must solve the equations $-2yx = 0$, $9y^2 - x^2 + 1 = 0$. From the first equation we see that $x = 0$ or $y = 0$. Substitution into the second equation gives $9y^2 + 1 = 0$ when $x = 0$. Since $9y^2 + 1 > 0$ for all y, $x = 0$ does not result in a solution to the second equation. When $y = 0$ the second equation becomes $-x^2 + 1 = 0$, hence $x = \pm 1$. Thus, the critical points are $(1,0)$ and $(-1,0)$. Since $f_{xx}(x,y) = -2y$, $f_{yy}(x,y) = 18y$, and $f_{xy}(x,y) = -2x$, we have

$f_{xy}^2(1,0) - f_{xx}(1,0) \, f_{yy}(1,0) = 4 - 0 > 0$ and so $(1,0)$ is a saddle point. Since $f_{xy}^2(-1,0) - f_{xx}(-1,0) \, f_{yy}(-1,0) = 4 - 0 > 0$, $(-1,0)$ is also a saddle point.

21. Since $f_x(x,y) = \dfrac{(x + y) - (x - y)}{(x + y)^2} = \dfrac{2y}{(x + y)^2}$ and $f_y(x,y) = \dfrac{-(x + y) - (x - y)}{(x + y)^2}$

$= \dfrac{-2x}{(x + y)^2}$, we must solve the equations $\dfrac{2y}{(x + y)^2} = 0$ and $\dfrac{2x}{(x + y)^2} = 0$. Thus, $(0,0)$ is the only possibility for a critical point. However, since $(0,0)$ is not in the domain of the function, $(0,0)$ is <u>not</u> a critical point. This function has no local extrema.

25. Here $f_x(x,y) = 1 + y \cos x$ and $f_y(x,y) = \sin x$. $f_y(x,y) = 0$ implies $x = k\pi$ where $k = 0, \pm 1, \pm 2, \ldots$ Since $\cos(k\pi) = (-1)^k$, we must have $f_x(k\pi,y) = 1 + (-1)^k y$. Thus, $f_x(k\pi,y) = 0$ implies $y = (-1)^{k+1}$. The critical points are $(k\pi,(-1)^{k+1})$, $k = 0, \pm 1, \pm 2, \ldots$ Since $f_{xx}(x,y) = -y \sin x$, $f_{xy}(x,y) = \cos x$, and $f_{yy}(x,y) = 0$, we have

$f_{xy}^2(x,y) - f_{xx}(x,y) \, f_{yy}(x,y) = \cos^2 x > 0$ where $x = k\pi$. Consequently, the

points $(k\pi, (-1)^{k+1})$, $k = 0, \pm1, \pm2, \ldots$ are all saddle points.

Exercises 15.6: Absolute Maxima and Minima (page 661)

1. Since $f_x(x,y) = -2$ and $f_y(x,y) = 1$ are never zero, the absolute maximum and absolute minimum must occur on the boundary of the rectangle $-2 \leq x \leq 0$, $1 \leq y \leq 2$. On the boundary we have

   $f(x,y) = y - 6 \qquad\qquad x = -2,\ 1 \leq y \leq 2,$

   $f(x,y) = y - 10 \qquad\quad x = 0,\ 1 \leq y \leq 2,$

   $f(x,y) = -2x - 9 \qquad\ y = 1,\ -2 \leq x \leq 0,$

   $f(x,y) = -2x - 8 \qquad\ y = 2,\ -2 \leq x \leq 0.$

   Since each of these restrictions is linear, its absolute extrema must occur at the endpoints of its interval of definition. Thus, $f(x,y)$ must attain its extreme values at $(-2,1)$, $(-2,2)$, $(0,1)$, or $(0,2)$. Since $f(-2,1) = -5$, $f(-2,2) = -4$, $f(0,1) = -9$, and $f(0,2) = -8$, $(-2,2,4)$ is the absolute maximum point and $(0,1,-9)$ is the absolute minimum point.

3. Since $f_x(x,y) = 3$ and $f_y(x,y) = 4$ are never zero, the absolute extrema must occur on the boundary of the region determined by $x \leq 1$, $y \geq 0$, $y \leq x$. This region along with the restrictions of $f(x,y)$ to the boundary are illustrated below.

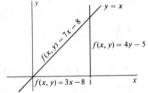

   Since the restrictions $f(x,y) = 7x - 8$, when $x = y$ and $0 \leq x \leq 1$,

   $\qquad\qquad\qquad\quad f(x,y) = 3x - 8$, when $y = 0$ and $0 \leq x \leq 1$,

   $\qquad\qquad\qquad\quad f(x,y) = 4y - 5$, when $x = 1$ and $0 \leq y \leq 1$

   are all linear, the absolute extrema can only occur at the endpoints $(0,0)$, $(1,0)$, $(1,1)$. Direct computation gives $f(0,0) = -8$, $f(1,0) = -5$, and $f(1,1) = -1$. Thus, $(1,1,-1)$ is the absolute maximum point and $(0,0,-8)$ is the absolute minimum point.

5. Since $f_x(x,y) = 2x$ and $f_y(x,y) = 2$ are never both zero, the absolute extrema must occur on the boundary of the rectangle. On the boundary we have

   $$f(x,y) = 1 + 2y,\ x = -1,\ -1 \leq y \leq 1,$$
   $$f(x,y) = 1 + 2y,\ x = 1,\ -1 \leq y \leq 1,$$
   $$f(x,y) = x^2 - 2,\ y = -1,\ -1 \leq x \leq 1,$$
   $$f(x,y) = x^2 + 2,\ y = 1,\ -1 \leq x \leq 1.$$

   The function $f(x,y) = 1 + 2y$ is linear and so must attain its extreme values at the endpoints of its interval of definition. Thus, $f(x,y)$ must attain its extreme values at $(-1,-1)$, $(-1,1)$, $(1,-1)$, or $(1,1)$. The function $f(x,y)$ $= x^2 - 2$, $y = -1$, $-1 \leq x \leq 1$ attains its extreme values either at $x = 0$ (where the derivative is zero) or at the endpoints $x = \pm 1$. The function $f(x,y)$ $= x^2 + 2$, $y = 1$, $-1 \leq x \leq 1$ attains its extreme values either at $x = 0$ (where the derivative is zero) or at the endpoints $x = \pm 1$. Consequently, $f(x,y)$ must attain its extreme values at $(-1,-1)$, $(-1,1)$, $(1,-1)$, $(1,1)$, $(0,1)$, or $(0,-1)$. Direct computation yields $f(-1,-1) = -1$, $f(-1,1) = 3$, $f(1,-1) = -1$, $f(1,1) = 3$, $f(0,1) = 2$, $f(0,-1) = -2$. Thus, $(1,1,3)$ and $(-1,1,3)$ are absolute maximum points and $(0,-1,-2)$ is the absolute minimum point.

9. Since $f_x(x,y) = 1 \neq 0$, $f(x,y)$ must attain its extreme values on the boundary of the rectangle. On the boundary we have

$f(x,y) = 7y^3 - 2$, $x = -2$, $-1 \leq y \leq 1$,

$f(x,y) = 7y^3 + 2$, $x = 2$, $-1 \leq y \leq 1$,

$f(x,y) = x + 7$, $y = 1$, $-2 \leq x \leq 2$,

$f(x,y) = x - 7$, $y = -1$, $-2 \leq x \leq 2$.

The function $f(x,y) = 7y^3 - 2$ attains its extreme values either at $y = 0$ (where the derivative is zero) or at the endpoints of the interval of definition. Consequently, $f(x,y)$ may attain its extreme values at $(-2,0)$, $(-2,-1)$, or $(-2,1)$. The function $f(x,y) = 7y^3 + 2$ attains its extreme values either at $y = 0$ (where the derivative is zero) or at the endpoints. Thus, the points $(2,0)$, $(2,-1)$, and $(2,1)$ must be considered. The functions $f(x,y) = x + 7$ and $f(x,y) = x - 7$ are linear and so may attain extreme values only at the endpoints of their intervals of definition. Consequently, we must consider the points $(-2,1)$, $(2,1)$, $(-2,-1)$, and $(2,-1)$. Direct computation gives $f(-2,1)$ $= 5$, $f(2,1) = 9$, $f(-2,-1) = -9$, $f(2,-1) = -5$, $f(2,0) = 2$, and $f(-2,0) = -2$. Thus, $(2,1,9)$ is the absolute maximum point and $(-2,-1,-9)$ is the absolute minimum point.

11. Here $f_x(x,y) = 2x - 4y + 6$ and $f_y(x,y) = 2y - 4x$. On setting $f_x(x,y) = 0$ and $f_y(x,y) = 0$ and solving for x and y, we obtain the point $(1,2)$. When $x = 2$, we have $f(x,y) = y^2 - 8y + 16$, $0 \leq y \leq 6$. Since this function has a zero derivative when $y = 4$, we must also test the point $(2,4)$. When $x = 0$, the function becomes $f(x,y) = y^2$. This function has zero derivative only at $(0,0)$. Finally, if $y = 3x$, we have $f(x,y) = -2x^2 + 6x$, a function which has a zero derivative at $(3/2,9/2)$. Inclusion of all corner points completes the list of points to be checked. Thus, we must check $(1,2)$, $(2,4)$, $(0,0)$, $(3/2,9/2)$, $(2,0)$, $(2,6)$. Direct computation shows that $(2,0,16)$ is the absolute maximum point and

(0,0,0) and (2,4,0) are absolute minimum points.

13. Since $f_x(x,y) = 2x$ and $f_y(x,y) = 2y$ are both zero at $(0,0)$, the absolute extrema must occur either at $(0,0)$ or on the boundary of the disk, $x^2 + y^2 = 1$.

Since $f(x,y) = 1 + 14 = 15$ when $x^2 + y^2 = 1$ and $f(0,0) = 14$, the absolute maximum value of 15 occurs at any boundary point and $(0,0,14)$ is the absolute minimum point.

17. Since $f_x(x,y) = 2x$ and $f_y(x,y) = -4$ are never both zero, the absolute extrema of $f(x,y)$ must occur on the boundary of the region, that is, where $x^2 + y^2 = 9$.

On the boundary $x^2 = 9 - y^2$, and $-3 \leq y \leq 3$. Thus, $f(x,y) = 2 - 9 + y^2 - 4y$ $= y^2 - 4y - 7$, $-3 \leq y \leq 3$ on the boundary, $x^2 + y^2 = 9$. The function

$f(x,y) = y^2 - 4y - 7$ attains its extreme values either where its derivative is zero, at $y = 2$, or at the endpoints $y = 3$ and $y = -3$. Since all these points are on the boundary where $x^2 = 9 - y^2$, we have $(0,3)$, $(0,-3)$, $(\sqrt{5},2)$, and

$(-\sqrt{5},2)$ as a complete list of points that could possibly determine absolute extrema of $f(x,y)$. Direct calculation gives $f(0,3) = -10$, $f(0,-3) = 14$,

$f(\sqrt{5}, 2) = -11$, and $f(-\sqrt{5},2) = -11$. Thus, $(0,-3,14)$ is the absolute maximum point and both $(\sqrt{5},2,-11)$ and $(-\sqrt{5},2,-11)$ are absolute minimum points.

19. Since $f_x(x,y) = 2$ and $f_y(x,y) = 8y$ are never both zero, the absolute extrema of $f(x,y)$ must occur on the boundary of the region, that is, where $x^2/4 + y^2$ $= 1$. On the boundary $y^2 = 1 - x^2/4$, and $-2 \leq x \leq 2$. Thus,

$f(x,y) = 4 - x^2 + 2x - 3 = -x^2 + 2x + 1$, $-2 \leq x \leq 2$, on the boundary $x^2/4 + y^2$ $= 1$. The function $f(x,y) = -x^2 + 2x + 1$, $-2 \leq x \leq 2$ attains its extreme values either where its derivative is zero, at $x = 1$, or at the endpoints $x = -2$ and

$x = 2$. Since all these points are on the boundary where $y^2 = 1 - x^2/4$, we have

$(1,\sqrt{3}/2)$, $(1,-\sqrt{3}/2)$, $(-2,0)$, $(2,0)$ as a complete list of points that could possibly determine absolute extrema of $f(x,y)$. Direct calculation gives $f(1,\sqrt{3}/2)$ $= 2$, $f(1,-\sqrt{3}/2) = 2$, $f(-2,0) = -7$, and $f(2,0) = 1$. Thus, both $(1,\sqrt{3}/2,2)$ and

$(1,-\sqrt{3}/2,2)$ are absolute maximum points and $(-2,0,-7)$ is the absolute minimum point.

21. Since $f_x(x,y) = 2x + 2y$ and $f_y(x,y) = -2y + 2x$ are only both 0 at $(0,0)$, this point is the only possible place where $f(x,y)$ could attain an absolute extreme. However, since $f_{xx}(x,y) = 2$, $f_{yy}(x,y) = -2$, and $f_{xy}(x,y) = 2$, we have

$f_{xy}^2(0,0) - f_{xx}(0,0) \, f_{yy}(0,0) = 4 + 4 > 0$. Thus, $(0,0)$ is a saddle point and cannot determine an extreme of $f(x,y)$.

23. Since $P_x(x,y) = 100 - 10y - 2x$ and $P_y(x,y) = 130 - 10x$, we find that both $P_x(x,y) = 0$ and $P_y(x,y) = 0$ only at $(13,7.4)$. Since $x \leq 10$ and $y \leq 5$, this point is not in the region specified by plant capacity. The restrictions of $P(x,y)$ to the boundary of the region $0 \leq x \leq 10$, $0 \leq y \leq 5$ are:

$P(x,y) = 130y - 13$ when $x = 0$ and $0 \leq y \leq 5$,

$P(x,y) = 30y + 887$ when $x = 10$ and $0 \leq y \leq 5$,

$P(x,y) = -x^2 + 100x - 13$ when $y = 0$ and $0 \leq x \leq 10$,

$P(x,y) = -x^2 + 50x - 637$ when $y = 5$ and $0 \leq x \leq 10$.

The first two restrictions are linear and so can attain their maximum only at the endpoints $(0,0)$, $(0,5)$, $(10,0)$, and $(10,5)$. The function $P(x,y) = -x^2 + 100x - 13$ has zero derivative at $x = 50$, and the function $P(x,y) = -x^2 + 50x + 637$ has a zero derivative at $x = 25$. Since we must have $x \leq 10$, these points are not in the region specified by plant capacity. Consequently, we need only consider the points $(0,0)$, $(0,5)$, $(10,0)$, and $(10,5)$. Direct calculation gives $P(0,0) = -13$, $P(0,5) = 637$, $P(10,0) = 887$, $P(10,5) = 1037$. Thus, the maximum profit occurs when 10 units of the first item are produced and 5 units of the second are produced.

25. The total revenue from the sale of x units at a price of $p(x)$ per unit and y units at a price of $q(y)$ per unit is $xp(x) + yq(y) = 5(10)^4 x - 5x^2 + 4(10)^4 y - 3y^2$. The cost of producing the $x + y$ units is $10^7 + 8(10)^3 (x + y) + 5(x + y)^2$. Thus, the profit is given by $P(x,y) = 5(10)^4 x - 5x^2 + 4(10)^4 y - 3y^2 - 10^7 - 8(10)^3 (x + y) - 5(x + y)^2$. Then $\frac{\partial P}{\partial x} = 5(10)^4 - 10x - 8(10)^3 - 10(x + y) = 5(10)^4 - 8(10)^3 - 20x - 10y$ and $\frac{\partial P}{\partial y} = 4(10)^4 - 6y - 8(10)^3 - 10(x + y) = 4(10)^4 - 8(10)^3 - 10x - 16y$. Solution of the equations $5(10)^4 - 8(10)^3 - 20x - 10y = 0$ and $4(10)^4 - 8(10)^3 - 10x - 16y = 0$ gives $x = 1600$ and $y = 1000$. Since $P_{xx} = -20$, $P_{yy} = -16$, and $P_{xy} = -10$,

$P_{xy}^2(1600,1000) - P_{xx}(1600,1000) P_{yy}(1600,1000) < 0$. Thus, $x = 1600$ and $y = 1000$ maximize the profit.

Technique Review Exercises, Chapter 15 (page 665)

1. $f_x(x,y) = \cos x + \tan(xy) + xy \sec^2(xy)$, $f_y(x,y) = x^2 \sec^2(xy)$,

$f_{xy}(x,y) = x \sec^2(xy) + x \sec^2(xy) + x^2 y \sec^2(xy) \tan(xy)$

$= 2x \sec^2(xy) + x^2 y \sec^2(xy) \tan(xy)$.

2. Since $f(X) = \frac{1}{|X|}$, we have $f(x_1, x_2, \ldots, x_n) = (x_1^2 + x_2^2 + \ldots + x_n^2)^{-1/2}$. Thus,

$\frac{\partial f}{\partial x_i} = -\frac{1}{2}(x_1^2 + x_2^2 + \ldots + x_n^2)^{-3/2} 2x_i = -\frac{x_i}{|X|^3}$ since $|X| = (x_1^2 + x_2^2$

$+ \ldots + x_n^2)^{1/2}$.

3. Substitution into $\frac{df}{dt} = \frac{\partial f}{\partial x}\frac{dx}{dt} + \frac{\partial f}{\partial y}\frac{dy}{dt}$ gives $\frac{df}{dt} = \frac{2x - 3y}{x^2 - 3xy} 2 + \frac{-3x}{x^2 - 3xy} (-\sin t)$.

Then since $x = 1$ and $y = 1$ when $t = 0$, we have $\frac{df}{dt} = 1 + 0 = 1$ when $t = 0$.

4. Substitution into $\frac{\partial f}{\partial u} = \frac{\partial f}{\partial x}\frac{\partial x}{\partial u} + \frac{\partial f}{\partial y}\frac{\partial y}{\partial u}$ gives $\frac{\partial f}{\partial u} = (y^2 - 3y)\cos v + (2xy - 3x)$

$\sin v$. Then since $x = -2$ and $y = 0$ when $u = 2$ and $v = \pi$, we have $\frac{\partial f}{\partial u} = 0$ when

$u = 2$ and $v = \pi$.

5. The vectors $\pm \nabla g(a,b,c)$ are normal to the surface $g(x,y,z) = k$ at (a,b,c).

Thus, if we let $g(x,y,z) = x^2 - 3xy + y^2 - z$, we have $\nabla g(x,y,z) = (2x - 3y,$

$-3x + 2y, -1)$. Thus, $\pm \nabla g(2,-1,11) = \pm(7,-8,-1)$ are both normal to the graph of

$x^2 - 3xy + y^2 - z = 0$ at $(2,-1,11)$. Since we want a unit vector, we divide by

$|\nabla g(2,-1,11)| = \sqrt{7^2 + 8^2 + 1} = \sqrt{114}$. Then $\pm \frac{1}{\sqrt{114}} (7,-8,-1)$ are unit vectors

normal to the graph of $x^2 - 3xy + y^2 = z$ at $(2,-1,11)$.

6. Since $f(x,y,z) = x(yz - z^2)$, we have $\nabla f(x,y,z) = (yz - z^2, xz, x(y - 2z))$, or

$\nabla f(1,0,-1) = (-1,-1,2)$. The direction derivative of $f(x,y,z)$ at the point

(a,b,c) in the direction of the unit vector U is given by $D_U f(a,b,c) = \nabla f(a,b,c)$

$\cdot U$. Since $U = \frac{1}{\sqrt{13}} (3,0,2)$ is the unit vector in the direction of $(3,0,2)$, the

desired directional derivative is $D_U f(1,0,-1) = (-1,-1,2) \cdot \frac{1}{\sqrt{13}} (3,0,2) = 1/\sqrt{13}$.

7. The gradient of $f(x,y,z)$ at $(1,0,-1)$, $\nabla f(1,0,-1)$, is a vector whose absolute

value is the maximal directional derivative of $f(x,y,z)$ at $(1,0,-1)$. Moreover,

this vector points in the direction of the maximal directional derivative.

Since $\nabla f(x,y,z) = (yz - z^2, xz, z(y - 2z))$, we have $\nabla f(1,0,-1) = (-1,-1,2)$.

Consequently, the maximal rate of change is $|\nabla f(1,0,-1)| = \sqrt{1 + 1 + 4} = \sqrt{6}$ and

this maximal rate of change occurs in the direction of the unit vector

$U = \frac{\nabla f(1,0,-1)}{|\nabla f(1,0,-1)|} = \frac{1}{\sqrt{6}} (-1,-1,2)$.

8. Since $f_x(x,y) = 3x - 3$ and $f_y(x,y) = 2y$ are both zero only at $(1,0)$, $(1,0)$ is

the only critical point to be considered. Since $f_{xx}(x,y) = 3$, $f_{yy}(x,y) = 2$, and $f_{xy}(x,y) = 0$, we have $f_{xy}^2(1,0) - f_{xx}(1,0) f_{yy}(1,0) = -6 < 0$. Then since $f_{xx}(1,0) = 3 > 0$, we know that $(1,0,-2)$ is a local minimum point of $f(x,y)$.

9. Since $f_x(x,y) = 2x - 12$ and $f_y(x,y) = 6y$ are only both zero at $(6,0)$, and since this point does not lie in the region $x^2 + y^2 \leq 25$, the absolute extrema must occur on the boundary of the region. The restriction of $f(x,y)$ to the boundary is, where $y^2 = 25 - x^2$, $f(x,y) = x^2 - 12x + 3(25 - x^2) = -2x^2 - 12x + 75$ for $-5 \leq x \leq 5$. The absolute extrema of this function must occur either where its derivative is zero, at $x = -3$, or at the endpoints $x = \pm 5$. Consequently, the absolute extrema occur at $(-3,4)$, $(-3,-4)$, $(5,0)$, or $(-5,0)$. Direct calculation gives $f(-3,4) = 93$, $f(-3,-4) = 93$, $f(5,0) = -35$, $f(-5,0) = 85$. Thus, $(-3,4,93)$ and $(-3,-4,93)$ are both absolute maximum points and $(5,0,-35)$ is the absolute minimum point.

Additional Exercises, Chapter 15 (page 665)

1. (a) $\dfrac{\partial f}{\partial x} = 2xy + 3y^2$, (b) $\dfrac{\partial f}{\partial y} = x^2 + 6xy$

3. (a) $\dfrac{\partial f}{\partial x} = \dfrac{(x + y) - (x - y)}{(x + y)^2} = \dfrac{2y}{(x + y)^2}$,

(b) $\dfrac{\partial f}{\partial y} = \dfrac{-(x + y) - (x - y)}{(x + y)^2} = \dfrac{-2x}{(x + y)^2}$

5. (a) $\dfrac{\partial f}{\partial x} = y^2 - 2xye^{-x^2} - e^z$, (b) $\dfrac{\partial f}{\partial y} = 2xy + e^{-x^2}$

7. $\dfrac{\partial^2 f}{\partial x^2} = \dfrac{\partial}{\partial x}\left(\dfrac{\partial f}{\partial x}\right) = \dfrac{\partial}{\partial x}(2xy + 3y^2) = 2y.$

$\dfrac{\partial^2 f}{\partial y^2} = \dfrac{\partial}{\partial y}\left(\dfrac{\partial f}{\partial y}\right) = \dfrac{\partial}{\partial y}(x^2 + 6xy) = 6x.$

9. $\dfrac{\partial^2 f}{\partial x \partial y} = \dfrac{\partial}{\partial x}\left(\dfrac{\partial f}{\partial y}\right) = \dfrac{\partial}{\partial x}(x^2 + 6xy) = 2x + 6y.$

$\dfrac{\partial^2 f}{\partial y \partial x} = \dfrac{\partial}{\partial y}\left(\dfrac{\partial f}{\partial x}\right) = \dfrac{\partial}{\partial y}(2xy + 3y^2) = 2x + 6y.$

11. Since $f(X) = |X|^4$, we have $f(x_1,\ldots,x_i,\ldots,x_n) = (x_1^2 + x_2^2 + \ldots + x_i^2 + \ldots + x_n^2)^2$. Thus, $f_{x_i}(x_1,\ldots,x_i,\ldots,x_n) = 4x_i(x_1^2 + x_2^2 + \ldots + x_i^2 + \ldots + x_n^2)$, or $f_{x_i}(X) = 4x_i|X|^2$.

13. Since $\dfrac{df}{dt} = \dfrac{\partial f}{\partial x}\dfrac{dx}{dt} + \dfrac{\partial f}{\partial y}\dfrac{dy}{dt}$, we have $\dfrac{df}{dt} = 2x \cos t - 2y \sin t$. Then since $x = \sin t$ and $y = \cos t$, $\dfrac{df}{dt} = 2 \sin t \cos t - 2 \cos t \sin t = 0$.

15. Since $\frac{df}{dt} = \frac{\partial f}{\partial x}\frac{dx}{dt} + \frac{\partial f}{\partial y}\frac{dy}{dt} + \frac{\partial f}{\partial z}\frac{dz}{dt}$, we have $\frac{df}{dt} = yz \sec(xyz) \tan(xyz)$

$+ e^t xz \sec(xyz) \tan(xyz) + 2t\, xy \sec(xyz) \tan(xyz)$. When $t = 0$ we have

$x = 0$, $y = 1$, and $z = 0$. Thus, when $t = 0$, $\frac{df}{dt} = 0$.

17. Since $\frac{\partial f}{\partial u} = \frac{\partial f}{\partial x}\frac{\partial x}{\partial u} + \frac{\partial f}{\partial y}\frac{\partial y}{\partial u}$, we have $\frac{\partial f}{\partial u} = (1 + 2x) \cos v + 2y \sin v$. Thus, since

$x = 0$ and $y = 1$ when $u = 1$ and $v = \pi/2$, we have $\frac{\partial f}{\partial u} = 2$ when $u = 1$ and $v = \pi/2$.

Similarly, $\frac{\partial f}{\partial v} = \frac{\partial f}{\partial x}\frac{\partial x}{\partial v} + \frac{\partial f}{\partial y}\frac{\partial y}{\partial v}$. Thus, $\frac{\partial f}{\partial v} = -(1 + 2x) u \sin v + 2yu \cos v$.

When $u = 1$ and $v = \pi/2$ we have $\frac{\partial f}{\partial v} = -1$.

19. Since $\frac{\partial f}{\partial u} = \frac{\partial f}{\partial x}\frac{\partial x}{\partial u} + \frac{\partial f}{\partial y}\frac{\partial y}{\partial u} + \frac{\partial f}{\partial z}\frac{\partial z}{\partial u}$, we have $\frac{\partial f}{\partial u} = (z \sec^2 x)(2u) + ze^y v$

$+ (\tan x + e^y) v^3$, or $\frac{\partial f}{\partial u} = 2uz \sec^2 x + vze^y + v^3(\tan x + e^y)$. Similarly,

$\frac{\partial f}{\partial v} = (z \sec^2 x)(0) + ze^y u + (\tan x + e^y)(3uv^2) = uze^y + 3uv^2 (\tan x + e^y)$.

21. Since $\nabla f(x,y) = (f_x(x,y), f_y(x,y))$, we have $\nabla f(x,y) = (3(x - y)^2, -3(x - y)^2)$.

Thus, $\nabla f(2,1) = (3,-3)$.

23. Since $\nabla f(x,y,z) = (f_x(x,y,z), f_y(x,y,z), f_z(x,y,z))$, we have

$\nabla f(x,y,z) = (\frac{z}{x}, \frac{z}{y}, \ln |xy|)$.

25. The vectors $\pm\nabla g(a,b,c)$ are both normal to the graph of $g(x,y,z) = 0$ at (a,b,c).

Thus, if we take $g(x,y,z) = 5x^2 + 4y^2 + z - 17$ we have $\nabla g(x,y,z) = (10x,8y,1)$.

Consequently, the vectors $\pm\nabla g(-1,1,8) = \pm(-10,8,1)$ are normal to the given

surface at $(-1,1,8)$. The desired unit vectors are thus $\pm \frac{1}{\sqrt{165}} (-10,8,1)$.

27. Since $D_{(\alpha)} f(a,b) = f_x(a,b) \cos \alpha + f_y(a,b) \sin \alpha$, we have

$D_{(\pi/6)} f(1,1) = 2 \cos \pi/6 + 2 \sin \pi/6 = \sqrt{3} + 1$.

29. Here $\nabla f(x,y) = (\ln (x^2 + y^2) + \frac{2x^2}{x^2 + y^2}, \frac{2xy}{x^2 + y^2})$ and so $\nabla f(1,0) = (2,0)$. Then

since

$D_{(U)} f(a,b) = \nabla f(a,b) \cdot U$, we get $D_{(U)} f(1,0) = (2,0) \cdot \frac{1}{\sqrt{10}} (3,-1) = 6/\sqrt{10}$.

31. In this case $\nabla f(X) = (2x_1, 2x_2, 2x_3, 2x_4) = 2X$. Thus, $\nabla f(2,2,2,2) = (4,4,4,4)$

and, $D_{(U)} f(2,2,2,2) = (4,4,4,4) \cdot \frac{1}{\sqrt{7}} (1,2,1,-1) = 12/\sqrt{7}$.

33. Since $f_x(x,y) = 4x$ and $f_y(x,y) = 2y$ are both zero only at $(0,0)$, $(0,0)$ is the

only critical point. Also, $f_{xx}(x,y) = 4$, $f_{yy}(x,y) = 2$, and $f_{xy}(x,y) = 0$.

Thus, $f_{xy}^2(0,0) - f_{xx}(0,0) f_{yy}(0,0) = -8 < 0$. Then since $f_{xx}(0,0) > 0$,

(0,0,0) is a local minimum point.

35. Since $f_x(x,y) = 2x$ and $f_y(x,y) = -2y$ are both zero only at $(0,0)$, $(0,0)$ is the only critical point. Also, $f_{xx}(x,y) = 2$, $f_{yy}(x,y) = -2$, and $f_{xy}(x,y) = 0$.

Thus, $f_{xy}{}^2(0,0) - f_{xx}(0,0)\, f_{yy}(x,y) = 4 > 0$ and $(0,0,0)$ is a saddle point.

37. In this case $f_x(x,y) = e^{-x(1+y)} - x(1 + y)\, e^{-x(1+y)} = e^{-x(1+y)}(1 - x - xy)$ and

$f_y(x,y) = -x^2\, e^{-x(1+y)}$. Note that $f_y(x,y) = 0$ only when $x = 0$. However, $f_x(0,y) = 1 \neq 0$ so the partial derivatives are not both zero at the same point. Consequently, there are no critical points.

39. Note that $f_x(x,y) = 2xy^3 + y^2 = y^2(2xy + 1)$ and $f_y(x,y) = 3x^2y^2 + 2xy + 1$. Use of the quadratic formula to find the values of xy for which $f_y(x,y) = 0$ gives

$xy = \dfrac{-2 \pm \sqrt{4 - 12}}{6}$. Consequently, the product xy is not real and so $f_y(x,y)$ is never zero. There are no critical points.

41. Here $f_x(x,y) = 12x - 24$ and $f_y(x,y) = 4y + 36$. Thus, both partial derivatives are zero only at $(2,-9)$. Since $f_{xx}(x,y) = 12$, $f_{yy}(x,y) = 4$, and $f_{xy}(x,y) = 0$,

we have $f_{xy}{}^2(2,-9) - f_{xx}(2,-9)\, f_{yy}(2,-9) = -48 < 0$. Then since $f_{xx}(2,-9) = 12 > 0$, $(2,-9,-184)$ is a local minimum point.

43. Since $f_x(x,y) = 3$ is never zero, the absolute maximum must occur on the boundary of the rectangle given by $1 \leq x \leq 4$ and $2 \leq y \leq 6$. Since the restrictions of $f(x,y) = 3x - 4y + 14$ to the boundary of the rectangle are all linear, the absolute extrema must occur at the endpoints of their intervals of definition. Consequently, we need only check the values of the function at $(1,2)$, $(1,6)$, $(4,2)$, and $(4,6)$. Direct calculation gives $f(1,2) = 9$, $f(1,6) = -7$, $f(4,2) = 18$, and $f(4,6) = 2$. Thus, the absolute maximum value is 18 and the absolute minimum value is -7.

45. Since $f_x(x,y) = 2x$ and $f_y(x,y) = 2y + 2$, the only point where both partial derivatives are zero is $(0,-1)$, a boundary point of the disk $x^2 + y^2 \leq 1$. The restriction of $f(x,y)$ to the boundary $x^2 + y^2 = 1$ is the linear function $f(x,y) = 2y - 9$, $-1 \leq y \leq 1$. Since the derivative of this function is never zero, it must attain its extreme values at $(0,-1)$ or $(0,1)$. Direct calculation gives $f(0,-1) = -11$ and $f(0,1) = -7$. Thus, -7 is the absolute maximum value and -11 is the absolute minimum value.

49. Since $f_x(x,y)$ is never zero, we know that the function $f(x,y)$ must attain its extreme values on the boundary of the disk, that is, where $x^2 + y^2 = 9$. The

restriction of $f(x,y)$ to the boundary is given by $f(x,y) = 9 - x^2 + 4x - 10$
$= -x^2 + 4x - 1$, $-3 \le x \le 3$. Since this function has a zero derivative at
$x = 2$, it must attain its extreme values at $(2,\sqrt{5})$, $(2,-\sqrt{5})$ or at the endpoints
$(-3,0)$ or $(3,0)$. Direct calculation gives $f(2,\sqrt{5}) = 3$, $f(2,-\sqrt{5}) = 3$, $f(-3,0)$
$= -22$, and $f(3,0) = 2$. Thus, the absolute maximum value is 3 and the absolute
minimum value is -22.

CHAPTER 16: MULTIPLE INTEGRATION

Exercises 16.1: Double Integration (page 675)

3. Since $f(x,y) = 6$ for all points of R, $f(x_i^*,y_i^*) = 6$ for all i. Also, each A_i is

the area of a rectangle of dimensions h and k where $h = \dfrac{10 - 3}{n} = \dfrac{7}{n}$ and

$k = \dfrac{9 - 4}{n} = \dfrac{5}{n}$, and $m = n^2$. Thus, $A_i = hk = \dfrac{35}{n^2}$, and $\displaystyle\int_R 6dA = \lim_{n \to \infty} \sum_{i=1}^{m}$

$(6 \cdot \dfrac{35}{n^2}) = \lim_{n \to \infty} (6 \cdot \dfrac{35}{n^2} \cdot n^2) = 210$ units^3.

5. For convenience, we take the point (x_i^*,y_i^*) in each subregion to be the lower

right corner. Since $h = \dfrac{5}{n}$ and $k = \dfrac{8}{n}$, $A_i = \dfrac{40}{n^2}$ for every i. R is a rectangle so

$m = n^2$. To calculate $\displaystyle\sum_{i=1}^{m} f(x_i^*,y_i^*) A_i$, we note that since $f(x,y) = x/2$ does not

depend on y, the part of the sum along each horizontal row of rectangles is
equal to the part of the sum along the bottom row. The part of the sum along

the bottom row is $(\dfrac{h}{2} + \dfrac{2h}{2} + \dfrac{3h}{2} + \ldots + \dfrac{nh}{2}) \dfrac{40}{2}$. Since there are n horizontal

rows of rectangles, we get $\displaystyle\sum_{i=1}^{m} f(x_i^*,y_i^*) A_i = n(\dfrac{h}{2} + \dfrac{2h}{2} + \ldots + \dfrac{nh}{2}) \dfrac{40}{n^2} = \dfrac{nh}{2}$

$(1 + 2 + \ldots + n) \dfrac{40}{n^2} = \dfrac{nh}{2} \cdot \dfrac{n(n + 1)}{2} \cdot \dfrac{40}{n^2} = 10h(n + 1)$. But $h = \dfrac{5}{n}$, so this

equals $10(\dfrac{5}{n})(n + 1) = 50(1 + \dfrac{1}{n})$. Thus, $\displaystyle\int_R f(x,y)dA = \lim_{n \to \infty} \sum_{i=1}^{m} (f(x_i^*,y_i^*) A_i$

$= \lim_{n \to \infty} 50(1 + \dfrac{1}{n}) = 50$ cubic units.

Exercises 16.2: Iterated Integrals; The Fundamental Theorem for Double Integrals (page 687)

1. $\displaystyle\int_{-1}^{3} 3xy^2 \Big|_{x-1}^{x^2} dx = \int_{-1}^{3} [3x^5 - 3x(x - 1)^2] dx = \int_{-1}^{3} (3x^5 - 3x^3 + 6x^2 - 3x) dx$

$= 348$

3. $\int_0^2 \left(\frac{x^3}{3} + xy^2\right) \Big|_y^3 \, dy = \int_0^2 \left(9 + 3y^2 - \frac{y^3}{3} - y^3\right) \, dy = 124$

5. $\int_0^{2\pi} \frac{r^4}{4} \Big|_0^2 \, d\theta = \int_0^{2\pi} 4\,d\theta = 8\pi$

7. $\int_0^{\pi} \frac{r^3}{3} \sin\theta \Big|_0^{1+\cos\theta} \, d\theta = \frac{1}{3} \int_0^{\pi} (1 + 3\cos^2\theta + 3\cos\theta + \cos^3\theta)\sin\theta \, d\theta = 4/3$

9. $\int_1^2 -r^2 \cos\theta \Big|_0^{\pi/2} \, dr = \int_1^2 r^2 \, dr = \frac{8}{3} - \frac{1}{3} = \frac{7}{3}$

11. $\int_1^2 6x^3 y \Big|_y^{\sqrt{y}} \, dy = 6\int_1^2 (y^{5/2} - y^4) \, dy = (98\sqrt{2} - 12)/7 - 186/5$

13. $\int_{-1}^2 \frac{1}{2}\left(\frac{2}{3}\right)(x^2 + y^2)^{3/2} \Big|_0^x \, dx = \frac{1}{3}\int_{-1}^2 [(2x^2)^{3/2} - (x^2)^{3/2}] \, dx$

$= \frac{1}{3}(2^{3/2} - 1)\int_{-1}^2 (x^2)^{3/2} \, dx.$ Since $(x^2)^{3/2} = -x^3$ for $x \le 0$, we get

$\int_{-1}^2 (x^2)^{3/2} \, dx = \int_{-1}^0 (-x^3) \, dx + \int_0^2 x^3 \, dx = -\frac{x^4}{4}\Big|_{-1}^0 + \frac{x^4}{4}\Big|_0^2 = \frac{1}{4} + 4 = 17/4$ so

the result is $\frac{1}{3}(2^{3/2} - 1)\left(\frac{17}{4}\right) = 17(2\sqrt{2} - 1)/12.$

15. $\int_1^2 \int_x^{x^2} \frac{1}{x^3 y^3} \, dy \, dx = \int_1^2 \int_x^{x^2} x^{-3}y^{-3} \, dy \, dx = \int_1^2 \left(-\frac{1}{2}x^{-3}y^{-2}\right)\Big|_x^{x^2} \, dx$

$= -\frac{1}{2}\int_1^2 (x^{-7} - x^{-5}) \, dx = \frac{9}{256}$

17. $\int_0^6 \int_0^{6-x} \frac{1}{2}(6 - x - y) \, dy \, dx = \frac{1}{2}\int_0^6 6y - xy - \frac{y^2}{3}\Big|_0^{6-x} \, dx$

$= \frac{1}{2}\int_0^6 \left[36 - 12x + x^2 - \frac{(6-x)^2}{2}\right] \, dx = 18$ cubic units

21. $V = \int_0^1 \int_{x^2}^x \frac{6 - x - y}{2} \, dy \, dx = \frac{17}{40}$ cubic units

25.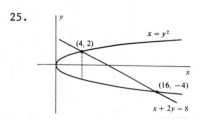

$y^2 + 2y = 8$

$y^2 + 2y - 8 = 0$

$(y - 2)(y + 4) = 0$

$y = 2, \ y = -4$

(a) $\int_{-4}^{2} \int_{y^2}^{8-2y} dx\, dy = 36$ square units.

(b) It takes 2 integrals.

$$\int_{0}^{4} \int_{-\sqrt{x}}^{\sqrt{x}} dy\, dx + \int_{4}^{16} \int_{-\sqrt{x}}^{\frac{8-x}{2}} dy\, dx = 36 \text{ square units.}$$

27. We integrate over the region $x^2 + y^2 - 5 = 4$, that is $x^2 + y^2 = 9$ in the plane $z = -5$. The height of the surface above each point in the region is $z + 5 = 4 - x^2 - y^2 + 5 = 9 - x^2 - y^2$. So we get

$$4 \int_{0}^{3} \int_{0}^{\sqrt{9-x^2}} (9 - x^2 - y^2)\, dy\, dx = \frac{8}{3} \int_{0}^{3} (9 - x^2)^{3/2}\, dx. \text{ Let } x = 3 \sin\theta, \text{ then}$$

$dx = \cos\theta\, d\theta$ and $(9 - x^2)^{3/2} = 27 \cos^3\theta$ so this equals $8 \cdot 27 \int_{0}^{\pi/2} \cos^4\theta\, d\theta$

$$= 216 \int_{0}^{\pi/2} [\frac{1 + \cos 2\theta}{2}]^2\, d\theta = \frac{81\pi}{2} \text{ cubic units.}$$

29. Let V_1 be the volume of the solid bounded by the paraboloid $z = 4 - 4x^2 - y^2$ and the plane $z = -12$, and let V_2 be the volume of the solid bounded by the paraboloid and the plane $z = 0$. Then the volume V we are to find is $V_1 - V_2$.

Since the plane $z = -12$ intersects the paraboloid in the ellipse $4x^2 + y^2 = 16$ or $y = \pm 2\sqrt{4 - x^2}$, we get (using the symmetry of the solid)

$$V_1 = 4 \int_{0}^{2} \int_{0}^{2\sqrt{4-x^2}} [4 - 4x^2 - y^2 - (-12)]\, dy\, dx = 4 \int_{0}^{2} \int_{0}^{2\sqrt{4-x^2}} (16 - 4x^2 - y^2)$$

dy dx. Since the plane $z = 0$ intersects the paraboloid in the ellipse $4x^2 + y^2 = 4$ or $y = \pm 2\sqrt{1 - x^2}$, we get (again using the symmetry)

$$V_2 = 4 \int_{0}^{1} \int_{0}^{2\sqrt{1-x^2}} (4 - 4x^2 - y^2)\, dy\, dx. \text{ Thus,}$$

$$V = V_1 - V_2 = 4 \int_{0}^{2} \int_{0}^{2\sqrt{4-x^2}} (16 - 4x^2 - y^2\, dy\, dx$$

$$- 4 \int_{0}^{1} \int_{0}^{2\sqrt{1-x^2}} (4 - 4x^2 - y^2)\, dy\, dx.$$

35. $\int_{R} f(x,y)\, dA = \int_{3}^{\infty} \int_{0}^{2} x^{-2}y\, dy\, dx = \int_{3}^{\infty} \frac{x^{-2}y^2}{2} \Big|_{0}^{2}\, dx = \int_{3}^{\infty} 2x^{-2}\, dx = \lim_{b \to \infty} \int_{3}^{b} 2x^{-2}\, dx$

$= \lim_{b \to \infty} (-2x^{-1}) \Big|_{3}^{b} = \lim_{b \to \infty} (-\frac{2}{b} + \frac{2}{3}) = \frac{2}{3}.$

37. $\int_1^\infty \int_0^{\frac{y}{2}} 24y^{-5}x^2 \, dx \, dy = \int_1^\infty 8y^{-5}x^3 \, \Big|_0^{\frac{y}{2}} \, dy = \int_1^\infty y^{-2} \, dy = \lim_{b \to \infty} \int_1^b y^{-2} \, dy$

$= \lim_{b \to \infty} (-y^{-1}) \, \Big|_1^b = \lim_{b \to \infty} (-\frac{1}{b} + 1) = 1$

39. Since the region of integration lies between $x = y^2$ and $x = 4$, and between $y = 0$ and $y = 4$, a sketch of the region shows that when the order is reversed, the region lies between $y = 0$ and $y = \sqrt{x}$; and $x = 0$ and $x = 4$. Thus, we get

$\int_0^4 \int_0^{\sqrt{x}} f(x,y) \, dy \, dx.$

47. $\int_2^\infty (4xy + 3y^2) \, \Big|_0^{x^{-4}} \, dx = \int_2^\infty (4x^{-3} + 3y^{-8}) \, dx = \lim_{b \to \infty} (-2b^{-2} - \frac{3}{7} b^{-7} + \frac{1}{2}$

$+ \frac{3}{7 \cdot 128}) = \frac{451}{896}$

49. $\int_0^\infty (\frac{x^2}{2} e^{-y^3}) \, \Big|_0^{2y} \, dy = \lim_{b \to \infty} \int_0^b 2y^2 \, e^{-y^3} \, dy = \lim_{b \to \infty} (-\frac{2}{3} e^{-y^3}) \, \Big|_0^b$

$= \lim_{b \to \infty} (-\frac{2}{3} e^{-b^3} + \frac{2}{3}) = \frac{2}{3}$

Exercises 16.3: Surface Area (page 693)

1. $z = \frac{3}{2} x - \frac{1}{2} y + \frac{7}{2}$; $S = \int_0^{\frac{7}{3}} \int_{3x-7}^0 \sqrt{1 + \frac{9}{4} + \frac{1}{4}} \, dy \, dx = \frac{49\sqrt{14}}{12}$ square units

3. Here $z = 21 - x^2 - y^2$, so $f_x = -2x$ and $f_y = -2y$. The region R is obtained from the intersection of $z = 5$ with the surface. We get $x^2 + y^2 = 16$, so R is bounded by the circle $x^2 + y^2 = 16$ in the xy-plane. Using the symmetry of $f(x,y)$ and R, we can take 4 times the area over the part of R in the first quadrent to get $4 \int_0^4 \int_0^{\sqrt{16-x^2}} \sqrt{1 + 4x^2 + 4y^2} \, dy \, dx.$

5. Here R is bounded by $x^2 + y^2 = 25 - 9 = 16$ and since $z = \sqrt{25 - x^2 - y^2}$,

$f_x = \frac{1}{2} (25 - x^2 - y^2)^{-1/2}(-2x) = \frac{-x}{\sqrt{25 - x^2 - y^2}}$, $f_y = \frac{-y}{\sqrt{25 - x^2 - y^2}}$, and

$\sqrt{1 + f_x^2 + f_y^2} = 5(25 - x^2 - y^2)^{-1/2}$. So $S = 4 \int_0^4 \int_0^{\sqrt{16-x^2}} 5(25 - x^2 - y^2)^{-1/2} \, dy \, dx.$

240

7. $z = \sqrt{9 - x^2}$, $f_x = \dfrac{-x}{\sqrt{9 - x^2}}$, $f_y = 0$. $\sqrt{1 + f_x^2 + f_y^2} = 3(9 - x^2)^{-1/2}$.

(a) $\displaystyle\int_0^2 \int_{2x}^{3x} 3(9 - x^2)^{-1/2} \, dy \, dx$.

(b) We need two integrals.

$$\int_0^4 \int_{y/3}^{y/2} 3(9 - x^2)^{-1/2} \, dx \, dy + \int_4^6 \int_{y/3}^{2} 3(9 - x^2)^{-1/2} \, dx \, dy.$$

9. R is bounded by $9x^2 + 9y^2 = 36$, or $x^2 + y^2 = 4$. Since $z = 3(x^2 + y^2)^{1/2}$,

$f_x = \dfrac{3x}{\sqrt{x^2 + y^2}}$, and $f_y = \dfrac{3y}{\sqrt{x^2 + y^2}}$, so $\sqrt{1 + f_x^2 + f_y^2} = \sqrt{10}$. Using the symmetry

we get $S = 4 \displaystyle\int_0^2 \int_0^{\sqrt{4-x^2}} \sqrt{10} \, dy \, dx = 4\sqrt{10} \int_0^2 (4 - x^2)^{1/2} \, dx$, let $x = 2 \sin \theta$,

then $dx = 2 \cos \theta \, d\theta$ and we get $S = 4\sqrt{10} \displaystyle\int_0^{\pi/2} 2^2 \cos^2 \theta \, d\theta = 4\sqrt{10}\pi$ units^2.

<u>Exercises 16.4</u>: Triple Integration (page 698)

1. $\displaystyle\int_{-1}^2 \int_1^3 6xy^2z^2 \Big|_2^5 \, dy \, dx = \int_{-1}^2 \int_1^3 6(21)xy^2 \, dy \, dx = \int_{-1}^2 42xy^3 \Big|_1^3 \, dx$

$= \displaystyle\int_{-1}^2 42(26)x \, dx = 1638$

3. $\displaystyle\int_{-2}^2 \int_{-\sqrt{4-x^2}}^{\sqrt{4-x^2}} (6xz + 4yz^2) \Big|_3^5 \, dy \, dx = \int_{-2}^2 \int_{-\sqrt{4-x^2}}^{\sqrt{4-x^2}} (12x + 64y) \, dy \, dx$

$= \displaystyle\int_{-2}^2 (12xy + 32y^2) \Big|_{-\sqrt{4-x^2}}^{\sqrt{4-x^2}} \, dx = \int_{-2}^2 24x \sqrt{4 - x^2} \, dx = 0$

5. $\displaystyle\int_{-2}^2 \int_0^{\sqrt{4-x^2}} (6xz + 4yz^2) \Big|_3^5 \, dy \, dx = \int_{-2}^2 \int_0^{\sqrt{4-x^2}} (12x + 64y) \, dy \, dx$

$= \displaystyle\int_{-2}^2 (12xy + 32y^2) \Big|_0^{\sqrt{4-x^2}} \, dx = \int_{-2}^2 [12x \sqrt{4 - x^2} + 128 - 32x^2] \, dx = 341\tfrac{1}{3}$

7. The ellipsoid and paraboloid intersect when $z^2 + z - 90 = 0$, giving $z = 9$ and $z = -10$. We reject $z = -10$ since it does not produce real values of x and y. Thus, the intersection is over the region R in the xy-plane bounded by the el-lipse $9x^2 + y^2 = 9$, or $y = \pm3\sqrt{1 - x^2}$. Since the solid T lies between $z = 9x^2 + y^2$ and $z = \sqrt{90 - 9x^2 + y^2}$, we obtain

$$\int_{-1}^{1} \int_{-3\sqrt{1-x^2}}^{3\sqrt{1-x^2}} \int_{9x^2+y^2}^{\sqrt{90-9x^2-y^2}} xz \; dz \; dy \; dx.$$

13. $\int_{1}^{2} \int_{1}^{x} 8xz \; \big|_{0}^{y} \; dy \; dx = \int_{1}^{2} \int_{1}^{x} 8xy \; dy \; dx = \int_{1}^{2} 4xy^2 \; \big|_{1}^{x} \; dx = \int_{1}^{2} (4x^3 - 4x) \; dx = 9$

15. $\int_{0}^{2} \int_{1}^{y} 5x^2y \; \big|_{1}^{z} \; dz \; dy = \int_{0}^{2} \int_{1}^{y} (5z^2y - 5y) \; dz \; dy = \int_{0}^{2} (\tfrac{5}{3} z^3y - 5yz) \; \big|_{1}^{y} \; dy$

$= \int_{0}^{2} (\tfrac{5}{3} y^4 - 5y^2 - \tfrac{5}{3} y + 5y) \; dy = 4$

17. $\int_{\pi/6}^{\pi/2} \int_{0}^{\sin \theta} 3r^2 \sin \theta \cos \theta \; dr \; d\theta = \int_{\pi/6}^{\pi/2} (r^3 \sin \theta \cos \theta) \; \big|_{0}^{\sin \theta} \; d\theta$

$= \int_{\pi/6}^{\pi/2} \sin^4 \theta \cos \theta \; d\theta = \frac{\sin^5 \theta}{5} \; \big|_{\pi/6}^{\pi/2} = \frac{31}{160}$

19. The density at any point is $3xy$ and we get $\int_{0}^{12} \int_{0}^{\frac{12-x}{2}} \int_{0}^{\frac{12-x-2y}{3}} 3xy \; dz \; dy \; dx.$

25. Using the symmetry we get $V = 4 \int_{0}^{2} \int_{0}^{\sqrt{4-x^2}} \int_{0}^{4-x^2-y^2} dz \; dy \; dx.$

<u>Exercises 16.5:</u> Transformation of Multiple Integrals (page 705)

1. $(3)(5) - (-4)(2) = 23$ 3. $2e^{2u}(3 \cos 3v) - 0 = 6e^{2u} \cos 3v$

5. $(2uv^3)(12u^3v^3) - (3u^2v^2)(9u^2v^4) = 24u^4v^6 - 27u^4v^6 = -3u^4v^6$

7. $(-v \cos^2 uv)(-u \sin uv) - (-u \csc^2 uv)(-v \sin uv) = 0$

9. $(1)(1) - (0)(0) = 1$ 11. $e^u e^v - (0)(0) = e^{u+v}$

13. $y = 2x$ transforms to $2v = 2u - 2v$, or $v = 1/2 \, u$,

   $x = 1$ transforms to $u - v = 1$,

   $y = 0$ transforms to $2v = 0$, $v = 0$.

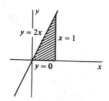 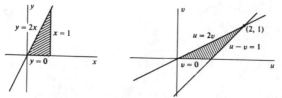

$|(1)(2) - (-1)(0)| = 2$ so we get $\int_{R} (x + y) \; dA = \int_{0}^{1} \int_{2v}^{v+1} (u - v + 2v)(2) \; du \; dv$

$= 2 \int_{0}^{1} \int_{2v}^{v+1} (u + v) \; du \; dv.$

15. $\dfrac{x^2}{16} + \dfrac{y^2}{9} \leq 1$ transforms into $u^2 + v^2 \leq 1$. $\big|(4)(3) - (0)(0)\big| = 12$ so we get

$$\int_R \left(\frac{x^2}{16} + \frac{y^2}{9}\right) dA = \int_{-1}^{1} \int_{-\sqrt{1-u^2}}^{\sqrt{1-u^2}} (u^2 + v^2)\, 12\ dv\ du.$$

17. The region in the xy-plane is bounded by $y = -x - 2$, $y = 0$, and $x = 0$ as shown. These transform to $u = -1$, $v = -u$, and $v = u$.

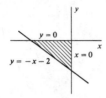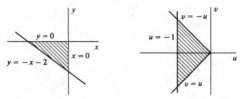

The absolute value of the Jacobian is $\big|(1)(1) = (-1)(1)\big| = 2$ and $xy + 3$

$= (u - v)(u + v) + 3 = u^2 - v^2 + 3$ so we get

$$2 \int_{-1}^{0} \int_{u}^{-u} (u^2 - v^2 + 3)\ dv\ du.$$

19.

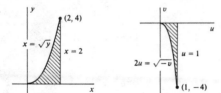

$x = \sqrt{y}$ transforms to $2u = \sqrt{-u}$,

$x = 2$ transforms to $u = 1$. We get $\displaystyle \int_{0}^{4} \int_{\sqrt{y}}^{2} y\ dy\ dx = - \int_{-4}^{0} \int_{\frac{1}{2}\sqrt{-v}}^{1} v\ du\ dv.$

Exercises 16.6: Double Integrals in Polar Coordinates (page 710)

1.

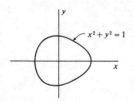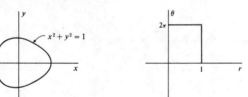

$$\int_R e^{-(x^2+y^2)}\ dA = \int_0^{2\pi} \int_0^1 e^{-r^2}\ r\ dr\ d\theta = -\frac{1}{2} \int_0^{2\pi} e^{-r^2}\ \Big|_0^1\ d\theta$$

243

$$= -\frac{1}{2} \int_0^{2\pi} (e^{-1} - 1) \, d\theta = (-\frac{1}{2})(e^{-1} - 1) \, 2\pi = \pi(1 - e^{-1})$$

3.

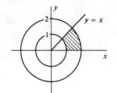

$$\int_R x(x^2 + y^2)^{1/2} \, dA = \int_0^{\pi/4} \int_1^2 r \cos\theta \; rr \, dr \, d\theta = \int_0^{\pi/4} \int_1^2 r^3 \cos\theta \, dr \, d\theta = \frac{15}{4\sqrt{2}}$$

5. $\displaystyle \int_0^{\pi/2} \int_0^6 r^2 \cdot r \, dr \, d\theta = 162\pi$

7. $y = \sqrt{6x - x^2}$ is the upper half of the circle $(x - 3)^2 + y^2 = 3^2$, or $x^2 + y^2 = 6x$ which in polar coordinates is $r^2 \cos^2\theta + r^2 \sin^2\theta = 6r \cos\theta$, or $r = 6 \cos\theta$.

$$\int_0^{\pi/2} \int_0^{6\cos\theta} r^2 \, r \, dr \, d\theta = \int_0^{\pi/2} \frac{r^4}{4} \Big|_0^{6\cos\theta} \, d\theta \int_0^{\pi/2} 324 \cos^4\theta \, d\theta$$

$$= 324 \int_0^{\pi/2} \left(\frac{1 + \cos 2\theta}{2}\right)^2 \, d\theta = 81 \int_0^{\pi/2} (1 + 2\cos 2\theta + \cos^2 2\theta) \, d\theta$$

$$= 81 \int_0^{\pi/2} \left(1 + 2\cos 2\theta + \frac{1 + \cos 4\theta}{2}\right) \, d\theta = \frac{243\pi}{4}.$$

9. The region is the upper semicircle of radius 2. $\displaystyle \int_0^{\pi} \int_0^2 e^{4-r^2} r \, dr \, d\theta$

$$= \frac{(e^4 - 1)\pi}{2}.$$

11. $\displaystyle \int_0^{\pi/2} \int_0^3 \sqrt{9 - r^2} \, r \, dr \, d\theta = \frac{9\pi}{2}$

13. $f_x = 2x$, $f_y = 2y$, and R is bounded by $x^2 + y^2 = 4$ so we get

$$S = 4 \int_0^2 \int_0^{\sqrt{4-x^2}} \sqrt{1 + 4x^2 + 4y^2} \, dy \, dx = 4 \int_0^{\pi/2} \int_0^2 \sqrt{1 + 4r^2} \, r \, dr \, d\theta$$

$$= \frac{\pi}{6} [17^{3/2} - 1] \text{ square units.}$$

15. $\displaystyle A = 4 \int_0^{\pi/2} \int_0^{\sin 2\theta} r \, dr \, d\theta = 2 \int_0^{\pi/2} \sin^2 2\theta \, d\theta = \int_0^{\pi/2} (1 + \cos 4\theta) d\theta = \frac{\pi}{2}$

square units

17. $2 \int_0^{\pi/2} \int_1^{1+\sin\theta} r\ dr\ d\theta = \int_0^{\pi/2} r^2 \Big|_1^{1+\sin\theta} d\theta = 2 + \frac{\pi}{4}$ units^2

19. $4 \int_0^2 \int_0^{\sqrt{4-x^2}} [2 - (-1)]\ dy\ dx = 4 \int_0^{\pi/2} \int_0^2 3r\ dr\ d\theta = 12\pi$ cubic units

21. $4 \int_0^3 \int_0^{\sqrt{9-x^2}} (x^2 + y^2)\ dy\ dx = 4 \int_0^{\pi/2} \int_0^3 r^2\ r\ dr\ d\theta = \frac{81\pi}{2}$ cubic units

Exercises 16.7: Transformation of Triple Integrals (page 715)

1. $\begin{vmatrix} a & 0 & 0 \\ 0 & b & 0 \\ 0 & 0 & c \end{vmatrix} = abc$

3. $\begin{vmatrix} \cos v & -u\sin v & 0 \\ \sin v & u\cos v & 0 \\ 0 & 0 & 1 \end{vmatrix} = \begin{vmatrix} \cos v & -u\sin v \\ \sin v & u\cos v \end{vmatrix} = u$

5. $\begin{vmatrix} v\sec^2 u & \tan u & 0 \\ 0 & w & v \\ e^u & 0 & 0 \end{vmatrix} = ve^u \tan u$

7. As in 1, the Jacobian is abc. The boundary $\frac{x^2}{a^2} + \frac{y^2}{b^2} + \frac{z^2}{c^2} = 1$ is transformed

   into $u^2 + v^2 + w^2 = 1$. We get

   $$\int_{-1}^1 \int_{-\sqrt{1-u^2}}^{\sqrt{1-u^2}} \int_{-\sqrt{1-u^2-v^2}}^{\sqrt{1-u^2-v^2}} (u^2 + v^2 + w^2)|abc|\ dw\ dv\ du.$$

9. The boundaries of T are transformed as follows: $x = 4$ gives $4 - u/3 = 4$, or
   $u = 0$; $y = x$ gives $4 - u/3 = 4 - u/3 - w/3$, or $w = 0$; $z = 0$ gives $v/3 = 0$, or
   $v = 0$; $z = y$ gives $v/3 = 4 - u/3 - w/3$, or $u + v + w = 12$. The function
   $3x + 6y$ is transformed to $12 - u + 24 - 2u - 2w = 36 - 3u - 2w$. The Jacobian

   is $\begin{vmatrix} -1/3 & 0 & 0 \\ -1/3 & 0 & -1/3 \\ 0 & 1/3 & 0 \end{vmatrix} = -\frac{1}{27}$. A sketch of the uvw region is

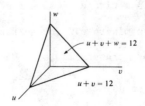

So $\int_T (3x + 6y)\ dV = \frac{1}{27} \int_0^{12} \int_0^{12-u} \int_0^{12-u-v} (36 - 3u - 2w)\ dw\ dv\ du.$

11. The boundaries of T are transformed as follows: x = 0 gives 2w = 0, or w = 0; y = 0 gives 2v - 2u = 0, or v = u; z = 0 gives 10 - 2v = 0, or v = 5; x + y + z = 10 gives 2w + 2v - 2u + 10 - 2v = 10, or w = u. The function xy + z is transformed to 4vw - 4uw - 2v + 10. The Jacobian is

$\begin{vmatrix} 0 & 0 & 2 \\ -2 & 2 & 0 \\ 0 & -2 & 0 \end{vmatrix} = 8.$ A sketch of the uvw region is

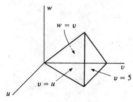

So $\int_T (xy + z)\ dV = 8 \int_0^5 \int_0^v \int_0^u (4vw - 4uw - 2v + 10)\ dw\ du\ dv.$

Exercises 16.8: Cylindrical Coordinates (page 719)

1. $\int_0^{2\pi} \int_0^{\sqrt{3}} \int_{-1}^{2} (zr)\ r\ dz\ dr\ d\theta = \int_0^{2\pi} \int_0^{\sqrt{3}} \frac{r^2 z^2}{2} \Big|_{-1}^{2}\ dr\ d\theta$

$= \int_0^{2\pi} \int_0^{\sqrt{3}} \frac{3r^2}{2}\ dr\ d\theta = \int_0^{2\pi} \frac{r^3}{2} \Big|_0^{\sqrt{3}}\ d\theta = \int_0^{2\pi} \frac{3^{3/2}}{2}\ d\theta = 3^{3/2}\pi$

3. $\int_0^{2\pi} \int_0^{\sqrt{3}} \int_{-1}^{2} \frac{zr}{r^2 + 1}\ dz\ dr\ d\theta = \int_0^{2\pi} \int_0^{\sqrt{3}} \frac{z^2 r}{2(r^2 + 1)} \Big|_{-1}^{2}\ dr\ d\theta$

$= \frac{3}{2} \int_0^{2\pi} \int_0^{\sqrt{3}} \frac{r}{r^2 + 1}\ dr\ d\theta = \frac{3}{4} \int_0^{2\pi} \ln |r^2 + 1| \ \Big|_0^{\sqrt{3}}\ d\theta = 3\pi \ln 2$

5. $(x^2 + y^2)^{1/2} = r$ when $r \geq 0$, so we get $\int_0^{2\pi} \int_0^{2} \int_0^{r} r \cdot r\ dz\ dr\ d\theta = 8\pi.$

7. In the xy-plane the region is a circle. Using the symmetry we can take twice the integral over the upper semicircle. The surface $z = x^2 + y^2$ is transformed to $z = r^2$. The upper boundary transforms to $r = 6 \cos \theta$. Thus, we get

$2 \int_0^{\pi} 2 \int_0^{6 \cos \theta} \int_0^{r^2} r(r)\ dz\ dr\ d\theta = 2 \int_0^{\pi/2} \int_0^{6 \cos \theta} r^4\ dr\ d\theta$

$= \frac{2}{5} \int_0^{\pi/2} 6^5 \cos^5 \theta\ d\theta = \frac{2}{5} (6)^5 \int_0^{\pi/2} (1 - \sin^2 \theta) \cos \theta\ d\theta$

$$= \frac{2}{5} (6)^5 \int_0^{\pi/2} (1 - 2 \sin^2 \theta + \sin^4 \theta) \cos \theta \, d\theta = \frac{41472}{25}.$$

9. $\displaystyle \int_{-1}^{1} \int_0^{\pi/2} \int_0^2 r \, dr \, d\theta \, dz = 2\pi$

11. $\displaystyle \int_{-2}^{0} \int_0^{\pi/4} \int_0^1 r \sin \theta \, r \cos \theta \, r \, dr \, d\theta \, dz = 1/8$

13. $\displaystyle \int_0^{2\pi} \int_0^3 \int_{r^2}^9 r \, dz \, dr \, d\theta = \frac{81\pi}{2}$ cubic units.

15. The smaller solid is bounded above by $z = (8 - x^2 - y^2)^{1/2}$, or $z = (8 - r^2)^{1/2}$, and below by $z = \frac{1}{2} (x^2 + y^2)$, or $z = \frac{r^2}{2}$. These two surfaces intersect at the positive root of $2z + z^2 = 8$, that is, at $z = 2$. Thus, the volume is over the region bounded by the circle $x^2 + y^2 = 4$, or $r = 2$. Thus, the volume is

$$\int_0^{2\pi} \int_0^2 \int_{\frac{r^2}{2}}^{\sqrt{8-r^2}} r \, dz \, dr \, d\theta = \frac{2\pi}{3} (8^{3/2} - 14) \text{ units}^3.$$

17. The solid is bounded above by $z^2 = x^2 + y^2$, or $z = r$, and below by $z = x^2 + y^2$, or $z = r^2$. These two surfaces intersect when $z = z^2$, i.e., at $z = 0$ and $z = 1$. Thus, the volume is over the region bounded by $x^2 + y^2 = 1$, or $r = 1$. Thus, the volume is $\displaystyle \int_0^{2\pi} \int_0^1 \int_{r^2}^r r \, dz \, dr \, d\theta = \frac{\pi}{6} \text{ units}^3.$

Exercises 16.9: Spherical Coordinates (page 725)

5. The cone $x^2 + y^2 = z^2$ is $\phi = \pi/4$ and the sphere $x^2 + y^2 + z^2 = 1$ is $\rho = 1$, so we get the set of all points (ρ, ϕ, θ) where $0 \leq \rho \leq 1$ and $0 \leq \phi \leq \pi/4$.

7. $\displaystyle \int_0^{2\pi} \int_0^{\pi} \int_0^1 \rho \cos \phi \, \rho^2 \sin \phi \, d\rho \, d\phi \, d\theta = \frac{1}{8} \int_0^{2\pi} \sin^2 \phi \, \Big|_0^{\pi} \, d\theta = 0$

9. $\displaystyle \int_0^{2\pi} \int_0^{\pi/4} \int_0^1 \frac{\rho^2 \sin \phi}{1 + \rho^2} \, d\rho \, d\phi \, d\theta = \int_0^{2\pi} \int_0^{\pi/4} \int_0^1 \sin \phi \left[1 - \frac{1}{1 + \rho^2}\right] d\rho \, d\phi \, d\theta$

$$= \int_0^{2\pi} \int_0^{\pi/4} \sin \phi \, [\rho - \arctan \rho] \, \Big|_0^1 \, d\phi \, d\theta = \int_0^{2\pi} \int_0^{\pi/4} \sin \phi \, (1 - \tfrac{\pi}{4}) \, d\phi \, d\theta$$

$$= (1 - \tfrac{\pi}{4}) \int_0^{2\pi} (-\cos \phi) \, \Big|_0^{\pi/4} \, d\theta = \pi(1 - \tfrac{\pi}{4})(2 - \sqrt{2})$$

11. $\displaystyle \int_0^{\pi/2} \int_0^{\pi/4} \int_0^{\sqrt{2}} \rho^2 \sin \phi \, d\rho \, d\phi \, d\theta = \frac{\pi}{3} (\sqrt{2} - 1)$

13.

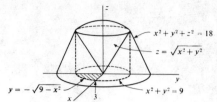

$$\int_{\frac{3\pi}{2}}^{2\pi} \int_0^{\pi/4} \int_0^{3\sqrt{2}} \rho^2 \sin\phi \, d\rho \, d\phi \, d\theta = 9\pi(\sqrt{2} - 1)$$

15. $$\int_0^{2\pi} \int_0^{\pi/3} \int_0^8 \rho^2 \sin\phi \, d\rho \, d\phi \, d\theta = \int_0^{2\pi} \int_0^{\pi/3} \frac{8^3}{3} \sin\phi \, d\phi \, d\theta$$

$$= \frac{512}{3} \int_0^{2\pi} (-\cos\phi) \Big|_0^{\pi/3} d\theta = \frac{512}{3} \int_0^{2\pi} (1 - \frac{1}{2}) \, d\theta = \frac{512\pi}{3} \text{ cubic units}$$

17. $$\int_0^{2\pi} \int_0^\pi \int_{10}^{11} \rho^2 \sin\phi \, d\rho \, d\phi \, d\theta = \frac{1324}{3} \pi \text{ cubic units}$$

19. $$\int_0^{2\pi} \int_0^\pi \int_0^2 3\rho^2 \cdot \rho^2 \sin\phi \, d\rho \, d\phi \, d\theta = \frac{384}{5} \pi \text{ units}$$

Technique Review Exercises, Chapter 16 (page 728)

1. Since $f(x,y) = 20$ for all points of R, $f(x_i^*, y_i^*) = 20$ for all i. Also, each

A_i = hk where h = $\frac{11 - 4}{n} = \frac{7}{n}$ and k = $\frac{7 - 3}{n} = \frac{4}{n}$. There are n^2 subregions so

$m = n^2$ and $A_i = hk = \frac{28}{n^2}$. Thus, the volume is

$$\int_R 20 \, dA = \lim_{n\to\infty} \sum_{i=1}^{n^2} 20 \, (\frac{28}{n^2}) = \lim_{n\to\infty} 20 \, (\frac{28}{n^2}) \, n^2 = \lim_{n\to\infty} 20(28) = 560 \text{ cubic units.}$$

2.

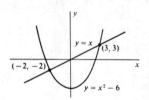

Intersection: $x^2 - 6 = x$

$x^2 - x - 6 = 0$

$(x - 3)(x + 2) = 0$

$x = 3, \ x = -2$

$$V = \int_{-2}^3 \int_{x^2-6}^x (8 - x + 2y) \, dy \, dx = 72 \frac{11}{12} \text{ cubic units.}$$

3.

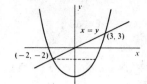

We need 2 integrals because the left boundary is $x = -\sqrt{y + 6}$ for $-6 \leq y \leq -2$ and $x = y$ for $-2 \leq x \leq 3$. The minimum point is at $(0,-6)$.

$$V = \int_{-6}^{-2} \int_{-\sqrt{y+6}}^{\sqrt{y+6}} (8 - x + 2y) \, dx \, dy + \int_{-2}^{3} \int_{y}^{\sqrt{y+6}} (8 - x + 2y) \, dx \, dy = 72\frac{11}{12}$$ cubic units.

4. Intersection: $14 - x^2 - y^2 = x^2 + y^2$, or $x^2 + y^2 = 7$.

$$V = 4 \int_{0}^{\sqrt{7}} \int_{0}^{\sqrt{7-x^2}} (14 - x^2 - y^2 - x^2 - y^2) \, dy \, dx$$

$$= 8 \int_{0}^{\sqrt{7}} \int_{0}^{\sqrt{7-x^2}} (7 - x^2 - y^2) \, dy \, dx$$

5. $z = \frac{9}{2} - \frac{5}{2} x + \frac{7}{2} y$, so $f_x = -\frac{5}{2}$, $f_y = \frac{7}{2}$, and $\sqrt{1 + f_x^2 + f_y^2} = \sqrt{1 + \frac{25}{4} + \frac{49}{4}} = \frac{\sqrt{78}}{2}$.

$$S = \int_{-\frac{7}{9}}^{0} \int_{0}^{\frac{9+7y}{5}} \frac{\sqrt{78}}{2} \, dx \, dy = \frac{81\sqrt{78}}{140}$$ square units.

6. Intersection points: $y^2 = 2y + 15$, $y = 5$, $y = -3$.

$$A = \int_{-3}^{5} \int_{y^2}^{2y+15} dx \, dy = 85\frac{1}{3}$$ square units.

7. The density at (x,y,z) is $z\sqrt{x^2 + y^2}$. The intersection: $x^2 + y^2 + 2x^2 + 2y^2 = 54$, or $x^2 + y^2 = 18 = (3\sqrt{2})^2$. Using the symmetry we get

$$4 \int_{0}^{3\sqrt{2}} \int_{0}^{\sqrt{18-x^2}} \int_{\sqrt{2x^2+2y^2}}^{\sqrt{54-x^2-y^2}} z\sqrt{x^2 + y^2} \, dz \, dy \, dx.$$

8.

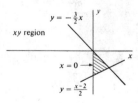

The boundaries are transformed as:

$y = -\frac{3}{2} x$ gives $\frac{v}{4} - \frac{u}{2} = -\frac{3v}{4}$, or

$u = 2v$; $x = 0$ gives $v = 0$; $y = \frac{x - 2}{2}$

gives $\frac{v}{4} - \frac{u}{4} = \frac{v - 4}{4}$, or $u = 2$; $12x - 8y$

gives $6v - 2v + 4u = 4v + 4u$. Jacobian

$$\begin{vmatrix} 0 & \frac{1}{2} \\ -\frac{1}{2} & \frac{1}{4} \end{vmatrix} = \frac{1}{4}.$$ We get

249

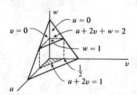

$$\int_0^2 \int_0^{u/2} (4v + 4u)\,\tfrac{1}{4}\,dv\,du$$

$$= \int_0^2 \int_0^{u/2} (v + u)\,dv\,du.$$

9. Use polar coordinates to get $\displaystyle\int_0^{\pi/2} \int_0^1 e^{r^2}\,dr\,d\theta = \frac{(e - 1)\pi}{4}$.

10. The boundaries of T are transformed as follows: $z = 0$ gives $u = 0$; $z = y$ gives $v = 0$; $x + 3y + z = 2$ gives $u + 2v + w = 2$; $x + y + z = 1$ gives $w = 1$. The function $xz + yz + z^2$ is transformed to $w\left(\frac{u}{2}\right) = \frac{uw}{2}$. The Jacobian is

$$\begin{vmatrix} -1 & -1 & 1 \\ \frac{1}{2} & 1 & 0 \\ \frac{1}{2} & 0 & 0 \end{vmatrix} = \frac{1}{2}\,(-1) = -\frac{1}{2}.$$ A sketch of the uvw region is

So we get $\displaystyle\int_0^{\frac{1}{2}} \int_0^{1-2v} \int_1^{2-u-2v} uw\,dw\,du\,dv$.

11. $\displaystyle\int_{-2}^3 \int_0^{2\pi} \int_0^3 (z^2 r)\,r\,dr\,d\theta\,dz = 210\pi$

Additional Exercises, Chapter 16 (page 729)

1. Since $f(x,y) = 9$ for all points of R, $f(x_i^*, y_i^*) = 9$ for all i. Also, $A_i = hk$

$= \left(\frac{3 + 4}{n}\right)\left(\frac{11 - 8}{n}\right) = \frac{21}{n^2}$, and $m = n^2$. Thus, the volume is

$$\int_R 9\,dA = \lim_{n\to\infty} \sum_{i=1}^{n^2} 9\,\left(\frac{21}{n^2}\right) = \lim_{n\to\infty} 9\,\left(\frac{21}{n^2}\right) n^2 = 189 \text{ cubic units.}$$

3. $\displaystyle\int_{-2}^1 \int_{x^2}^x x^2 y\,dy\,dx = \int_{-2}^1 \frac{x^2 y^2}{2}\Big|_{x^2}^x dx = \frac{1}{2}\int_{-2}^1 (x^4 - x^6)\,dx = -\frac{207}{35}$

5. $\displaystyle\int_2^4 \int_0^y x(x^2 + y^2)^2\,dx\,dy = \int_2^4 \frac{1}{2}(x^2 + y^2)^3\left(\frac{1}{3}\right)\Big|_0^y dy = \frac{1}{6}\int_2^4 7y^6\,dy = 2709\,1/3$

7. We use the minus since the solid is below the xy-plane.

$$V = -\int_0^2 \int_0^{\frac{10-5y}{2}} \frac{2x + 5y - 10}{2}\,dx\,dy = 25/3 \text{ units}^3.$$

11. $\int_1^\infty \int_0^{y-3} (4x + y)\ dx\ dy = \int_1^\infty (x^2 + xy)\ \Big|_0^{y-3}\ dy = \lim_{b\to\infty} \int_1^b (2y^2 - 9y + 9)\ dy$

$= \lim_{b\to\infty} (\dfrac{2b^3}{3} - \dfrac{9}{2} b^2 + 9b - \dfrac{2}{3} + \dfrac{9}{2} - 9)$ does not exist.

17. $\int_{-3}^3 \int_{-\sqrt{9-x^2}}^{\sqrt{9-x^2}} \int_{-2}^7 10yz\ dz\ dy\ dx = \int_{-3}^3 \int_{-\sqrt{9-x^2}}^{\sqrt{9-x^2}} 225y\ dy\ dx = 0$

19. $\begin{vmatrix} 2u & -2v \\ 2uv^3 & 3u^2v^2 \end{vmatrix} = 6u^3v^2 + 4uv^4$

21.

$\qquad\qquad\qquad\qquad$ Jacobian $\begin{vmatrix} 1 & 1 \\ 1 & -1 \end{vmatrix} = -2$

The boundaries are transformed as follows: $y = 3 - x$ gives $u - v = 3 - u - v$, or $u = 3/2$; $x = 0$ gives $u + v = 0$, or $v = -u$; $y = 0$ gives $u - v = 0$, or $v = u$. Also, xy gives $u^2 - v^2$.

We get $2 \int_0^{3/2} \int_{-u}^u (u^2 - v^2)\ dv\ du.$

23. $\int_{-\pi/2}^{\pi/2} \int_0^3 r^5 \cdot r\ dr\ d\theta = \dfrac{3^7\pi}{7}$

25. $\int_0^{\pi/2} \int_0^3 r^4 r\ dr\ d\theta = \dfrac{243\pi}{4}$

27. $A = 2 \int_0^\pi \int_0^{3+\cos\theta} r\ dr\ d\theta = \int_0^\pi (9 + 6\cos\theta + \cos^2\theta)$

$= \int_0^\pi (9 + 6\cos\theta + \dfrac{1 + \cos 2\theta}{2})\ d\theta = (9\theta + 6\sin\theta + \dfrac{\theta}{2} + \dfrac{\sin 2\theta}{2})$

$= \dfrac{19\pi}{2}$ square units.

29. The upper boundary is $z = x^2 + y^2$, or $z = r^2$, and the solid is above the region bounded by $x^2 + y^2 = 36$, or $r = 6$. Take 4 times the volume over a quarter of the region to get

$V = 4 \int_0^{\pi/2} \int_0^6 r^2 \cdot r\ dr\ d\theta = 648\pi$ cubic units.

31. $\begin{vmatrix} 1 & -2 & 1 \\ 2 & 3 & 2 \\ 1 & 5 & -3 \end{vmatrix} = \begin{vmatrix} 3 & 2 \\ 5 & -3 \end{vmatrix} + 2 \begin{vmatrix} 2 & 2 \\ 1 & -3 \end{vmatrix} + \begin{vmatrix} 2 & 3 \\ 1 & 5 \end{vmatrix} = -28$

33. The boundaries are transformed as follows: $x = 0$ gives $u = 0$; $y = 0$ gives $v = 0$; $z = 0$ gives $w = v$; $2x + y + z = 2$ gives $u + w = 2$. Also, $8xy (x + z)$

gives $4uvw$. Jacobian: $\begin{vmatrix} \frac{1}{2} & 0 & 0 \\ 0 & 1 & 0 \\ 0 & -1 & 1 \end{vmatrix} = \frac{1}{2}$.

A sketch of the region is

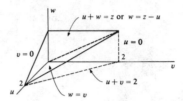

Intersection of $w = v$ and $u + w = 2$ is $u + v = 2$. We get

$$\int_0^2 \int_0^{2-u} \int_v^{2-u} 4uvw \, \frac{1}{2} \, dw \, dv \, du = 2 \int_0^2 \int_0^{2-u} \int_v^{2-u} uvw \, dw \, dv \, du.$$

35. $\displaystyle\int_0^{\pi/2} \int_0^5 \int_{-2}^3 r \, dz \, dr \, d\theta = \int_0^{\pi/2} \int_0^5 5r \, dr \, d\theta = \frac{125\pi}{4}$

37. $\displaystyle\int_0^{2\pi} \int_0^2 \int_{-2}^3 z^2 r \cdot r \, dz \, dr \, d\theta = \frac{560\pi}{9}$

39. $x^2 + y^2 = 8y$ gives $r = 8 \sin \theta$ and $z = (x^2 + y^2)^{1/2}$ gives $z = r$, so we get

$$\int_0^\pi \int_0^{8 \sin \theta} \int_0^r r \cdot r \, dz \, dr \, d\theta = \int_0^\pi \int_0^{8 \sin \theta} r^3 \, dr \, d\theta = \int_0^\pi \frac{r^4}{4} \Big|_0^{8 \sin \theta} d\theta$$

$$= 1024 \int_0^\pi \sin^4 \theta \, d\theta = 1024 \int_0^\pi \left(\frac{1 + \cos 2\theta}{2}\right)^2$$

$$= 256 \int_0^\pi (1 + 2 \cos 2\theta + \cos^2 2\theta) \, d\theta = 256 \int_0^\pi \left(1 + 2 \cos \theta + \frac{1 + \cos 2\theta}{2}\right) d\theta$$

$$= 384\pi.$$

41. The upper boundary is $z^2 = 4x^2 + 4y^2$, or $z^2 = 4r^2$, or $z = 2r$. The lower boundary is $z = x^2 + y^2$, or $z = r^2$. The surfaces intersect when $z^2 = 4z$, that is, at $z = 0$ and $z = 4$. Thus, the volume lies over the region bounded by $x^2 + y^2 = 4$, or $r = 2$. Using the symmetry we take 4 times the volume over one quarter of the region to get $V = 4 \displaystyle\int_0^{\pi/2} \int_0^2 \int_{r^2}^{2r} r \, dz \, dr \, d\theta = \frac{8\pi}{3}$ cubic units.

47. $\int_0^{2\pi} \int_0^{\pi} \int_0^1 \frac{\rho^2 \cos^2 \phi}{\rho} \rho^2 \sin \phi \, d\rho \, d\phi \, d\theta = \int_0^{2\pi} \int_0^{\pi} \int_0^1 \rho^3 \cos^2 \phi \sin \phi \, d\rho \, d\phi \, d\theta$

$= \int_0^{2\pi} \int_0^{\pi} \frac{1}{4} \cos^2 \phi \sin \phi \, d\phi \, d\theta = \frac{\pi}{3}$

49. $\int_0^{2\pi} \int_{\pi/3}^{\pi/2} \int_0^6 \rho^2 \sin \phi \, d\rho \, d\phi \, d\theta = 72\pi$ cubic units

CHAPTER 17: VECTOR CALCULUS

Exercises 17.1: Work and Line Integrals (page 737)

1. By definition $\int_C f(x,y) \, dx = \lim_{n \to \infty} \sum_{i=1}^n f(x_i^*, y_i^*)(x_i - x_{i-1})$ where (x_i, y_i),

(x_{i-1}, y_{i-1}), and (x_i^*, y_i^*) are all on C. Then since $f(x_i^*, y_i^*) = 5$, we have

$\int_C f(x,y) \, dx = \lim_{n \to \infty} \sum_{i=1}^n (x_i - x_{i-1}) = \lim_{n \to \infty} 5(x_n - x_0)$. But since $(0,2)$ is the

initial point, $x_0 = 0$ and since $(1,1)$ is the terminal point, $x_n = 1$. Thus,

$\int_C f(x,y) \, dx = \lim_{n \to \infty} 5 = 5$.

3. By definition $\int_C f(x,y) \, dy = \lim_{n \to \infty} \sum_{i=1}^n f(x_i^*, y_i^*)(y_i - y_{i-1})$ where (x_i, y_i),

(x_{i-1}, y_{i-1}), and (x_i^*, y_i^*) are all on C. Note that $f(x,y) = x^2 - 2xy + y^2$

$= (x - y)^2$ and on C, $x - y = 1$. Thus, $f(x,y) = (-1)^2 = 1$ for all points (x,y)

on C. Thus, $\int_C f(x,y) \, dy = \lim_{n \to \infty} \sum_{i=1}^n (y_i - y_{i-1}) = \lim_{n \to \infty} (y_n - y_0)$. Now since the

initial point is $(0,1)$, we have $y_0 = 1$, and since the terminal point is $(1,2)$,

we have $y_n = 2$. Thus, $\int_C f(x,y) \, dy = \lim_{n \to \infty} (2 - 1) = 1$.

Exercises 17.2: Computation of Line Integrals (page 742)

1. If we let $r(t) = t^2$ and $s(t) = 5t$, we can apply Theorem 16.2.2 to obtain

$\int_C (x + y^2) \, dx + xy \, dy = \int_1^3 (t^2 + (5t)^2) \, 2t \, dt + \int_1^3 t^2 (5t) \, 5dt$. Thus,

$\int_C (x + y^2) \, dx + xy \, dy = 2 \int_1^3 (t^3 + 25t^3) \, dt + 25 \int_1^3 t^3 \, dt = 77 \int_1^3 t^3 \, dt$

$= 77 \left(\frac{1}{4} t^4 \right) \Big|_1^3 = 1540.$

3. By Theorem 16.2.2 we have $\int_C \cos y\, dx + e^x\, dy = \int_0^1 \cos(2t)\, 3dt + \int_0^1 e^{3t}\, 2dt$

$= \dfrac{3}{2} \sin 2t \Big|_0^1 + \dfrac{2}{3} e^{3t} \Big|_0^1 = \dfrac{3}{2} \sin 2 + \dfrac{2}{3}(e^3 - 1).$

5. By Theorem 16.2.2 we have $\int_C \dfrac{dx}{x^2 + y^2} + \dfrac{dy}{x^2 + y^2} = \int_0^\pi \dfrac{-\sin t\, dt}{\cos^2 t + \sin^2 t}$

$+ \int_0^\pi \dfrac{\cos t\, dt}{\cos^2 t + \sin^2 t} = -\int_0^\pi \sin t\, dt + \int_0^\pi \cos t\, dt = \cos t \Big|_0^\pi + \sin t \Big|_0^\pi$

$= -1 - 1 = -2.$

7. The line segment from $(0,0)$ to $(2,4)$ has the parametric equation

$$\begin{aligned} x &= 2t \\ y &= 4t \end{aligned} \qquad 0 \le t \le 1.$$

Thus, $X(t) = (2t, 4t)$, $0 \le t \le 1$ is an equation of the line segment and so,

$\int_C 2xy\, dx + x^2\, dy = \int_0^1 2(2t)(4t)\, 2dt + \int_0^1 (2t)^2\, 4dt = 32 \int_0^1 t^2\, dt + 16 \int_0^1 t^2\, dt$

$= 16.$

9. If we let $x = t$, we have $y = 2t$. Thus, $(X)t = (t, 2t)$, $2 \le t \le 3$ is an equation

for C. Consequently, $\int_C (x^2 + y)\, dx = \int_2^3 (t^2 + 2t)\, dt = (\dfrac{1}{3} t^3 + t^2) \Big|_2^3 = 34/3.$

11. If we let $y = t$, then $x = \sin t$ and $X(t) = (\sin t, t)$, $0 \le t \le \pi$ is an equation

of C. Thus, $\int_C y\, dx + x\, dy = \int_0^\pi t \cos t\, dt + \int_0^\pi \sin t\, dt.$ An application of

integration by parts gives $\int t \cos t\, dt = t \sin t - \int \sin t\, dt = t \sin t + \cos$

$t + C.$ Consequently, $\int_C y\, dx + x\, dy = (t \sin t + \cos t - \cos t) \Big|_0^\pi = 0.$

13. The curve C is split into two parts C_1 and C_2. If we let $x = t$, C_1 and C_2 have

equations $X_1(t)$ and $X_2(t)$ respectively where

$$X_1(t) = (t, t^{1/2}) \qquad t_I = 0,\ t_T = 1$$

$$X_2(t) = (t, t^2) \qquad t_I = 1,\ t_T = 0.$$

Thus, since $\int_C (x + y)\, dx + dy = \int_{C_1} (x + y)\, dx + dy + \int_{C_2} (x + y)\, dx + dy$

$= \int_0^1 (t + t^{1/2})\, dt + \int_0^1 \dfrac{1}{2} t^{-1/2}\, dt + \int_1^0 (t + t^2)\, dt + \int_1^0 2t\, dt$

$= (\dfrac{1}{2} t^2 + \dfrac{2}{3} t^{3/2} + t^{1/2}) \Big|_0^1 + (\dfrac{1}{2} t^2 + \dfrac{1}{3} t^3 + t^2) \Big|_1^0 = \dfrac{1}{3}.$

17. (a) In this case $\int_C (x + y)\, dx + (x - y)\, dy = \int_1^4 (t + 2t^{1/2})\, dt$

$+ \int_1^4 (t - 2t^{1/2})\, t^{-1/2}\, dt = (\frac{1}{2} t^2 + \frac{4}{3} t^{3/2})\, \Big|_1^4 + (\frac{2}{3} t^{3/2} - 2t)\, \Big|_1^4$

$= (8 + \frac{32}{3} - \frac{1}{2} - \frac{4}{3}) + (\frac{16}{3} - 8 - \frac{2}{3} + 2) = 31/2.$

(b) Here $\int_C (x + y)\, dx + (x - y)\, dy = \int_1^2 (t^2 + 2t)\, 2t\, dt + \int_1^2 (t^2 - 2t)\, 2dt$

$= (\frac{2}{4} t^4 + \frac{4}{3} t^3)\, \Big|_1^2 + 2 (\frac{1}{3} t^3 - t^2)\, \Big|_1^2 = (8 + \frac{32}{3} - \frac{2}{4} - \frac{4}{3}) + 2 (\frac{8}{3} - 4 - \frac{1}{3} + 1)$

$= 31/2.$

Exercises 17.3: Green's Theorem (page 749)

1. Since $P(x,y) = xy$ and $\dot{Q}(x,y) = x + y$, we have by Green's Theorem that

$\int_C xy\, dx + (x + y)\, dy = \int_R (1 - x)\, dA.$ Thus, $\int_C xy\, dx + (x + y)\, dy$

$= \int_{-1}^0 \int_0^1 (1 - x)\, dy\, dx = \int_{-1}^0 (1 - x)\, dx = x - \frac{1}{2} x^2 \Big|_{-1}^0 = -(-1 - \frac{1}{2}) = 3/2.$

3. In this case $P(x,y) = 2xy$ and $Q(x,y) = 0$. Thus, by Green's Theorem

$\int_C 2xy\, dx = - \int_R 2x\, dx$ where R is the rectangular region with vertices $(3,1)$,

$(3,3)$, $(5,3)$, $(5,1)$. Thus, $\int_C 2xy\, dx = - \int_3^5 \int_1^3 2x\, dy\, dx = - \int_3^5 4x\, dx$

$= -2x^2 \Big|_3^5 = -32.$

5. In this case $P(x,y) = e^x \sin y$ and $Q(x,y) = e^x \cos y$. Thus, $\frac{\partial P}{\partial y} = e^x \cos y$ and

$\frac{\partial Q}{\partial x} = e^x \cos y$. Then by Green's Theorem $\int_C e^x \sin y\, dx + e^x \cos y\, dy$

$= \int_R (e^x \cos y - e^x \cos y)\, dA = 0.$

7. A sketch of the curve C, and the region R bounded by C is given below.

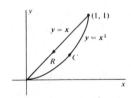

Since $P(x,y) = -y^3$ and $Q(x,y) = x^3$, we have $\int_C x^3\, dy - y^3\, dx = \int_R (3x^2 + 3y^2)\, dA$

$$= \int_0^1 \int_{x^3}^x (3x^2 + 3y^2) \, dy \, dx = \int_0^1 (3x^2 y + y^3) \Big|_{x^3}^x \, dx = \int_0^1 (4x^3 - 3x^5 - x^9) \, dx$$

$$= (x^4 - \frac{1}{2} x^6 - \frac{1}{10} x^{10}) \Big|_0^1 = 2/5.$$

9. In this case $P(x,y) = y$ and $Q(x,y) = x$. Thus, $P_y(x,y) = 1$ and $Q_x(x,y) = 1$. Consequently, from Green's Theorem $\int_C y \, dx + x \, dy = \int_R (1 - 1) \, dA = 0$.

11. Here $P(x,y) = 0$ and $Q(x,y) = x^3 - y$. Thus, by Green's Theorem

$\int_C (x^3 - y) \, dy = \int_R 3x^2 \, dA$ where R is the region bounded by the circle $x^2 + y^2$

$$= 1. \quad \text{Thus,} \quad \int_C (x^2 - y^2) \, dx = 3 \int_{-1}^1 \int_{-\sqrt{1-x^2}}^{\sqrt{1-x^2}} x^2 \, dy \, dx = 6 \int_{-1}^1 x^2 \sqrt{1 - x^2} \, dx.$$

On substitution of $x = \sin \theta$, we obtain $\int_C (x^2 - y^2) \, dx = 6 \int_{-\pi/2}^{\pi/2} \sin^2 \theta \cos^2 \theta$

$$d\theta = -\frac{3}{4} (\frac{1}{4} \sin 4\theta - \theta) \Big|_{-\pi/2}^{\pi/2} = -\frac{3}{4} (-\frac{\pi}{2} - \frac{\pi}{2}) = 3\pi/4.$$

13. In this case $P(x,y) = y^3$ and $Q(x,y) = y - x^3$. Thus, $P_y(x,y) = 3y^2$ and $Q_x(x,y)$

$= -3x^2$. Consequently, from Green's Theorem $\int_C y^3 \, dx + (y - x^3) \, dy$

$= \int_R (-3x^2 - 3y^2) \, dA$ where R is the disk bounded by $x^2 + y^2 = 1$. This last

integral is most easily evaluated by a change to polar coordinates. Then

$$\int_C y^3 \, dx + (y - x^3) \, dy = -3 \int_0^{2\pi} \int_0^1 r^2 \, r \, dr \, d\theta = -3 \int_0^{2\pi} \frac{r^4}{4} \Big|_0^1 \, d\theta = -\frac{3}{2} \pi.$$

15. Since $P(x,y) = x^3 y^4$ and $Q(x,y) = x^4 y^3$, $P_y(x,y) = 4x^3 y^3$ and $Q_x(x,y) = 4x^3 y^3$.

Thus, from Green's Theorem $\int_C x^3 y^4 \, dx + x^4 y^3 \, dy = \int_R (4x^3 y^3 - 4x^3 y^3) \, dA = 0$.

17. The hypothesis of Green's Theorem demands that the partial derivatives $Q_x(x,y)$

and $P_y(x,y)$ be continuous in the region S bounded by C. In this case $P(x,y)$

$= \dfrac{y}{x^2 + y^2}$ and $Q(x,y) = -\dfrac{x}{x^2 + y^2}$ have partial derivatives that are discontinu-

ous only at $(0,0)$. Thus,

(a) Green's Theorem may not be used since $(0,0)$ is in the region bounded by C.

(b) Green's Theorem may be used since the specified triangle does not enclose $(0,0)$.

19. By Exercise 12 the area A enclosed by C is given by $A = \int_C x\,dy$ where C is the

circle $x^2 + y^2 = r^2$. Thus, using the parametric equation $X(t) = (r\cos t,$

$r\sin t)$ with $t_I = 0$, $t_T = 2\pi$, we have $A = \int_0^{2\pi} r\cos t\,(r\cos t)\,dt$

$$= r^2 \int_0^{2\pi} \cos^2 t\,dt = r^2 \int_0^{2\pi} \frac{1 + \cos 2t}{2}\,dt = \frac{r^2}{2}(2\pi + 0) = \pi r^2.$$

Exercises 17.4: Independence of Path (page 755)

1. Since $P(x,y) = 3x^2 y^2$ and $Q(x,y) = 2x^3 y$, we have $P_y(x,y) = 6x^2 y = Q_x(x,y)$.

Thus, the integral is independent of path in the entire plane. Thus, we may

use the polygonal path consisting of the line segments C_1 from $(0,1)$ to $(1,1)$

and C_2 from $(1,1)$ to $(1,3)$. Then

$$\int_{(0,1)}^{(1,3)} 3x^2 y^2\,dx + 2x^3 y\,dy = \int_{C_1} 3x^2 y^2\,dx + 2x^3 y\,dy + \int_{C_2} 3x^2 y^2\,dx + 2x^3 y\,dy$$

$$= \int_0^1 3t^2\,dt + \int_1^3 2t\,dt = t^3 \Big|_0^1 + t^2 \Big|_1^3 = 1 + 9 - 1 = 9.$$

3. Since $P(x,y) = 2xe^y$ and $Q(x,y) = x^2 e^y$, we have $P_y(x,y) = 2xe^y = Q_x(x,y)$. Thus,

the integral is independent of path and we may use the polygonal path consist-

ing of the line segments C_1 from $(1,1)$ to $(2,1)$ and C_2 from $(2,1)$ to $(2,2)$.

Then $\displaystyle\int_{(1,1)}^{(2,2)} 2xe^y\,dx + x^2 e^y\,dy = \int_{C_1} 2xe^y\,dx + x^2 e^y\,dy + \int_{C_2} 2xe^y\,dx + x^2 e^y\,dy$

$$= \int_1^2 2te\,dt + \int_1^2 4e^t\,dt = et^2 \Big|_1^2 + 4e^t \Big|_1^2 = 3e + 4e^2 - 4e = 4e^2 - e.$$

5. Since $P(x,y) = e^x y^2$ and $Q(x,y) = 2ye^x$, we have $P_y(x,y) = 2ye^x = Q_x(x,y)$. Thus,

the integral is independent of path and we may use the polygonal path consist-

ing of the line segments C_1 from $(0,1)$ to $(0,0)$ and C_2 from $(0,0)$ to $(-1,0)$.

Then $\displaystyle\int_{(0,1)}^{(-1,0)} e^x y^2\,dx + 2ye^x\,dy = \int_{C_1} e^x y^2\,dx + 2ye^x\,dy + \int_{C_2} e^x y^2\,dx + 2ye^x\,dy$

$$= \int_1^0 2t\,dt = -1.$$

7. Since $P(x,y) = \cos x \sin y$ and $Q(x,y) = \sin x \cos y$, we have $P_y(x,y) = \cos x$

$\cos y = Q_x(x,y)$. Thus, the integral is independent of path and we may use the

polygonal path consisting of the line segments C_1 from $(0,0)$ to $(\pi,0)$ and C_2

from $(\pi,0)$ to (π,π). Then $\int_{(0,0)}^{(\pi,\pi)} \cos x \sin y \, dx + \sin x \cos y \, dy$

$= \int_{C_1} \cos x \sin y \, dx + \sin x \cos y \, dy + \int_{C_2} \cos x \sin y \, dx + \sin x \cos y \, dy$

$= \int_0^\pi \cos t \sin 0 \, dt + \int_0^\pi \sin \pi \cos t \, dt = 0.$

9. Since $P(x,y) = 2xe^y - e^x y^2$ and $Q(x,y) = x^2 e^y - 2ye^x$, we have $P_y(x,y) = 2xe^y - 2ye^x = Q_x(x,y)$. Thus, the integral is independent of path and we may use the polygonal path consisting of the line segments C_1 from $(0,1)$ to $(0,0)$ and C_2 from $(0,0)$ to $(1,0)$.

Then $\int_{(0,1)}^{(1,0)} (2xe^y - e^x y^2) \, dx + (x^2 e^y - 2ye^x) \, dy = \int_{C_1} (2xe^y - e^x y^2) \, dx$

$+ (x^2 e^y - 2ye^x) \, dy + \int_{C_2} (2xe^y - e^x y^2) \, dx + (x^2 e^y - 2ye^x) \, dy$

$= \int_1^0 (-2t) \, dt + \int_0^1 2t \, dt = 4 \int_0^1 t \, dt = 2t^2 \Big|_0^1 = 2.$

11. In this case $P(x,y) = e^x \sin y - e^y \sin x$ and $Q(x,y) = e^y \cos x + e^x \cos y$. Thus, $P_y(x,y) = e^x \cos y - e^y \sin x = Q_x(x,y)$, and the integral is independent of path. Thus, we may use the polygonal path consisting of the line segments C_1 from $(0,\pi/2)$ to $(\pi/4,\pi/2)$ and C_2 from $(\pi/4,\pi/2)$ to $(\pi/4,\pi)$. Then

$\int_{(0,\pi/2)}^{(\pi/4,\pi)} (e^x \sin y - e^y \sin x) \, dx + (e^y \cos x + e^x \cos y) \, dy$

$= \int_{C_1} (e^x \sin y - e^y \sin x) \, dx + (e^y \cos x + e^x \cos y) \, dy$

$+ \int_{C_2} (e^x \sin y - e^y \sin x) \, dx + (e^y \cos x + e^x \cos y) \, dy$

$= \int_0^{\pi/2} (e^t \sin \pi/2 - e^{\pi/2} \sin t) \, dt + \int_{\pi/2}^\pi (e^t \cos \pi/4 + e^{\pi/4} \cos t) \, dt$

$= \int_0^{\pi/2} (e^t - e^{\pi/2} \sin t) \, dt + \int_{\pi/2}^\pi (\frac{1}{\sqrt{2}} e^t + e^{\pi/4} \cos t) \, dt$

$= (e^t + e^{\pi/2} \cos t) \Big|_0^{\pi/2} + (\frac{1}{\sqrt{2}} e^t + e^{\pi/4} \sin t) \Big|_{\pi/2}^\pi$

$$= e^{\pi/2} - 1 - e^{\pi/2} + \frac{1}{\sqrt{2}} e^{\pi} - \frac{1}{\sqrt{2}} e^{\pi/2} - e^{\pi/4} = \frac{1}{\sqrt{2}} e^{\pi} - \frac{1}{\sqrt{2}} e^{\pi/2} - e^{\pi/4} - 1.$$

13. Since $P(x,y) = e^x \sin y$ and $Q(x,y) = e^x \cos y$, we have $P_y(x,y) = e^x \cos y$

$= Q_x(x,y)$. Thus, this integral is independent of path and the specified path

from $X(0) = (0,1)$ to $X(\pi) = (0,-1)$ may be replaced by the line segment $X(t)$

$= (0,t)$, $t_I = 1$, $t_T = -1$. Then

$$\int_C e^x \sin y \, dx + e^x \cos y \, dy = \int_1^{-1} \cos t \, dt = \sin t \Big|_1^{-1} = \sin(-1) - \sin 1$$

$= -2 \sin 1$.

15. Since $P(x,y) = \dfrac{y+1}{(x-1)^2 + (y+1)^2}$ and $Q(x,y) = \dfrac{1-x}{(x-1)^2 + (y+1)^2}$, we have

$$P_y(x,y) = \frac{[(x-1)^2 + (y+1)^2] - (y+1)2(y+1)}{[(x-1)^2 + (y+1)^2]^2} = \frac{(x-1)^2 - (y+1)^2}{[(x-1)^2 + (y+1)^2]^2}$$

and $Q_x(x,y) = \dfrac{-[(x-1)^2 + (y+1)^2] - (1-x)2(x-1)}{[(x-1)^2 + (y+1)^2]} = \dfrac{(x-1)^2 - (y+1)^2}{[(x-1)^2 + (y+1)^2]^2}$.

Consequently, $P_y(x,y) = Q_x(x,y)$ except at $(1,-1)$ where neither partial deriva-

tive exists. Thus, the integral is independent of path in any simply connected

region not containing $(1,-1)$.

Exercises 17.5: Line Integrals in Three Dimensions (page 761)

1. $\displaystyle\int_C y \, dx + (x+z) \, dy + z \, dz = \int_0^1 t^2 \, dt + \int_0^1 (t+t^3) \, 2t \, dt + \int_0^1 t^3 \, 3t^2 \, dt$

$= \dfrac{1}{3} t^3 \Big|_0^1 + (\dfrac{2}{3} t^3 + \dfrac{2}{5} t^5) \Big|_0^1 + \dfrac{3}{6} t^6 \Big|_0^1 = 19/10.$

3. $\displaystyle\int_C xy \, dx + yz \, dy + z^2 \, dz = \int_{\pi/2}^0 \cos t \sin t (-\sin t) \, dt + \int_{\pi/2}^0 \sin^3 t \cos t \, dt$

$+ \displaystyle\int_{\pi/2}^0 \sin^4 t \, (2 \sin t \cos t) \, dt = - \int_{\pi/2}^0 \sin^2 t \, d \sin t + \int_{\pi/2}^0 \sin^3 t \, d \sin t$

$+ 2 \displaystyle\int_{\pi/2}^0 \sin^5 t \, d \sin t = - \dfrac{1}{3} \sin^3 t \Big|_{\pi/2}^0 + \dfrac{1}{4} \sin^4 t \Big|_{\pi/2}^0 + \dfrac{1}{3} \sin^6 t \Big|_{\pi/2}^0$

$= 1/3 - 1/4 - 1/3 = -1/4.$

5. $\displaystyle\int_C \sin x \, dx + x \, dy + e^z \, dz = \int_0^1 \sin(t^3)(3t^2) \, dt + \int_0^1 t^3 \, (2t) \, dt + \int_0^1 e^t \, dt$

$= -\cos t^3 \Big|_0^1 + \dfrac{2}{5} t^5 \Big|_0^1 + e^t \Big|_0^1 = -\cos 1 + 1 + \dfrac{2}{5} + e - 1 = \dfrac{2}{5} + e - \cos 1.$

7. The line segment in question has parametric equations

$$x = -5t + 3$$
$$y = 2t + 4 \quad \text{for } 0 \le t \le 1.$$
$$z = t + 1$$

Thus, $\int_C y^2 \, dx + (x + y) \, dy + (x + z) \, dz = \int_0^1 (2t + 4)^2(-5) \, dt$

$+ \int_0^1 (-3t + 7) \, 2dt + \int_0^1 (-4t + 4) \, dt = -\frac{5}{6} (2t + 4)^3 \Big|_0^1 + 2(-\frac{3}{2} t^2 + 7t) \Big|_0^1$

$+ (-2t^2 + 4t) \Big|_0^1 = -5(36) + \frac{5}{6} 4^3 - 2(\frac{3}{2} - 7) + (-2 + 4) = -341/3.$

9. The circle C has equation $X(t) = (\cos t, \sin t, -2)$, $t_I = 0$, $t_T = 2\pi$. Hence,

$$\int_C x \, dx + y \, dy + z \, dz = \int_0^{2\pi} \cos t \, (-\sin t) \, dt + \int_0^{2\pi} \sin t \cos t \, dt = 0.$$

11. Since $P(x,y,z) = yz$, $Q(x,y,z) = xz$, and $R(x,y,z) = xy$, we have

$$P_y(x,y,z) = z = Q_x(x,y,z),$$
$$Q_z(x,y,z) = x = R_y(x,y,z), \text{ and}$$
$$R_x(x,y,z) = y = P_z(x,y,z).$$

Thus, the integral is independent of path in R_3 and we may use the polygonal path consisting of the line segment C_1 from $(-1,3,2)$ to $(2,3,2)$, C_2 from $(2,3,2)$ to $(2,1,2)$, and C_3 from $(2,1,2)$ to $(2,1,-1)$. Then

$\int_C yz \, dz + xz \, dy + xy \, dz = \int_{C_1} yz \, dx + xz \, dy + xy \, dz + \int_{C_2} yz \, dx + xz \, dy$

$+ xy \, dz + \int_{C_3} yz \, dx + xz \, dy + xy \, dz = \int_{-1}^2 6dt + \int_3^1 4dt + \int_2^{-1} 2dt = 4.$

13. Since $P(x,y,z) = 2xy$, $Q(x,y,z) = x^2 + 2yz$, and $R(x,y,z) = y^2 + 1$, we have

$$P_y(x,y,z) = 2x = Q_x(x,y,z),$$
$$Q_z(x,y,z) = 2y = R_y(x,y,z), \text{ and}$$
$$R_x(x,y,z) = 0 = P_z(x,y,z).$$

Thus, the integral is independent of path in R_3 and we may use the polygonal path consisting of the line segments C_1 from $(0,1,-1)$ to $(1,1,-1)$, C_2 from $(1,1,-1)$ to $(1,2,-1)$, and C_3 from $(1,2,-1)$ to $(1,2,0)$. Then

$\int_C 2xy \, dx + (x^2 + 2yz) \, dy + (y^2 + 1) \, dz = \int_{C_1} 2xy \, dx + (x^2 + 2yz) \, dy$

$+ (y^2 + 1)$ dz $+ \int_{C_2}$ 2xy dx $+ (x^2 + 2yz)$ dy $+ (y^2 + 1)$ dz $+ \int_{C_3}$ 2xy dx

$+ (x^2 + 2yz)$ dy $+ (y^2 + 1)$ dz $= \int_0^1$ 2t dt $+ \int_1^2 (1 - 2t)$ dt $+ \int_{-1}^0$ 5 dt

$= 1 - 2 + 5 = 4.$

15. Since $P(x,y,z) = y^2 + z^2 + yz$, $Q(x,y,z) = 2yx + xz$, and $R(x,y,z) = 2xz + xy$,

we have $\qquad P_y(x,y,z) = 2y + z = Q_x(x,y,z),$

$\qquad\qquad Q_z(x,y,z) = x = R_y(x,y,z),$ and

$\qquad\qquad R_x(x,y,z) = 2z + y = P_z(x,y,z).$

Thus, the integral is independent of path and we can use any path from
$X(1/2) = (1,0,1/2)$ to $X(1) = (0,-1,1)$. Thus, if C_1 denotes the line segment
from $(1,0,1/2)$ to $(0,0,1/2)$, C_2 the line segment from $(0,0,1/2)$ to $(0,-1,1/2)$,
and finally, C_3 the line segment from $(0,-1,1/2)$ to $(0,-1,1)$, we have

$\int_C$ P dx + Q dy + R dz $= \int_{C_1}$ P dx + Q dy + R dz $+ \int_{C_2}$ P dx + Q dy + R dz

$+ \int_{C_3}$ P dx + Q dy + R dz. Thus, $\int_C (y^2 + z^2 + yz)$ dx $+ (2yx + xz)$ dy

$+ (2xz + xy)$ dz $= \int_1^0 \frac{1}{4}$ dt $+ \int_0^{-1}$ 0 dt $+ \int_{1/2}^1$ 0 dt $= -\frac{1}{4}.$

17. (a) If $x = 3$, $z = t$, and $y = \sqrt{4 - t^2}$, $-2 \le t \le 2$, we have

$\int_C$ y dx + dy + x dz $= -\int_{-2}^2 t(4 - t^2)^{-1/2}$ dt $+ \int_{-2}^2$ 3 dt $= \frac{1}{2} \cdot 2 (4 - t^2)^{1/2} \Big|_{-2}^2$

$+ 3t \Big|_{-2}^2 = 12.$

(b) If $x = 3$, $z = -2 \cos t$, $y = 2 \sin t$, $0 \le t \le \pi$, we have

$\int_C$ y dx + dy + x dz $= \int_0^\pi (2 \cos t)$ dt $+ \int_0^\pi 3(2 \sin t)$ dt

$= 2 \sin t \Big|_0^\pi - 6 \cos t \Big|_0^\pi = -6(-1 - 1) = 12.$

19. The work is given by $W = \int_C f_1(x,y,z)$ dx $+ f_2(x,y,z)$ dy $+ f_3(x,y,z)$ dz

$= \int_C ye^{xy}$ dx $+ xe^{xy}$ dy $+ 2z$ dz. Thus, if $P(x,y,z) = ye^{xy}$, $Q(x,y,z) = xe^{xy}$, and

$R(x,y,z) = 2z$, we have $\qquad P_y(x,y,z) = e^{xy} + xye^{xy} = Q_x(x,y,z),$

$$Q_z(x,y,z) = 0 = R_y(x,y,z), \text{ and}$$
$$R_x(x,y,z) = 0 = P_z(x,y,z).$$

Consequently, the integral is independent of path and may be evaluated along any path with initial point $(1,2,3)$ and terminal point $(7,-2,-2)$. Thus, if we take C_1 to be the line segment from $(1,2,3)$ to $(7,2,3)$, C_2 to be the line segment from $(7,2,3)$ to $(7,-2,3)$, and C_3 to be the line segment from $(7,-2,3)$ to $(7,-2,-2)$, we have

$$W = \int_{C_1} ye^{xy} \, dx + xe^{xy} \, dy + 2z \, dz + \int_{C_2} ye^{xy} \, dx + xe^{xy} \, dy + 2z \, dz$$

$$+ \int_{C_3} ye^{xy} \, dx + xe^{xy} \, dy + 2z \, dz. \text{ Thus, } W = \int_1^7 2e^{2t} \, dt + \int_2^{-2} 7e^{7t} \, dt + \int_3^{-2}$$

$$2t \, dt = e^{2t} \Big|_1^7 + e^{7t} \Big|_2^{-2} + t^2 \Big|_3^{-2} = e^{-14} - e^2 - 5.$$

Exercises 17.6: The Surface Integral (page 769)

1. In this case $g(x,y) = 2x - 4y$ and R is the region pictured.

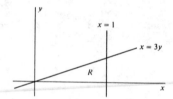

Thus, since $g_x(x,y) = 2$ and $g_y(x,y) = -4$, we have

$$\int_S (x + y + z) \, dS = \int_R (x + y + 2x - 4y) \sqrt{1 + 4 + 16} \, dA$$

$$= \sqrt{21} \int_0^1 \int_0^{x/3} (3x - 3y) \, dy \, dx = \sqrt{21} \int_0^1 (3xy - \frac{3}{2} y^2) \Big|_0^{x/3} \, dx$$

$$= \sqrt{21} \int_0^1 (x^2 - \frac{1}{6} x^2) \, dx = \frac{5\sqrt{21}}{6 \cdot 3} x^3 \Big|_0^1 = \frac{5\sqrt{21}}{18}.$$

3. The region R is as pictured above in Exercise 1. Then since $g(x,y) = 2x - 4y$, we have $\int_S (-4x^2 - 16y^2 + z^2) \, dS = \int_R (-4x^2 - 16y^2 + (2x - 4y)^2) \sqrt{21} \, dA$

$$= \sqrt{21} \int_0^1 \int_0^{x/3} (-16xy) \, dy \, dx = -8\sqrt{21} \int_0^1 xy^2 \Big|_0^{x/3} \, dx = -\frac{8\sqrt{21}}{9} \int_0^1 x^3 \, dx$$

$$= -\frac{2\sqrt{21}}{9} x^4 \Big|_0^1 = -\frac{2\sqrt{21}}{9}.$$

5. Here $g(x,y) = 1 - 3y - 2x$ so $g_x(x,y) = -2$ and $g_y(x,y) = -3$. Thus,

$$\int_S \cos z \, dS = \int_R (\cos(1 - 3y - 2x))\sqrt{1 + 4 + 9}\, dA = \sqrt{14} \int_0^1 \int_{-1}^2$$

$$\cos(1 - 3y - 2x)\, dy\, dx = -\frac{\sqrt{14}}{3} \int_0^1 \sin(1 - 3y - 2x)\Big|_{-1}^2 \, dx$$

$$= -\frac{\sqrt{14}}{3} \int_0^1 (\sin(-5 - 2x) - \sin(4 - 2x))\, dx = \frac{\sqrt{14}}{6}(-\cos(-5 - 2x)$$

$$+ \cos(4 - 2x))\Big|_0^1 = \frac{\sqrt{14}}{6}(-\cos(-7) + \cos(2) + \cos(-5) - \cos(4))$$

$$= \frac{\sqrt{14}}{6}(-\cos 7 + \cos 2 + \cos 5 - \cos 4).$$

7. Here $g(x,y) = 1 - x - y$, so $g_x(x,y) = -1 = g_y(x,y)$. Thus, $\int_S y \, dS$

$$= \int_R y\sqrt{1 + 1 + 1}\, dA \text{ where } R \text{ is the disk in the xy-plane bounded by } x^2 + y^2$$

$$= 1. \text{ Consequently, } \int_S y \, dS = \sqrt{3} \int_{-1}^1 \int_{-\sqrt{1-x^2}}^{\sqrt{1-x^2}} y\, dy\, dx = \frac{\sqrt{3}}{2} \int_{-1}^1 y^2 \Big|_{-\sqrt{1-x^2}}^{\sqrt{1-x^2}} \, dx$$

$$= \frac{\sqrt{3}}{2} \int_{-1}^1 ((1 - x^2) - (1 - x^2))\, dx = 0.$$

9. Here $g(x,y) = 4 - x - y$, so $g_x(x,y) = -1 = g_y(x,y)$. Thus,

$$\int_S (x^2 - \sin y + z)\, dS = \int_R (x^2 - \sin y + 4 - x - y)\sqrt{1 + 1 + 1}\, dA \text{ where } R \text{ is}$$

the region in the xy-plane bounded by the coordinate axes and the lines $x + y = 4$ and $x + y = 5$. R is pictured as

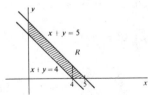

Then $\int_S (x^2 - \sin y + z)\, dS = \sqrt{3} \int_0^4 \int_{4-x}^{5-x} (x^2 - \sin y + 4 - x - y)\, dy\, dx$

$$+ \sqrt{3} \int_4^5 \int_0^{5-x} (x^2 - \sin y + 4 - x - y)\, dy\, dx = \sqrt{3} \int_0^4 (x^2 y + \cos y + 4y - xy$$

$$- \frac{1}{2} y^2)\Big|_{4-x}^{5-x} \, dx + \sqrt{3} \int_4^5 (x^2 y + \cos y + 4y - \frac{1}{2} y^2)\Big|_0^{5-x} \, dx = \sqrt{3} \int_0^4 (x^2 + \cos$$

$(5 - x) - \cos (4 - x) - \frac{1}{2})\, dx + \sqrt{3} \int_4^5 (-x^3 + \frac{11}{2} x^2 - 4x + \frac{15}{2} + \cos (5 - x))\, dx$

$= \sqrt{3}\, (\frac{1}{3} x^3 - \sin (5 - x) + \sin (4 - x) - \frac{1}{2} x) \Big|_0^4 + \sqrt{3}\, (-\frac{1}{4} x^4 + \frac{11}{6} x^3 - 2x^2$

$+ \frac{15}{2} x - \sin (5 - x) \Big|_4^5 = \sqrt{3}\, (\frac{64}{3} - \sin 1 - 2 + \sin 5 - \sin 4) + \sqrt{3}\, (- \frac{625}{4}$

$+ \frac{1375}{6} - 50 + \frac{75}{2} + 64 - \frac{704}{6} + 32 - 30 + \sin 1) = \sqrt{3}\, (\frac{341}{12} + \sin 5 - \sin 4).$

13. The tetrahedron S is the union of subsurfaces S_1, S_2, S_3, and S_4 as indicated where S_4 is a portion of the plane $2x + 4y + z = 8$.

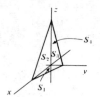

Thus, $\int_S z\, dS = \int_{S_1} z\, dS + \int_{S_2} z\, dS + \int_{S_3} z\, dS + \int_{S_4} z\, dS.$ On S_1, $z = 0$, so $\int_{S_1} z\, dS = 0.$ S_2 has equation $g(x,z) = 0$, so $g_x(x,z) = g_z(x,z) = 0$ and $\int_{S_2} z\, dS = \int_R z\, dA$ where R is the

triangle in the xz-plane bounded by $x = 0$, $z = 0$, and $2x + z = 8$. Consequently,

$\int_{S_2} z\, dS = \int_0^4 \int_0^{8-2x} z\, dz\, dx = \frac{1}{2} \int_0^4 (8 - 2x)^2\, dx = -\frac{1}{12} (8 - 2x)^3 \Big|_0^4 = \frac{8^3}{12}.$ The

surface S_3 has equation $g(x,z) = 0$. Thus, $g_y(y,z) = g_z(y,z) = 0$ and $\int_{S_3} z\, dS$

$= \int_R z\, dA$ where R is the triangle in the yz-plane bounded by $y = 0$, $z = 0$, and

$4y + z = 8$. Thus, $\int_{S_3} z\, dS = \int_0^2 \int_0^{8-4y} z\, dz\, dy = \frac{1}{2} \int_0^2 (8 - 4y)^2\, dy$

$= -\frac{1}{24} (8 - 4y)^3 \Big|_0^2 = \frac{8^3}{24}.$ The surface S_4 has equation $g(x,y) = 8 - 2x - 4y.$

Thus, $g_x(x,y) = -2$ and $g_y(x,y) = -4$ and $\int_{S_4} z\, dS = \int_R \sqrt{21}\, (8 - 2x - 4y)\, dA$

where R is the triangle in the xy-plane bounded by $x = 0$, $y = 0$, and $x + 2y$

$= 4$. Then, $\int_{S_4} z\, dS = \sqrt{21} \int_0^2 \int_0^{4-2y} (8 - 2x - 4y)\, dx\, dy = \sqrt{21} \int_0^2 (8x - x^2$

$+ 4xy) \Big|_0^{4-2y}\, dy = \sqrt{21} \int_0^2 (-12y^2 + 16y + 16)\, dy = \sqrt{21}\, (-4y^3 + 8y^2 + 16y) \Big|_0^2$

$= 32\sqrt{21}$. Consequently, $\int_S z \, dS = \frac{8^3}{12} + \frac{8^3}{24} + 32\sqrt{21} = 32(2 + \sqrt{21})$.

15. In this case $g(x,y) = (x^2 + y^2)^{1/2}$, so $g_x(x,y) = x(x^2 + y^2)^{-1/2}$ and $g_y(x,y)$

$= y(x^2 + y^2)^{-1/2}$. Thus, $\int_S e^{x^2+y^2} \, dS = \int_R e^{x^2+y^2} \sqrt{1 + \frac{x^2}{x^2 + y^2} + \frac{y^2}{x^2 + y^2}} \, dA$

$= \sqrt{2} \int_R e^{x^2+y^2} \, dA$ where R is the quarter disk determined by $x^2 + y^2 \leq 1$,

$x \geq 0$, $y \leq 0$. A transformation to polar coordinates will aid in the evaluation

of this integral. Then $\int_S e^{x^2+y^2} \, dS = \sqrt{2} \int_{-\pi/2}^{0} \int_0^1 e^{r^2} r \, dr \, d\theta = \frac{\sqrt{2}}{2} \int_{-\pi/2}^{0}$

$e^{r^2} \Big|_0^1 \, d\theta = \frac{\sqrt{2}}{2} (e - 1) \Big|_{-\pi/2}^{0} \, d\theta = \frac{\sqrt{2}\pi}{4} (e - 1)$.

Exercises 17.7: Steady State Fluid Flows (page 775)

1. Here $g(x,y,z) = x + y + z$ and since $N(x,y,z) = \pm \frac{\nabla g(x,y,z)}{|\nabla g(x,y,z)|}$, we have $N(x,y,z)$

$= \pm \frac{(1,1,1)}{\sqrt{3}}$. Thus, since we are interested in the upwardly directed normal, we

select $N(x,y,z) = \frac{1}{\sqrt{3}} (1,1,1)$. Then the flux across S is given by

$F = \int_S G(x,y,z) \cdot N(x,y,z) \, dS = \frac{1}{\sqrt{3}} \int_S (x + y + z) \, dS$, where S is the surface

given by $x + y + z = 3$, $x \geq 0$, $y \geq 0$. Thus, $F = \frac{1}{\sqrt{3}} \int_R 3\sqrt{3} \, dA = 3 \int_R dA$ where

R is the triangle in the xy-plane bounded by the coordinate axes and $x + y = 3$.

Thus, $F = 3 \cdot \frac{9}{2} = 13.5$ cubic units per unit time.

3. $2x + 2y + z = 6$, $x > 0$, $y > 0$. Here $g(x,y,z) = 2x + 2y + z$ and since

$N(x,y,z) = \pm \frac{\nabla g(x,y,z)}{|\nabla g(x,y,z)|}$, we have $N(x,y,z) = \pm \frac{(2,2,1)}{3}$. Thus, since we are

interested in the upwardly directed normal, we select $N(x,y,z) = \frac{1}{3} (2,2,1)$.

Then the flux across S is given by $F = \int_S G(x,y,z) \cdot N(x,y,z) \, dS = \int_S \frac{1}{3} (4x$

$- 2x^2 + z - 2x + 2y) \, dS = \frac{1}{3} \int_S (-2x^2 + 2x + 2y + z) \, dS$ where S is the surface

given by $2x + 2y + z = 6$, $x > 0$, $y > 0$. Thus, since $z = 6 - 2x - 2y$,

$F = \frac{1}{3} \int_R (-2x^2 + 2x + 2y + 6 - 2x - 2y) \sqrt{1 + 4 + 4} \, dA$ where R is the triangle

in the xy-plane bounded by the coordinate axes and the line $x + y = 3$. Thus,

265

$$F = \int_0^3 \int_0^{3-x} (-2x^2 + 6) \, dy \, dx = \int_0^3 (-2x^2 (3 - x) + 6 (3 - x)) \, dx$$

$$= \int_0^3 (2x^3 - 6x^2 - 6x + 18) \, dx = (\tfrac{1}{2} x^4 - 2x^3 - 3x^2 + 18x) \Big|_0^3 = 13.5 \text{ cubic units}$$

per unit time.

5. Here $g(x,y,z) = z - (x^2 + y^2)^{1/2}$ and since $N(x,y,z) = \pm \dfrac{\nabla g(x,y,z)}{|\nabla g(x,y,z)|}$, we have

$$N(x,y,z) = \pm \dfrac{(\dfrac{-x}{\sqrt{x^2 + y^2}}, \dfrac{-y}{\sqrt{x^2 + y^2}}, 1)}{\sqrt{2}}.$$ Since we are interested in the upwardly

directed normal, we take $N(x,y,z) = \dfrac{1}{\sqrt{2}} (\dfrac{-x}{\sqrt{x^2 + y^2}}, \dfrac{-y}{\sqrt{x^2 + y^2}}, 1)$. Then the flux

across S is given by $F = \dfrac{1}{\sqrt{2}} \int_S (y,-x,xy) \cdot (\dfrac{-x}{\sqrt{x^2 + y^2}}, \dfrac{-y}{\sqrt{x^2 + y^2}}, 1) \, dS$

$= \dfrac{1}{\sqrt{2}} \int_S xy \, dS$ where S is the surface given by $z = (x^2 + y^2)^{1/2}$, $x^2 + y^2 \le 1$.

Thus, $F = \dfrac{1}{\sqrt{2}} \int_R xy \sqrt{2} \, dA = \int_R xy \, dA$ where R is the disk $x^2 + y^2 \le 1$. Thus, in

polar coordinates $F = \int_0^{2\pi} \int_0^1 r \cos \theta \, r \sin \theta \, r \, dr \, d\theta = \dfrac{1}{4} \int_0^{2\pi} \cos \theta \sin \theta \, d\theta$

$= \dfrac{1}{8} \sin^2 \theta \Big|_0^{2\pi} = 0$ cubic units per unit time.

7. Here $g(x,y,z) = x^2 + y^2 + z^2$ and since $N(x,y,z) = \pm \dfrac{\nabla g(x,y,z)}{|\nabla g(x,y,z)|}$, we have

$$N(x,y,z) = \pm \dfrac{(2x,2y,2z)}{\sqrt{4x^2 + 4y^2 + 4z^2}} = \pm \dfrac{(x,y,z)}{\sqrt{x^2 + y^2 + z^2}}.$$ Since we are interested in

the upwardly directed normal, we take $N(x,y,z) = \dfrac{(x,y,z)}{\sqrt{x^2 + y^2 + z^2}}$. Then, the

flux across S is given by $F = \int_S (x,y,z) \cdot \dfrac{(x,y,z)}{\sqrt{x^2 + y^2 + z^2}} \, dS = \int_S \sqrt{x^2 + y^2 + z^2} \, dS$

dS where S is the upper hemisphere $x^2 + y^2 + z^2 = 1$, $z > 0$. That is, $z = \sqrt{1 - x^2 - y^2}$. Hence, $F = \int_R \sqrt{x^2 + y^2 + 1 - x^2 - y^2}$

$\sqrt{1 + \dfrac{x^2}{1 - x^2 - y^2} + \dfrac{y^2}{1 - x^2 - y^2}} \, dA$ or, $F = \int_R \dfrac{1}{\sqrt{1 - x^2 - y^2}} \, dA$ where R is the

disk given by $x^2 + y^2 \le 1$. On transforming to polar coordinates we have

$$F = \int_0^{2\pi} \int_0^1 \frac{1}{\sqrt{1-r^2}} \; r \; dr \; d\theta = - \int_0^{2\pi} (1-r^2)^{1/2} \Big|_0^1 \; d\theta = \int_0^{2\pi} d\theta = 2\pi \text{ cubic units}$$

per unit time.

9. The tetrahedron is the union of subsurfaces S_1, S_2, S_3, and S_4 where S_4 is the portion of the plane $x + y + z = 3$ for $x \geq 0$, $y \geq 0$, $z \geq 0$. The tetrahedron is indicated below.

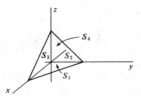

Thus, the total flux in the outward direction is given by

$$F = \int_{S_1} G \cdot N \; dS + \int_{S_2} G \cdot N \; dS + \int_{S_3} G \cdot N \; dS + \int_{S_4} G \cdot N \; dS.$$

In Exercise 1 we found that $\int_{S_4} G \cdot N \; dS = 13.5$ cubic units per time. On S_1 the outwardly directed normal is $(0,0,-1)$. Thus, $\int_{S_1} G \cdot N \; dS = \int_{S_1} (-z) \; dS$.

But, since $z = 0$ on S_1, we have $\int_{S_1} G \cdot N \; dS = 0$. On S_2 the outwardly directed normal is $(-1,0,0)$. Thus, $\int_{S_2} G \cdot N \; dS = \int_{S_2} (-x) \; dS$. Then since $x = 0$ on S_2,

$\int_{S_2} G \cdot N \; dS = 0$. On S_3 the outwardly directed normal is $(0,-1,0)$. Thus,

$\int_{S_3} G \cdot N \; dS = \int_{S_3} (-y) \; dS$. Since $y = 0$ on S_3, $\int_{S_3} G \cdot N \; dS = 0$. Thus, the total flux in the outward direction is given by

$$F = \int_{S_1} G \cdot N \; dS + \int_{S_2} G \cdot N \; dS + \int_{S_3} G \cdot N \; dS + \int_{S_4} G \cdot N \; dS = 13.5 \text{ cubic}$$

units per unit time.

11. The tetrahedron is the union of subsurfaces S_1, S_2, S_3, and S_4 where S_4 is the portion of the plane $2x + 2y + z = 6$, $x > 0$, $y > 0$, $z > 0$. The tetrahedron is indicated below.

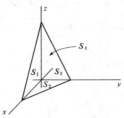

Thus, the total flux in the outward direction is given by

$$F = \int_{S_1} G \cdot N \, dS + \int_{S_2} G \cdot N \, dS + \int_{S_3} G \cdot N \, dS + \int_{S_4} G \cdot N \, dS \text{ where N is taken}$$

as the outwardly directed normal in each case. In Exercise 3 we found that $\int_{S_4} G \cdot N \, dS = 13.5$ cubic units per unit time. On S_1 the outwardly directed

unit normal is $(0,-1,0)$. Thus, $\int_{S_1} G \cdot N \, dS = \int_{S_1} x^2 \, dS = \int_0^3 \int_0^{6-2x} x^2 \, dz \, dx$

$= \int_0^3 (6x^2 - 2x^3) \, dx = 2x^3 - \frac{1}{2} x^4 \Big|_0^3 = 13.5$ cubic units per unit time. On S_2

the outwardly directed normal is $(0,0,-1)$. Thus, $\int_{S_2} G \cdot N \, dS = \int_{S_2}$

$(-z + 2x - 2y) \, dS$. On S_2, $z = 0$ so $\int_{S_2} G \cdot N \, dS = 2 \int_0^3 \int_0^{3-x} (x - y) \, dy \, dx$

$= 2 \int_0^3 (x(3 - x) - \frac{1}{2} (3 - x)^2) \, dx = \int_0^3 (-3x^2 + 12x - 9) \, dx = -x^3 + 6x^2 - 9x \Big|_0^3$

$= 0$. On S_3 the outwardly directed normal is $(-1,0,0)$. Thus, $\int_{S_3} G \cdot N \, dS$

$= \int_{S_3} (-2x) \, dS = 0$ since $x = 0$ on S_3. Consequently, the total flux $F = 13.5$

$+ 13.5 = 27$ cubic units per unit time.

13. The surface S is the union of two subsurfaces S_1 and S_2 where S_1 is a portion

of the cone $z = (x^2 + y^2)^{1/2}$ and S_2 is a portion of the plane $z = 1$. The total

flux across S is given by $\int_S G \cdot N \, dS = \int_{S_1} G \cdot N \, dS + \int_{S_2} G \cdot N \, dS$. In Exer-

cise 5 we found that $\int_{S_1} G \cdot N \, dS = 0$. On S_2 the outwardly directed normal is

$(0,0,1)$. Thus, $\int_{S_2} G(x,y,z) \cdot N \, dS = \int_{S_2} xy \, dS = \int_R xy \, dS$ where R is the disk

$x^2 + y^2 \leq 1$. Then using polar coordinates we have $\int_{S_2} G(x,y,z) \cdot N \, dS$

$= \int_0^{2\pi} \int_0^1 r \cos \theta \, r \sin \theta \, r \, dr \, d\theta = \frac{1}{4} \int_0^{2\pi} \cos \theta \sin \theta \, d\theta = \frac{1}{8} \sin^2 \Big|_0^{2\pi} = 0$.

Consequently, the total flux across S is 0 cubic units per unit time.

15. The total flux is $F = \int_{S_1} G \cdot N \, dS + \int_{S_2} G \cdot N \, dS$ where S_1 is the upper hemis-

phere and S_2 is the lower. In Exercise 7 we found that $\int_{S_1} G \cdot N \, dS = 2\pi$ cubic units per unit time. Thus, we need only compute $\int_{S_2}$ G • N dS where S_2 is the

surface $x^2 + y^2 + z^2 = 1$, $z < 0$. On S_2, $g(x,y,z) = x^2 + y^2 + z^2$ so $N(x,y,z)$

$= \dfrac{\nabla g(x,y,z)}{|\nabla g(x,y,z)|} = \pm \dfrac{(x,y,z)}{\sqrt{x^2 + y^2 + z^2}}$. Since we are interested in the outwardly di-

rected normal, we select $N(x,y,z) = \dfrac{(x,y,z)}{\sqrt{x^2 + y^2 + z^2}}$. Then $\int_{S_2}$ G • N dS

$= \int_{S_2} \sqrt{x^2 + y^2 + z^2} \, dS = \int_R \sqrt{x^2 + y^2 + 1 - x^2 - y^2} \dfrac{1}{\sqrt{1 - x^2 - y^2}} \, dA$

$= \int_R \dfrac{dA}{\sqrt{1 - x^2 - y^2}}$ where R is the disk $x^2 + y^2 \leq 1$. As in Exercise 7,

$\int_{S_2}$ G • N dS = 2π cubic units per unit time. Thus, the total flux is

$F = \int_{S}$ G • N dS = 4π cubic units per unit time.

<u>Exercises 17.8</u>: The Divergence Theorem (page 780)

1. Since $G(X) = (x + y, y + z, z + x)$, div G = 1 + 1 + 1 = 3. Thus, $\int_S$ G(X) •

N(X) dS = $\int_R$ 3 dV where R is the region bounded by $x^2 + y^2 + z^2 = 4$. Thus,

$\int_S$ G(X) • N(X) dS = 3 $\int_R$ dV. However, since $\int_R$ dV is simply the volume of the

sphere bounded by $x^2 + y^2 + z^2 = 4$, $\int_R$ dV = $\dfrac{4}{3} \pi (2)^3 = \dfrac{32}{3} \pi$. Thus, $\int_S$ G(X) •

N(X) dS = 32π.

3. Since $G(X) = kX = (kx, ky, kz)$, div G = k + k + k = 3k. Thus, $\int_S$ G(X) • N(X) dS

$= \int_R$ 3k dV = 3k $\int_R$ dV where R is the region bounded by $x^2 + y^2 + z^2 = a^2$.

Since $\int_R$ dV is the volume of a sphere of radius a, $\int_S$ G(X) • N(X) dS

$= 3k \left(\dfrac{4}{3} \pi a^3\right) = 4ka^3\pi$.

5. Since $g(x,y,z) = (x,y,z)$, div G = 1 + 1 + 1 = 3. Thus, the total flux across
    the tetrahedron is given by $\int_S$ G(X) • N(X) dS = $\int_R$ 3 dV where R is the region

bounded by the tetrahedron. Thus, $\int_R$ dV = volume of the tetrahedron = 9/2.

Hence, the total flux is F = $\int_S$ G(X) • N(X) dS = $3 \cdot \dfrac{9}{2}$ = 13.5 cubic units per

unit time.

7. Since $G(x,y,z) = (2x, -x^2, z - 2x + 2y)$, div $G = 2 + 0 + 1 = 3$. Thus, the total flux across the tetrahedron is given by $\int_S G(X) \cdot N(X)\, dS = \int_R 3\, dV$ where R is the region bounded by the tetrahedron. Thus, $\int_R dV$ = volume of the tetrahedron = 9. Hence, the total flux is $F = \int_S G(X) \cdot N(X) = 3 \int_R dV = 27$ cubic units per unit time.

9. Since $G(x,y,z) = (y,-x,xy)$, div $G = 0 + 0 + 0 = 0$. Thus, $\int_S G(X) \cdot N(X)\, dS = \int_R \text{div } G\, dV = 0$.

11. Since $G(x,y,z) = (x,y,z)$, div $G = 1 + 1 + 1 = 3$. Thus, the total flux across S is given by $F = \int_S G(X) \cdot N(X)\, dS = \int_R \text{div } G\, dV = 3 \int_R dV$ where R is the volume bounded by $x^2 + y^2 + z^2 = 1$. Hence, $\int_R dV = \frac{4}{3}\pi$ and so $F = \int_S G(X) \cdot N(X)\, dS = 3\left(\frac{4}{3}\pi\right) = 4\pi$ cubic units per unit time.

13. Let $G(x,y,z) = (C_1,C_2,C_3)$ where C_1, C_2, and C_3 are constants. Then div $G = 0$, so $\int_S G \cdot N\, dS = \int_R \text{div } G\, dV = 0$ as was to be shown.

Exercises 17.9: Stokes' Theorem (page 787)

1. The circle bounds the surface S given by $x^2 + y^2 \le 4$, $z = 2$. Thus, $N(X) = (0,0,1)$. Then since $P(x,y,z) = x + z$, $Q(x,y,z) = y + z$, and $R(x,y,z) = \sin z$, we have $\nabla \times F = \begin{vmatrix} i & j & k \\ \frac{\partial}{\partial x} & \frac{\partial}{\partial y} & \frac{\partial}{\partial z} \\ x + z & y + z & \sin z \end{vmatrix} = (-1,1,0)$. Thus, by Stokes' Theorem

$$\int_C (x + z)\, dx + (y + z)\, dy + \sin z\, dz = \int_S (-1,1,0) \cdot (0,0,1)\, dS = \int_S 0\, dS = 0.$$

3. The circle formed by the intersection of the cylinder $x^2 + y^2 = 1$ and the hemisphere $z = \sqrt{10 - x^2 - y^2}$ lies in the plane $z = 3$. Thus, the circle of intersection bounds the surface S given by $z = 3$, $x^2 + y^2 \le 1$. On this surface $N(X) = (0,0,1)$. Then since $P(x,y,z) = 2y + 3x$, $Q(x,y,z) = \sin y + z$, and $R(x,y,z) = \cos z + x$, we have $\nabla \times F = \begin{vmatrix} i & j & k \\ \frac{\partial}{\partial x} & \frac{\partial}{\partial y} & \frac{\partial}{\partial z} \\ 2y + 3x & \sin y + z & \cos z + x \end{vmatrix}$

$= (-1,-1,-2)$. Thus, by Stokes' Theorem, $\int_C (2y + 3x)\, dx + (\sin y + z)\, dy + (\cos z + x)\, dz = \int_S (-1,-1,-2) \cdot (0,0,1)\, dS = -2 \int_S dS$. Since $\int_S dS$ is the area of the surface S, $\int_S dS = \pi$. Hence, $\int_C (2y + 3x)\,dx + (\sin y + z)\,dy$

$+ (\cos z + x)\,dz = -2\pi$.

5. Since the triangle lies in the plane given by $x + z = 1$, we may take

$N = (1,0,1)/\sqrt{2}$. Then since $F(X) = (y,z,x)$, $\nabla \times F = \begin{vmatrix} i & j & k \\ \dfrac{\partial}{\partial x} & \dfrac{\partial}{\partial y} & \dfrac{\partial}{\partial z} \\ y & z & x \end{vmatrix} = (-1,-1,-1)$.

Thus, by Stokes' Theorem $\int_C y\,dx + z\,dy + x\,dz = \int_S (-1,-1,-1) \cdot (1,0,1)/\sqrt{2}\ dS$

$= -\sqrt{2} \int_S dS$. Since $\int_S dS$ is the area of the triangle bounded by C, $\int_S dS$

$= \dfrac{1}{2} \cdot 1 \cdot \sqrt{2} = 1/\sqrt{2}$. Thus, $\int_C y\,dx + z\,dy + x\,dz = -\sqrt{2}\,(1/\sqrt{2}) = -1$.

7. The curve is the circle of $x^2 + y^2 = 4$ in the plane $z = 1$. Thus, C bounds the

surface S given by $z = 1$, $x^2 + y^2 \le 4$. On this surface $N(X) = (0,0,1)$. Then

since $P(x,y,z) = -z - y$, $Q(x,y,z) = x + z$, and $R(x,y,z) = x - y$, we have

$\nabla \times F = \begin{vmatrix} i & j & k \\ \dfrac{\partial}{\partial x} & \dfrac{\partial}{\partial y} & \dfrac{\partial}{\partial z} \\ -z - y & x + z & x - y \end{vmatrix} = (-2,-2,2)$. Thus, by Stokes' Theorem

$\int_C (-z - y)\,dx + (x + z)\,dy + (x - y)\,dz = \int_S (-2,-2,2) \cdot (0,0,1)\ dS = 2 \int_S dS$.

Since $\int_S dS$ is the area of the surface S, $\int_S dS = 4\pi$. Hence, $\int_C (-z - y)\,dx$

$+ (x + z)\,dy + (x - y)\,dz = 8\pi$.

9. The surface S is bounded by the curve in the xy-plane with equation $x^2 + y^2$

$= 4$. Thus, since this curve has parametric equation $X(t) = (2 \cos t, 2 \sin t,$

$0)$, $0 \le t \le 2\pi$, we have $\int_C z\,dx + y\,dy - x\,dz = \int_0^{2\pi} 2 \sin t\,(2 \cos t)\,dt$

$= 2 \sin^2 t \,\Big|_0^{2\pi} = 0$. Since $F(X) = (z,y,-x)$, $\nabla \times F = \begin{vmatrix} i & j & k \\ \dfrac{\partial}{\partial x} & \dfrac{\partial}{\partial y} & \dfrac{\partial}{\partial z} \\ z & y & -x \end{vmatrix} = (0,2,0)$.

Since the surface S is given by $x^2 + y^2 + z^2 = 4$, $z \ge 0$, the upwardly directed

normal to S is given by $N(X) = \dfrac{(x,y,z)}{\sqrt{x^2 + y^2 + z^2}}$. Thus, $(\nabla \times F) \cdot N = (0,2,0) \cdot$

$\dfrac{(x,y,z)}{\sqrt{x^2 + y^2 + z^2}} = \dfrac{2y}{\sqrt{x^2 + y^2 + z^2}}$. Then $\int_S (\nabla \times F) \cdot N(X)\ dS = \int_S \dfrac{2y}{\sqrt{x^2 + y^2 + z^2}}\ dS$

$= \int_R \dfrac{2y}{\sqrt{x^2 + y^2 + 4 - x^2 - y^2}}\sqrt{\dfrac{4}{4 - x^2 - y^2}}\ dA = 2 \int_R \dfrac{y}{\sqrt{4 - x^2 - y^2}}\ dA$. Since R is

271

bounded by $x^2 + y^2 = 4$ we get $2 \int_{-2}^{2} \int_{-\sqrt{4-x^2}}^{\sqrt{4-x^2}} (4 - x^2 - y^2)^{-1/2} y \, dy \, dx$

$= -2 \int_{-2}^{2} (4 - x^2 - y^2)^{1/2} \Big|_{-\sqrt{4-x^2}}^{\sqrt{4-x^2}} dx = -2 \int_{-2}^{2} 0 \, dx = 0.$

Technique Review Exercises, Chapter 17 (page 791)

1. Use of Theorem 16.2.2 gives $\int_{C} (x + y) \, dx + x^2 \, dy = \int_{0}^{2} ((t^2 + t^3) \, 2t + t^4$

   $(3t^2)) \, dt = \int_{0}^{2} (3t^6 + 2t^4 + 2t^3) \, dt = \frac{3}{7} t^7 + \frac{2}{5} t^5 + \frac{1}{2} t^4 \Big|_{0}^{2} = \frac{2648}{35}.$

2. C has the parametric equation $X(t) = (t^2, t)$. Thus, $\int_{C} e^{\sqrt{x}} y \, dx + e^y x \, dy$

   $= \int_{1}^{2} (e^t t (2t) + e^t t^2) \, dt = 3 \int_{1}^{2} t^2 e^t \, dt.$ In order to evaluate this integral we will use integration by parts with $u = t^2$ and $dv = e^t \, dt$. Then $du = 2t \, dt$ and $v = e^t$ so $\int_{C} e^{\sqrt{x}} y \, dx + e^y x \, dy = 3 \left[t^2 e^t \Big|_{1}^{2} - 2 \int_{1}^{2} t e^t \, dt \right]$

   $= 12e^2 - 3e - 6 \int_{1}^{2} t e^t \, dt.$ A second integration by parts with $u = t$ and $dv = e^t$ gives $\int_{C} e^{\sqrt{x}} y \, dx + e^y x \, dy = 12e^2 - 3e - 6 \left[t e^t \Big|_{1}^{2} - \int_{1}^{2} e^t \, dt \right]$

   $= 12e^2 - 3e - 6(2e^2 - e - e^2 + e) = 6e^2 - 3e.$

3. By Green's Theorem $\int_{C} P(x,y) \, dx + Q(x,y) \, dy = \int_{R} (Q_x(x,y) - P_y(x,y)) \, dA$ where R is the region bounded by the simple closed curve C. Thus, $\int_{C} x^2 y \, dx + xy^2 \, dy$

   $= \int_{R} (y^2 - x^2) \, dA = \int_{-1}^{1} \int_{-1}^{1} (y^2 - x^2) \, dy \, dx = \int_{-1}^{1} (\frac{2}{3} - 2x^2) \, dx = \frac{2}{3} x - \frac{2}{3} x^3 \Big|_{-1}^{1}$
   $= 0.$

4. In this case $P(x,y,z) = yz$, $Q(x,y,z) = xz$, and $R(x,y,z) = xy$. Thus,
   $$P_y(x,y,z) = z = Q_x(x,y,z),$$
   $$Q_z(x,y,z) = x = R_y(x,y,z), \text{ and}$$
   $$R_x(x,y,z) = y = P_z(x,y,z).$$

   Consequently, the given integral is independent of path in R^3.

5. Use of Theorem 16.5.1 gives $\int_C (x + z)\, dx + (x + z)\, dy + (y + z)\, dz$

$$= \int_0^{\pi/4} [(\sin^2 t + \tan^3 t)\, 2 \sin t \cos t - (\sin^2 t + \tan^3 t)\, 2 \cos t \sin t$$

$$+ (\cos^2 t + \tan^3 t)\, 3 \tan^2 t \sec^2 t]\, dt = 3 \int_0^{\pi/4} (\tan^2 t + \tan^5 t \sec^2 t)\, dt$$

$$= 3 \int_0^{\pi/4} (\sec^2 t - 1 + \tan^5 t \sec^2 t)\, dt = 3\, [\tan t - t + \frac{1}{6} \tan^6 t] \,\Big|_0^{\pi/4}$$

$$= 3\,(1 - \pi/4 + 1/6] = 7/2 - 3\pi/4.$$

6. The surface S is not the graph of a single function. Rather, S is the union

of two surfaces S_1 and S_2 where S_1 is the graph of $z = -\sqrt{1 - y^2}$, $1 \le y \le 1$,

$0 \le x \le 1$ and S_2 is the graph of $z = \sqrt{1 - y^2}$, $-1 \le y \le 1$, $0 \le x \le 1$. Then

$\int_S z\, dS = \int_{S_1} z\, dS + \int_{S_2} z\, dS$. Since $z_x = 0$ and $z_y = y/\sqrt{1 - y^2}$ on S_1, and

$z_x = 0$ and $z_y = -y/\sqrt{1 - y^2}$ on S_2, $\int_S z\, dS = -\int_R \sqrt{1 - y^2}\, \sqrt{1 + \dfrac{y^2}{1 - y^2}}\, dA$

$+ \int_R \sqrt{1 - y^2}\, \sqrt{1 + \dfrac{y^2}{1 - y^2}}\, dA$ where R is the rectangle determined by $0 \le x \le 1$,

$-1 \le y \le 1$. Thus, $\int_S z\, dS = 0$.

7. Since $V(X) = aX = (ax, ay, az)$, div $V = 3a$. Thus, by the Divergence Theorem, the
outward flux is given by $F = \int_V \text{div } V\, dV = 3a \int_V dV = 24a$ cubic units per unit
time since $\int_V dV$ is the volume of the cube V.

8. Here unit normals to the plane are given by $N(x,y,z) = \pm \dfrac{\nabla g(x,y,z)}{|\nabla g(x,y,z)|}$. Thus,
since $g(x,y,z) = x + 2y + z$, $N(x,y,z) = \pm (1,2,1)/\sqrt{6}$. Since we are interested
in the flow in the downward direction, we select $N(x,y,z) = -(1,2,1)/\sqrt{6}$. Then
the flux in the downward direction across S is given by $F = \int_S V(x,y,z) \cdot$

$N(x,y,z)\, dS = -\int_S a(x,y,z) \cdot (1,2,1)/\sqrt{6}\, dS = -\dfrac{a}{\sqrt{6}} \int_S (x + 2y + z)\, dS$ where S

is the surface given by $z = 4 - x - 2y$ for $x > 0$, $y > 0$, $z > 0$. Thus,

$F = -\dfrac{a}{\sqrt{6}} \int_R (x + 2y + 4 - x - 2y)\, \sqrt{1 + 1 + 4}\, dA = -4a \int_R dA$ where R is the

triangle bounded by the coordinate axes and the line $x + 2y = 4$. Then since
$\int_R dA$ is the area of the triangle R, $F = -4a \int_R dA = -16a$ cubic units per unit
time.

9. The curve C bounds the disk S given by $x^2 + y^2 = 4$, $z = 4$. Since the surface S is horizontal, the upwardly directed unit normal is $(0,0,1)$. In this case $F(x,y,z) = (-2y, 2x, z^2 x)$. Thus,

$$\nabla \times F = \begin{vmatrix} i & j & k \\ \frac{\partial}{\partial x} & \frac{\partial}{\partial y} & \frac{\partial}{\partial z} \\ -2y & 2x & z^2 z \end{vmatrix} = (0, -z^2, 4).$$ Use of Stokes' Theorem then gives

$$\int_C -2y\, dx + 2x\, dy + z^2 x\, dz = \int_S (0, -z^2, 4) \cdot (0,0,1)\, dS = 4 \int_S dS = 16\pi.$$

Additional Exercises, Chapter 17 (page 792)

1. By definition $\int_C f(x,y)\, dx = \lim\limits_{n \to \infty} \sum\limits_{i=1}^{n} f(x_i^*, y_i^*)(x_i - x_{i-1})$ where (x_i, y_i),

(x_{i-1}, y_{i-1}), and (x_i^*, y_i^*) all lie on C. Since $f(x,y) = x^2 + 2xy + y^2 = (x + y)^2$ and $x + y = 5$ on C, we have that $f(x,y) = 25$ for all (x,y) on C. Thus,

$$\int_C f(x,y)\, dx = \lim_{n \to \infty} \sum_{i=1}^{n} 25\,(x_i - x_{i-1}) = 25 \lim_{n \to \infty} \sum_{i=1}^{n} (x_i - x_{i-1}) = 25 \lim_{n \to \infty}$$

$(x_n - x_0)$. Since C has the initial point $(1,4)$, we have $x_0 = 1$. Since C has the terminal point $(0,5)$, $x_n = 0$. Thus, $\int_C f(x,y)\, dx = 25 \lim\limits_{n \to \infty} (0 - 1) = -25$.

3. $\int_C (xy + y^2)\, dx + y^2\, dy = \int_1^2 (t^3 + t^4)\, dt + \int_1^2 t^4\, 2t\, dt$

$= \frac{1}{4} t^4 + \frac{1}{5} t^5 + \frac{2}{6} t^6 \Big|_1^2 = 30 \frac{19}{20}.$

5. The line segment in question has the parametric equation $X(t) = (-5t + 3, t + 1)$, $0 \le t \le 1$. Thus, $\int_C y^2\, dx + 2xy\, dy = \int_0^1 (t + 1)^2 (-5)\, dt + \int_0^1 2(-5t + 3)(t + 1)$

$dt = -\frac{5}{3} (t + 1)^3 \Big|_0^1 + 2(-\frac{5}{3} t^3 - t^2 + 3t) \Big|_0^1 = -\frac{5}{3} (8 - 1) + 2(-\frac{5}{3} - 1 + 3)$

$= -11.$

7. The curve C has the parametric equation $X(t) = (t, \tan t)$, $0 \le t \le \pi/4$. Thus,

$$\int_C y\, dx + y^2\, dy = \int_0^{\pi/4} \tan t\, dt + \int_0^{\pi/4} \tan^2 t \sec^2 t\, dt = -\ln |\cos t| \Big|_0^{\pi/4}$$

$+ \frac{1}{3} \tan^3 t \Big|_0^{\pi/4} = -\ln (1/\sqrt{2}) + \frac{1}{3} = \frac{1}{3} + \frac{1}{2} \ln 2.$

9. Let C_1 denote the portion of C that follows $y = -x$ and C_2 denote the portion of C that follows $y = x^2$. Then $\int_C y^2 \, dx + (x + y) \, dy = \int_{C_1} y^2 \, dx + (x + y) \, dy$ $+ \int_{C_2} y^2 \, dx + (x + y) \, dy$. The curve C_1 has the parametric equation $X(t) = (t, -t)$ with initial point $X(0)$ and terminal point $X(-1)$. The curve C_2 has the parametric equation $X(t) = (t, t^2)$ with initial point $X(-1)$ and terminal point $X(0)$. Thus, $\int_C y^2 \, dx + (x + y) \, dy = \int_0^{-1} t^2 \, dt + \int_0^{-1} 0 \, dt + \int_{-1}^0 t^4 \, dt$ $+ \int_{-r}^0 (t + t^2) \, 2t \, dt = \frac{1}{3} t^3 \Big|_0^{-1} + \frac{1}{5} t^5 \Big|_{-1}^0 + (\frac{2}{3} t^3 + \frac{1}{2} t^4) \Big|_{-1}^0 = -\frac{1}{3} + \frac{1}{5} + \frac{2}{3}$ $- \frac{1}{2} = \frac{1}{30}$.

11. Since $P(x,y) = (x - y)^2$ and $Q(x,y) = (x + y)^2$, we have $P_y(x,y) = -2(x - y)$ and $Q_x(x,y) = 2(x + y)$. Thus, by Green's Theorem $\int_C (x - y)^2 \, dx + (x + y)^2 \, dy$ $= \int_R 2(x + y) + 2(x - y) \, dA = \int_R 4x \, dA$ where R is the region in the xy-plane bounded by $y = x^3$ and $y = x$ between $x = 0$ and $x = 1$. Thus, $\int_C (x - y)^2 \, dx$ $+ (x + y)^2 \, dy = 4 \int_0^1 \int_{x^3}^x x \, dy \, dx = 4 \int_0^1 (x^2 - x^4) \, dx = 4(\frac{1}{3} x^3 - \frac{1}{5} x^5) \Big|_0^1$ $= 4(\frac{1}{3} - \frac{1}{5}) = \frac{8}{15}$.

13. In this case $P(x,y) = 0$ and $Q(x,y) = x^3 - 3y^2$, thus, $P_y(x,y) = 0$ and $Q_x(x,y) = 3x^2$. Thus, by Green's Theorem $\int_C (x^3 - 3y^2) \, dy = \int_R 3x^2 \, dA$ where R is the disk $x^2 + y^2 \le 4$. Then in polar coordinates we have $\int_C (x^3 - 3y^2) \, dy = 3 \int_0^{2\pi}$ $\int_0^2 r^2 \cos^2 \theta \, r \, dr \, d\theta = 12 \int_0^{2\pi} \cos^2 \theta \, d\theta = 6 \int_0^{2\pi} (1 + \cos 2\theta) \, d\theta$ $= (6\theta + 3 \sin 2\theta) \Big|_0^{2\pi} = 12\pi$.

15. Since $P(x,y) = \sin y$ and $Q(x,y) = \sin x$, $P_y(x,y) = \cos y$ and $Q_x(x,y) = \cos x$. Thus, by Green's Theorem $\int_C \sin y \, dx + \sin x \, dy = \int_R (\cos x - \cos y) \, dA$ where R is the triangle bounded by $y = -x + 1$, $x = 0$ and $y = 0$. Thus,

$$\int_C \sin y \, dx + \sin x \, dy = \int_0^1 \int_0^{-x+1} (\cos x - \cos y) \, dy \, dx = \int_0^1 (y \cos x - \sin y) \Big|_0^{-x+1}$$

$$dx = \int_0^1 ((-x + 1) \cos x - \sin (-x + 1)) \, dx = -x \sin x \Big|_0^1 - \cos x \Big|_0^1$$

$$+ \sin x \Big|_0^1 - \cos (-x + 1) \Big|_0^1 = -\sin 1 - \cos 1 + 1 + \sin 1 - 1 + \cos 1 = 0.$$

17. Since $P(x,y) = \cos y$ and $Q(x,y) = -x \sin y$, we have $P_y(x,y) = -\sin y = Q_x(x,y)$.

Thus, the integral is independent of path and we can evaluate the integral along the polygonal **path** consisting of the line segments C_1 from $(1,-1)$ to $(-1,-1)$ and C_2 from $(-1,-1)$ to $(-1,1)$. Then $\int_{(1,-1)}^{(-1,1)} \cos y \, dx - x \sin y \, dy$

$$= \int_{C_1} \cos y \, dx - x \sin y \, dy + \int_{C_2} \cos y \, dx - x \sin y \, dy = \int_1^{-1} \cos (-1) \, dx$$

$$+ \int_{-1}^1 \sin y \, dy = -2 \cos (-1) - \cos y \Big|_{-1}^1 = -2 \cos 1 - \cos 1 + \cos (-1)$$

$$= -2 \cos 1.$$

21. Since $P(x,y) = \ln y$ and $Q(x,y) = x/y$, we have $P_y(x,y) = 1/y$ and $Q_x(x,y) = 1/y$.

Then $P_y(x,y)$ and $Q_x(x,y)$ are continuous and equal except along the line $y = 0$ where they do not exist. Consequently, the given integral is independent of path in any simply connected region that does not intersect the line $y = 0$.

23. The curve C has the parametric equation $X(t) = (t,t,t)$ with initial point $X(0)$ and terminal point $X(1)$. Thus, $\int_C (x^2 - z) \, dx + (y^2 - x) \, dy + (z^2 - y) \, dz$

$$= \int_0^1 (t^2 - t) \, dt + \int_0^1 (t^2 - t) \, dt + \int_0^1 (t^2 - t) \, dt = 3 \int_0^1 (t^2 - t) \, dt$$

$$= (t^3 - \frac{3}{2} t^2) \Big|_0^1 = -\frac{1}{2}.$$

25. $\int_C z \sin (xz) \, dx + xz \cos (xyz) \, dy + x \sin (xz) \, dz = \int_0^1 t^2 \sin t^3 \, dt$

$$+ \int_0^1 t^3 \cos (t^7)(4t^3) \, dt + \int_0^1 t \sin (t^3)(2t) \, dt = \int_0^1 t^2 \sin t^3 \, dt$$

$$+ 4 \int_0^1 t^6 \cos t^7 \, dt + 2 \int_0^1 t^2 \sin t^3 \, dt = -\frac{1}{3} \cos t^3 \Big|_0^1 + \frac{4}{7} \sin t^7 \Big|_0^1$$

$$- \frac{2}{3} \cos t^3 \Big|_0^1 = -\cos t^3 \Big|_0^1 + \frac{4}{7} \sin t^7 \Big|_0^1 = 1 - \cos 1 + \frac{4}{7} \sin 1.$$

27. Since $P(x,y,z) = 2x$, $Q(x,y,z) = z^2$, and $R(x,y,z) = 2yz$, we have

$$P_y(x,y,z) = 0 = Q_x(x,y,z),$$

$$Q_z(x,y,z) = 2z = R_y(x,y,z), \text{ and}$$

$$R_x(x,y,z) = 0 = P_z(x,y,z).$$

Thus, the integral is independent of path.

29. Since $g(x,y) = x + y$, $g_x(x,y) = g_y(x,y) = 1$. Thus, $\int_S (x^2y - z) \, dS$

$= \int_R \sqrt{3} \, (x^2y - x - y) \, dA$ where R is the region in the xy-plane bounded by

$x + y = 1$, $x = 0$, and $y = 0$. Thus, $\int_S (x^2y - z) \, dS = \sqrt{3} \int_0^1 \int_0^{1-x} (x^2y - x - y)$

$dy \, dx = \sqrt{3} \int_0^1 (\frac{1}{2} x^2y^2 - xy - \frac{1}{2} y^2) \big|_0^{1-x} dx = \sqrt{3} \int_0^1 (\frac{1}{2} x^4 - x^3 + x^2 - \frac{1}{2}) \, dx$

$= \sqrt{3} \, (\frac{1}{10} - \frac{1}{4} + \frac{1}{3} - \frac{1}{2}) = -19\sqrt{3}/60$.

31. $\int_S (x^2 + y^2) \, dS = \sqrt{3} \int_R (x^2 + y^2) \, dA = \sqrt{3} \int_0^1 \int_0^{1-x} (x^2 + y^2) \, dy \, dx$

$= \sqrt{3} \int_0^1 (x^2y + \frac{1}{3} y^3) \big|_0^{1-x} = \sqrt{3} \int_0^1 (x^2 - x^3 + \frac{1}{3} (1 - x)^3) \, dx$

$= \sqrt{3} \, (\frac{1}{3} - \frac{1}{4} + \frac{1}{12}) = \sqrt{3}/6$.

33. Since $g(x,y) = 2 - (x^2 - y^2)$, we have $g_x(x,y) = -2x$ and $g_y(x,y) = 2y$. Thus,

$\int_S dS = \int_R \sqrt{1 + 4x^2 + 4y^2} \, dA$ where R is the disk $x^2 + y^2 \le 2$. Thus, in polar

coordinates we have $\int_S dS = \int_0^{2\pi} \int_0^{\sqrt{2}} \sqrt{1 + 4r^2 \cos^2 \theta + 4r^2 \sin^2 \theta} \, r \, dr \, d\theta$

$= \int_0^{2\pi} \int_0^{\sqrt{2}} \sqrt{1 + 4r^2} \, r \, dr \, d\theta = \frac{1}{8} \int_0^{2\pi} \frac{2}{3} (1 + 4r^2)^{3/2} \big|_0^{\sqrt{2}} \, d\theta = \frac{1}{12} \int_0^{2\pi} (27 - 1) \, d\theta$

$= 13\pi/3$.

35. Here $g(x,y,z) = z - 1$, so $N(x,y,z) = \pm \frac{\nabla g(x,y,z)}{|\nabla g(x,y,z)|} = \pm (0,0,1)$. Since we are

interested in the flow in the upward direction, we select $N(x,y,z) = (0,0,1)$.

Then the flux across S in the upward direction is given by $F = \int_S (zx^2, -z^2y,$

$x - 2z) \cdot (0,0,1) \, dS = \int_S (x - 2z) \, dS$ where S is the surface of the plane $z = 1$

where $0 \le x \le 1$ and $0 \le y \le 1$. Then $F = \int_0^1 \int_0^1 (x - 2) \, dy \, dx = \int_0^1 (x - 2) \, dx$

$= \frac{1}{2} - 2 = -3/2$ cubic units per unit time.

37. Here $g(x,y,z) = x^2 + y^2 - z^2$, so $N(x,y,z) = \pm \dfrac{\nabla g(x,y,z)}{|\nabla g(x,y,z)|} = \pm \dfrac{(2x, 2y, -2z)}{2\sqrt{x^2 + y^2 + z^2}}$.

Since we are interested in the flow in the upward direction, we select the unit normal with a positive z component. Thus, $N(x,y,z) = \dfrac{(-2x, -2y, 2z)}{2\sqrt{x^2 + y^2 + z^2}}$

$= \dfrac{(-x, -y, z)}{\sqrt{x^2 + y^2 + z^2}}$. Then the flux across S in the upward direction is given by

$F = \int_S \dfrac{(-x, -y, z)}{\sqrt{x^2 + y^2 + z^2}} \cdot (0, 0, z) \, dS = \int_S \dfrac{z^2}{\sqrt{x^2 + y^2 + z^2}} \, dS$ where S is the surface

given by $z = (x^2 + y^2)^{1/2}$, $x^2 + y^2 \le 2$, $z \ge 0$. Then $F = \int_R \dfrac{x^2 + y^2}{\sqrt{2x^2 + 2y^2}}$

$\sqrt{1 + \dfrac{x^2}{x^2 + y^2} + \dfrac{y^2}{x^2 + y^2}} \, dA = \int_R (x^2 + y^2)^{1/2} \, dA$ where R is the disk $x^2 + y^2 \le 2$.

In polar coordinates we have $F = \int_0^{2\pi} \int_0^{\sqrt{2}} r \, r \, dr \, d\theta = \dfrac{2\sqrt{2}}{3} \int_0^{2\pi} d\theta = \dfrac{4\pi\sqrt{2}}{3}$ cubic

units per unit time.

39. Since $G(X) = (x + z, x + y, y + z)$, div $G = 1 + 1 + 1 = 3$. Thus,

$\int_S G(X) \cdot N(X) \, dS = \int_R 3 \, dV = 3 \int_R dV$, where R is the solid sphere $x^2 + y^2 + z^2$

≤ 4. Then $\int_R dV$ = volume of the sphere $= \dfrac{4}{3} \pi \, 2^3 = \dfrac{32\pi}{3}$. Consequently,

$\int_S G(X) \cdot N(X) \, dS = 32\pi$.

41. Since $G(X) = (xy^2 + 2zy, yz^2 - x^3, x^2 z)$, div $G = y^2 + z^2 + x^2$. Thus,

$\int_S G(X) \cdot N(X) \, dS = \int_R (x^2 + y^2 + z^2) \, dV$ where R is the solid sphere

$x^2 + y^2 + z^2 \le 1$. Then in spherical coordinates we have $\int_S G(X) \cdot N(X) \, dS$

$= \int_0^{2\pi} \int_0^{\pi} \int_0^1 \rho^2 \, \rho^2 \sin\phi \, d\rho \, d\phi \, d\theta = \dfrac{1}{5} \int_0^{2\pi} \int_0^{\pi} \sin\phi \, d\phi \, d\theta = \dfrac{2}{5} \int_0^{2\pi} d\theta = 4\pi/5$.

43. The flux is given by $F = \int_S G(X) \cdot N(X) \, dS$ where S is the surface of the

tetrahedron and N(X) is the outwardly directed normal. Since $G(x,y,z) = (zx^2, -z^2 y, x - 2z)$, we have div $G = 2xz - z^2 - 2$ and by the Divergence Theorem

$F = \int_R 2xz - z^2 - 2 \, dV$ where R is the region bounded by $x + y + z = 1$ and the

coordinate planes. Thus, $F = \int_0^1 \int_0^{1-x} \int_0^{1-x-y} (2xz - z^2 - 2) \, dz \, dy \, dx$

$$= \int_0^1 \int_0^{1-x} (x(1-x-y)^2 - \frac{1}{3}(1-x-y)^3 - 2(1-x-y)) \; dy \; dx = \int_0^1 -\frac{x}{3}$$

$$(1-x-y)^3 + \frac{1}{12}(1-x-y)^4 + (1-x-y)^2 \Big|_0^{1-x} dx = \int_0^1 (\frac{x}{3}(1-x)^3$$

$$-\frac{1}{12}(1-x)^4 - (1-x)^2) \; dx = \frac{1}{3}\int_0^1 (x - 3x^2 + 3x^3 - x^4) \; dx + [\frac{1}{60}(1-x)^5$$

$$+\frac{1}{3}(1-x)^3] \Big|_0^1 = \frac{1}{3}[\frac{1}{2} - 1 + \frac{3}{4} - \frac{1}{5}] - \frac{1}{60} - \frac{1}{3} = \frac{1}{60} - \frac{1}{60} - \frac{1}{3} = -\frac{1}{3} \text{ cubic units}$$

per unit time.

45. The circle C bounds the disk $x^2 + y^2 \leq 9$ in the xy-plane. Thus, $N(X) = (0,0,1)$. Then since $P(x,y,z) = 2y$, $Q(x,y,z) = 2x$, and $R(x,y,z) = -z^2$, we have

$$\nabla \times F = \begin{vmatrix} i & j & k \\ \frac{\partial}{\partial x} & \frac{\partial}{\partial y} & \frac{\partial}{\partial z} \\ 2y & 3x & -z^2 \end{vmatrix} = (0,0,1). \text{ Thus, by Stokes' Theorem } \int_C 2y \; dx + 3x \; dy$$

$$- z^2 \; dz = \int_S (0,0,1) \cdot (0,0,1) \; dS = \int_S dS \text{ where S is the disk } x^2 + y^2 \leq 9 \text{ in}$$

the xy-plane. Since $\int_S dS = $ area of the disk, $\int_C 2y \; dx + 3x \; dy - z^2 \; dz = 9\pi$.

47. Since the triangle lies in the plane given by $x - z = 0$, the upwardly directed unit normal vector is $N = (-1,0,1)/\sqrt{2}$. Then since $F(x,y,z) = (x^2 - y + z, \sin y + x, x - e^z)$, we have

$$\nabla \times F = \begin{vmatrix} i & j & k \\ \frac{\partial}{\partial x} & \frac{\partial}{\partial y} & \frac{\partial}{\partial z} \\ x^2 - y + z & \sin y + x & x - e^2 \end{vmatrix} = (0,0,2). \text{ Thus, by Stokes' Theorem,}$$

$$\int_C (x^2 - y + z) \; dz + (\sin y + x) \; dy + (x - e^z) \; dz = \int_S (0,0,2) \cdot \frac{1}{\sqrt{2}}(-1,0,1) \; dS$$

$$= \sqrt{2} \int_S dS \text{ where S is the triangle bounded by C. Since } \int_S dS \text{ is the area of}$$

the triangle bounded by C, we have $\int_C (x^2 - y + z) \; dx + (\sin y + x) \; dy$

$$+ (x - e^z) \; dz = \sqrt{2} \cdot \frac{1}{2} \cdot \sqrt{2} = 1. \text{ (Area of triangle is } \frac{1}{2}|A \times B| \text{ where A}$$

$= (1,1,1)$ and $B = (0,1,0)$. $A \times B = (-1,0,1)$, so the area is $\frac{1}{2}\sqrt{2}$.)